Bioprospecting of Natural Sources for Cosmeceuticals

# *Bioprospecting of Natural Sources for Cosmeceuticals*

Edited by

**Deepika Kathuria**
*Chandigarh University, India*
*Email: niperdeepika12@gmail.com*

**Ajay Sharma**
*Chandigarh University, India*
*Email: sharmaajay9981@gmail.com*

**Meenakshi Verma**
*Chandigarh University, India*
*Email: meenakshi.e8491@cumail.in*

and

**Gulzar Ahmad Nayik**
*Government Degree College Shopian, India*
*Email: gulzarnaik@gmail.com*

ROYAL SOCIETY OF **CHEMISTRY**

Print ISBN: 978-1-83767-046-8
PDF ISBN: 978-1-83767-228-8
EPUB ISBN: 978-1-83767-229-5

A catalogue record for this book is available from the British Library



The Royal Society of Chemistry is a charity, registered in England and Wales, Number 207890, and a company incorporated in England by Royal Charter (Registered No. RC000524), registered office: Burlington House, Piccadilly, London W1J 0BA, UK, Telephone: +44 (0) 20 7437 8656.

Visit our website at books.rsc.org

Printed in the United Kingdom by CPI Group (UK) Ltd, Croydon, CR0 4YY, UK

# *Dedication*

Dr Deepika Kathuria dedicates this book to
My Parents
(Mr Mange Ram Kathuria and Mrs Janak Dulari)
and
My Beloved Husband
(Mr Sanjeev Kumar)

Bioprospecting of Natural Sources for Cosmeceuticals
Edited by Deepika Kathuria, Ajay Sharma, Meenakshi Verma and Gulzar Ahmad Nayik

Published by the Royal Society of Chemistry, www.rsc.org

# Editor Biographies

**Deepika Kathuria** is a medicinal chemist with a key knowledge of natural products. Presently, she is an Assistant Professor at the University Center for Research and Development, Chandigarh University, Gharuan, Punjab, India. She graduated with a B Pharm from Pandit Bhagwat Dayal Sharma Post Graduate Institute of Medical Sciences, Rohtak. She completed her Master's and Doctorate in pharmaceutical sciences at the National Institute of Pharmaceutical Education and Research, Mohali, India. She has published over 25 peer-reviewed articles in reputed international journals. She has various honors/fellowships/awards to her credit *i.e.*, an INSPIRE fellowship (JRF) from DST, an INSPIRE fellowship (SRF) from DST, a Bristol Myers Squibb fellowship (Merit Cum Topper scholarship during her M Pharm), a Newton Bhabha placement, and the DST–DAAD exchange programme. She has knowledge of medicinal plants, pharmaceuticals, and cosmeceuticals, and thus it is very interesting to bring these three topics together to provide an enthralling book. Currently, she is editing several book projects.

**Ajay Sharma** is a natural product chemist with a strong organic chemistry, phytochemistry, biochemistry, and ethnobotany background. Presently, he is working as an Assistant Professor at the University Centre for Research and Development, Department of Chemistry, Chandigarh University, Mohali, India. He graduated (BSc Medicine) from the Government College Hoshiarpur (affiliated with Panjab University, Chandigarh), and obtained his Master's degree in organic chemistry (natural products) from Punjab Agriculture University Ludhiana and his PhD in organic chemistry (natural products) from SLIET (Sant Longowal Institute of Engineering & Technology, Deemed University) Punjab. He has published over 58 peer-reviewed papers and book chapters in reputed international/national journals and books. He has also published three edited books and presently, his three edited books are under production in different international publication houses.

Bioprospecting of Natural Sources for Cosmeceuticals
Edited by Deepika Kathuria, Ajay Sharma, Meenakshi Verma and Gulzar Ahmad Nayik

Published by the Royal Society of Chemistry, www.rsc.org

He has various honors/fellowships/awards to his credit *i.e.*, a Roll of honor from PAU Ludhiana (2011), a Merit Cum Topper scholarship during his MSc (PAU Ludhiana) (2009–10), an MHRD fellowship award for his PhD (2012), an INSPIRE fellowship (JRF, 2013) from DST, an INSPIRE fellowship (SRF, 2015) from DST, the 1st poster prize award at a national conference (NICS, 2016) and the best poster award at TSFS, 2016. He carries out research in organic chemistry, natural product chemistry, phytochemistry, and food chemistry. His research activities are focused on the isolation and characterization of bioactive extracts from traditional medicinal plants and evaluation of their biological potential. Currently, his research group at Chandigarh University is investigating the field of more environmentally friendly syntheses of metal nanoparticles and carbon nanodots using plant extracts/plant-based biomass, followed by assessing their biological, catalytic, and sensing potential. To date, he has successfully supervised six MSc dissertation students.

**Meenakshi Verma** obtained her PhD from Thapar University, Patiala. She has completed her National Postdoctoral fellowship at the Indian Institute of Technology, Ropar (IIT Ropar) and her WISTEMM Fellowship at the University of Wisconsin, Green Bay, USA. She joined the University Centre for Research & Development, Chandigarh University, Mohali in 2019 and is currently working as an Assistant Professor. Her research interests include organic and supramolecular chemistry.

**Gulzar Ahmad Nayik** completed his Master's degree in food technology at the Islamic University of Science & Technology, Awantipora, Jammu and Kashmir, India, and his PhD at the Sant Longowal Institute of Engineering & Technology, Sangrur, Punjab, India. He has published over 80 peer-reviewed research and review papers and more than 40 book chapters, and he has edited 13 books. Dr Nayik has also published a textbook on food chemistry and nutrition and has delivered presentations at various national and international conferences, seminars, workshops and webinars. Dr Nayik was shortlisted twice for the prestigious INSPIRE Faculty Award, from the Indian National Science Academy, New Delhi, India, in 2017 and 2018. He was nominated for India's prestigious national award (the Indian National Science Academy Medal for Young Scientists 2019–2020). Dr Nayik also fills the roles of editor, associate editor, assistant editor and reviewer for many food science and technology journals. He has received many awards, appreciations and recognitions and holds memberships of various international societies and organizations. Dr Nayik is currently editing several book projects.

# *Preface*

Natural biodiversity, such as medicinal plants, microbes, fungi and algae, has been acting as a rich resource of several therapeutic agents used in traditional medicine systems from prehistoric times to cure different kinds of ailments and diseases. These natural resources are the store house of a variety of bioactive secondary metabolites. These secondary metabolites are well-known for their wide range of applications in the food, pharmaceutical, cosmetics and personal care product industries. Owing to the widespread applications of these natural resources, nowadays they are extensively used as a reservoir of bioactive raw materials for personal care products and cosmetics. The concept of beauty or cosmetics is as old as the evolution of mankind. For the purpose of looking beautiful, charming and young, people are using a variety of beauty products. These beauty products include various types of secondary metabolites obtained from herbs, plants, fungi, lichens, microbes, *etc.* Herbal cosmetics are an invaluable gift of nature and their demand has been growing day-by-day in the world market.

This book presents a comprehensive guide to metabolites obtained from traditional medicinal plants, microbes, fungi, lichens, algae, *etc.*, which are used in the field of cosmetics. The isolation, characterization and authentication of the materials used in cosmeceuticals are very crucial and hence all of the related aspects will be discussed in this book. Lots of debate is occurring regarding the regulatory aspects of the natural products in cosmeceuticals. Therefore, this topic has been discussed as part of a chapter. All of the chapters cover the key components from each resource, such as plants, algae, bacteria and lichens, as well as clay resources, and their extraction, cosmeceutical applications, natural cosmetic formulations, allergic responses, and toxicology. The impact of nanotechnology will be discussed in various cosmetic products, such as skin care and hair care products, *etc.* This book represents a valuable asset for researchers and graduate students of chemistry,

Bioprospecting of Natural Sources for Cosmeceuticals
Edited by Deepika Kathuria, Ajay Sharma, Meenakshi Verma and Gulzar Ahmad Nayik

Published by the Royal Society of Chemistry, www.rsc.org

botany, biotechnology, microbiology, cosmetology and the pharmaceutical sciences. This book will make a substantial contribution to the knowledge of people from every field. Also, this book will be very useful to scientists, graduates and undergraduates, along with researchers in the fields of natural products, herbal medicines, ethnobotany, pharmacology, chemistry and biology. Furthermore, it is an equally significant resource for people working in different traditional medicinal systems, including doctors (especially those engaged in Ayurveda and the Chinese traditional medicinal system), the pharmaceutical industry (for cosmetic formulation design), microbiologists, and scientists working in biotechnology and nanoscience.

Deepika Kathuria
Ajay Sharma
Meenakshi Verma
Gulzar Ahmad Nayik

# Contents

**Chapter 1 Natural Products for Cosmeceuticals** 1
*Kashvi C. Shah, Kashish Verma, Jeel D. Trivedi, Shreeja H. Mehta, Sanjeev Kumar, Tejas M. Dhameliya and Deepika Kathuria*

1.1 Introduction 1
1.2 History of Natural Product-based Cosmeceuticals 3
1.3 Cosmeceutical Properties 3
1.4 Economic Overview of Green Cosmetics 5
1.5 Natural Cosmetic-based Formulations 6
1.5.1 Skin Care 6
1.5.2 Hair Care 9
1.5.3 Toiletries 11
1.5.4 Nail Care 12
1.5.5 Lip Care 12
1.5.6 Perfumes 14
1.6 Limitations of Natural Cosmeceuticals 14
1.7 Conclusion 14
References 15

**Chapter 2 Natural Products as Cosmeceuticals in Different Countries: A Regulatory Perspective** 18
*Avantika Agrawal and Sumit S. Chourasiya*

2.1 Introduction 18
2.2 Regulations in Different Countries 21
2.2.1 United States of America 21
2.2.2 Canada 26

Bioprospecting of Natural Sources for Cosmeceuticals
Edited by Deepika Kathuria, Ajay Sharma, Meenakshi Verma and Gulzar Ahmad Nayik

Published by the Royal Society of Chemistry, www.rsc.org

2.2.3 Japan 28
2.2.4 China 29
2.2.5 European Union 31
2.2.6 India 36
2.2.7 ISO 16128-1:2016 and ISO 16128-2:2017/Amd 1:2022(en) 41
2.3 Conclusion 41
List of Abbreviations 46
References 47

**Chapter 3 Methods of Isolation, Characterization and Authentication of Essential Oils for Cosmetics** 53
*Nibedita Nath, Subhendu Chakroborty and Kaushik Pal*

3.1 Introduction 53
3.2 Essential Oils 54
3.3 Chemistry of Essential Oils 55
3.4 Methods of Isolation 56
3.5 Methods of Characterization 63
3.6 Methods of Authentication 64
3.7 Application of Essential Oils in Cosmetics 65
3.8 Conclusions and Future Prospects 67
References 67

**Chapter 4 Toxic and Allergic Responses Caused by Secondary Metabolites Used in Cosmetic Formulations** 73
*Swati Gajbhiye and Kaushik Pal*

4.1 Introduction 73
4.2 Toxic and Allergenic Plant Secondary Metabolites Used in Cosmetics 76
4.2.1 Terpene Secondary Metabolites 76
4.2.2 Phenolic Secondary Metabolites 81
4.2.3 Alkaloid Secondary Metabolites 86
4.3 Secondary Metabolites Derived from Microorganisms and Animals That Cause Toxic and Allergic Responses 91
4.3.1 Carminic Acid 93
4.3.2 Bee Products 94
4.3.3 Toxicological Profile 95
4.4 Conclusions and Future Perspectives 95
References 96

**Chapter 5 The Wonders of Plant Secondary Metabolites as Cosmeceuticals 105**
*P. Shah, Y. Soni, M. Raval and N. Lunagariya*

5.1 Introduction 105
5.2 Cultural History of Natural Product-based Cosmeceuticals 106
5.3 Major Secondary Metabolites with Cosmetic Applications 106
5.3.1 Curcuminoids 107
5.3.2 Caffeine 109
5.3.3 Capsaicin 111
5.3.4 Boswellic Acids 114
5.3.5 Resveratrol 116
5.3.6 Bakuchiol 118
5.3.7 Other Natural Products as Wonders in Cosmeceuticals 120
5.4 Conclusion and Future Prospects 123
Acknowledgements 124
References 124

**Chapter 6 Applications of Mushrooms in Cosmeceuticals 129**
*Chasheen Fizza, Waseem Khalid, Safura Kausar, Zahra Maqbool, Shanza Mukhtar, Eman Ali, Blessing Titylayo Ajala, Muhammad Zeeshan Ahmad and Muhammad Irfan*

6.1 Introduction 129
6.2 Types of Mushrooms 131
6.2.1 Poisonous Mushrooms 131
6.2.2 Medicinal Mushrooms 132
6.2.3 Edible Mushrooms 133
6.3 Constituents of Mushrooms 134
6.3.1 Carbohydrates 134
6.3.2 Proteins 135
6.3.3 Minerals 135
6.3.4 Vitamins 136
6.3.5 Lipids 136
6.4 Phytochemicals 136
6.5 Cosmetics Applications of Mushrooms 137
6.5.1 Skincare 138
6.5.2 Hair Care 140
6.6 Conclusion 140
References 140

**Chapter 7 Algal and Microalgal Compounds in Cosmeceuticals** **144**
*Tridip Boruah, Barsha Devi, Nibedita Gogoi and Chiranjib Mili*

7.1 Introduction 144
7.2 Cultural and Historical Background 146
7.3 Use of Algal Secondary Metabolites in Cosmeceuticals 147
7.4 Use of Microalgal Secondary Metabolites in Cosmeceuticals 147
7.5 Traditional Uses of Bioactive Cosmeceuticals 150
7.6 Methods of Instrumental Extraction of Algal-based Raw Materials (Biometabolites) for Cosmeceuticals 152
7.6.1 Conventional Methods 152
7.6.2 Novel Extraction Methods 154
7.7 Cosmeceutical Activities 158
7.7.1 Anti-aging 158
7.7.2 Skin Whitening 159
7.7.3 Moisturizing 160
7.7.4 Antioxidant 160
7.7.5 Healing and Therapeutic Agent 161
7.8 Cosmeceutical-based Formulations and Marketed Products 161
7.9 Conclusions and Future Prospects 162
References 165

**Chapter 8 Cosmetic Applications of Lichens** **172**
*Yingshu Zhao, Baojun Xu and Jessica Pandohee*

8.1 Introduction 172
8.2 Lichen Morphology, Symbiosis, and Diversity 173
8.3 Overview of Biochemical Metabolites in Lichens 175
8.4 Pharmacological and Cosmeceutical Applications of Lichens 176
8.4.1 Antimicrobial Agents 176
8.4.2 Antibacterial Properties 178
8.4.3 Antifungal Properties 179
8.4.4 Antioxidants 180
8.4.5 Sunscreen Agents 183
8.4.6 Anti-inflammatory Agents 186
8.4.7 Skin Lightening and Depigmenting Agents 188
8.4.8 Moisturizers 189
8.4.9 Use in Perfume and Other Products 190
8.5 Conclusion 191
References 192

**Chapter 9 Bacteria as a Source of Bioactive Cosmeceuticals** 198
*Md. Anisur Rahman Mazumder, Md. Fahad Jubayer, Md. Shahidullah Kayshar, Dayanand Peter and Thottiam Vasudevan Ranganathan*

9.1 Introduction 198
9.2 Cultural and Historical Background 200
9.3 Use of Secondary Metabolites in Cosmeceuticals 203
9.3.1 Bacterial Cellulose 203
9.3.2 Secondary Metabolites from Cyanobacteria 205
9.3.3 Bacterial Polysaccharides 206
9.3.4 Oligosaccharides 209
9.3.5 Bio-surfactants 209
9.4 Traditional Uses of Bioactive Cosmeceuticals 209
9.5 Pharmacological Activities 210
9.6 Cosmeceutical-based Formulations and Market Products 216
9.7 Conclusions and Future Prospects 217
References 217

**Chapter 10 Marine Natural Products as a Bioresource for Cosmeceuticals** 223
*Yousef Rasmi, Kevser Kübra Kirboğa, Tazeen Rao, Mazhar Ali and Muhammad Zeeshan Ahmed*

10.1 Introduction 223
10.2 Discovery of Bioresource Marine Natural Products and Their Biodiversity 224
10.3 Developing Marine Natural Products: Challenges and Opportunities 225
10.3.1 Biodiversity Challenges 226
10.3.2 Supply and Technical Challenges 227
10.3.3 Market Challenges 229
10.4 Bioresources of Cosmeceuticals from Marine Environments 230
10.4.1 Fungi-derived Cosmeceuticals 231
10.4.2 Bacteria-derived Cosmeceuticals 231
10.4.3 Macroalgal-derived Cosmeceuticals 232
10.4.4 Microalgal-derived Cosmeceuticals 233
10.4.5 Marine Invertebrate-derived Cosmeceuticals 234
10.5 Properties of Marine Natural Products for Cosmeceutical Applications 235
10.6 Secondary Metabolite Marine Natural Products for Cosmeceutical Applications 235

10.6.1 Carbohydrates 236
10.6.2 Marine Phytochemical Compounds 239
10.6.3 Mycosporines 241
10.6.4 Neo Fatty (Carboxylic) Acid Derivatives 242
10.7 Applications of Marine Natural Products for Cosmeceuticals 243
10.7.1 Antiaging Activities 243
10.7.2 Antimelanogenic Activities 244
10.7.3 Anti-acne Activities 245
10.7.4 Antioxidant Activities 246
10.7.5 Sunscreens 246
10.7.6 Wound Healing, Anti-inflammatory, and Analgesic Activities 247
10.7.7 Moisturizing Activities 248
10.7.8 Repairing Activities 250
10.7.9 Skin Lipid Activities 250
10.7.10 Antitumor Activities 250
10.7.11 Antiviral Activities 253
10.8 Side Effects of Cosmeceutical Marine Natural Products 254
10.9 Conclusion and Future Perspectives 254
References 255

**Chapter 11 Cosmeceutical Applications of Clay Minerals** **276**
*Arunima Kalita, Arup Giri, Nitish Kumar and Pushpender Bhardwaj*

11.1 Introduction 276
11.2 History of Clay Mineral Use as Cosmetics 278
11.3 Uses of Clay Minerals in Cosmeceuticals 280
11.4 Cosmeceutical-based Formulations and Marketed Products that Use Clay Minerals 283
11.5 Conclusions and Future Perspectives 285
References 286

**Chapter 12 Nanotechnology for Cosmetics** **292**
*Gagandeep Kaur, Mandeep Kaur, Kirtanjot Kaur, Shweta Sareen and Vishal Mutreja*

12.1 Introduction 292
12.2 Nanomaterials in Cosmetics 293
12.2.1 Inorganic Nanomaterials 294
12.2.2 Organic Nanomaterials 295
12.3 Nanomaterials Used in Commercialized Products 299
12.4 Conclusions 302
References 303

**Subject Index** **306**

CHAPTER 1

# *Natural Products for Cosmeceuticals*

KASHVI C. SHAH[a], KASHISH VERMA[b], JEEL D. TRIVEDI[a], SHREEJA H. MEHTA[a], SANJEEV KUMAR[c], TEJAS M. DHAMELIYA[a,*†] AND DEEPIKA KATHURIA[c*]

[a]L. M. College of Pharmacy, Navrangpura, Ahmedabad 380 009, Gujarat, India; [b]Department of Biosciences, Chandigarh University, Gharuan, Punjab 140 413, India; [c]University Center for Research and Development, Chandigarh University, Gharuan, Punjab 140 413, India
*E-mail: tmdhameliya@gmail.com, tejasm.dhameliya@nirmauni.ac.in, niperdeepika12@gmail.com

## 1.1 Introduction

Natural cosmetics are made up of natural products containing only natural raw materials that are directly obtained from nature, like ingredients from plants, animals, mineral resources and renewable natural resources.[1] The raw materials undergo very little processing, due to which they possess minimal impact on humans, animals and the Earth. Thus, these are produced in an ecological, sustainable and ethical way. Each raw material required for the formation of natural cosmetics has a rich composition and various properties, like anti-oxidant, anti-aging, anti-inflammatory, anti-wrinkle, acne reducing, skin dehydration, cleansing, skin lightening, emulsification, exfoliation and surfactant effects.[2–4]

†Present Address: Department of Pharmaceutical Chemistry, Institute of Pharmacy, Nirma University, Ahmedabad 382 481, Gujarat, India.

Bioprospecting of Natural Sources for Cosmeceuticals
Edited by Deepika Kathuria, Ajay Sharma, Meenakshi Verma and Gulzar Ahmad Nayik

Published by the Royal Society of Chemistry, www.rsc.org

Products that can act as both cosmetics and medicine are called cosmeceuticals. They can be combined as cosmetic products with pharmaceuticals. In essence, they are biological active ingredients that possess medical benefits. The active substance helps in beautifying, cleansing, promoting attractiveness, or altering the appearance without affecting the structure or function of cosmeceuticals. The term cosmeceutical (Figure 1.1) is less common than the term cosmetics.[5] Although both of them are associated with beauty and are categorized under beauty products, they are different from each other. Cosmetics work on the outer layer of the skin and beautify the skin externally, but do not cure the skin internally. As a result, the effect is temporary and immediate. Meanwhile, cosmeceuticals are a combination of cosmetics and natural ingredients that have healing properties, which cure skin-related problems internally. When the product is applied on the skin, it penetrates from the epidermis and works on the dermis area. Cosmetics do not have rejuvenating properties, while cosmeceuticals do have this property.[6]

Synthetic products are mainly used nowadays in the production of cosmetics because they are cheap in price compared to other natural cosmetics and they also have a longer shelf life. However, many of the synthetic cosmetic formulations contain paraben, toluene, oxybenzone, artificial fragrances or phthalates, which have carcinogenic potential and also have a negative impact on the environment. In contrast, natural cosmetics do not contain any synthetically created ingredients. A natural product is extracted or produced in a sustainable way, but the products used in synthetic cosmetics are engineered to mimic natural ones. The natural products used in cosmetics are synthetic chemical-free, cause less irritation, are eco-friendly and have long term benefits.[7]

There are various sectors in which cosmetic products are used, such as in hair care, skin care, lip care, nail care, toiletries and perfumery. An outline of the different types of cosmetic products used in the field of cosmetics has been provided in Figure 1.2.

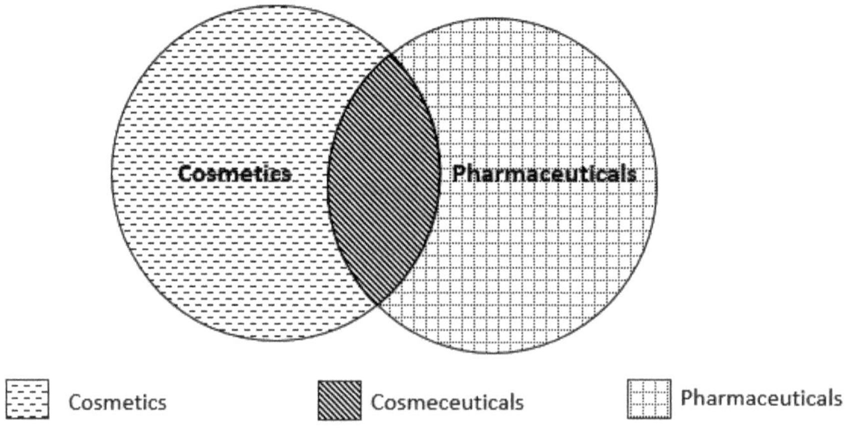


**Figure 1.1** General representation of cosmeceuticals.

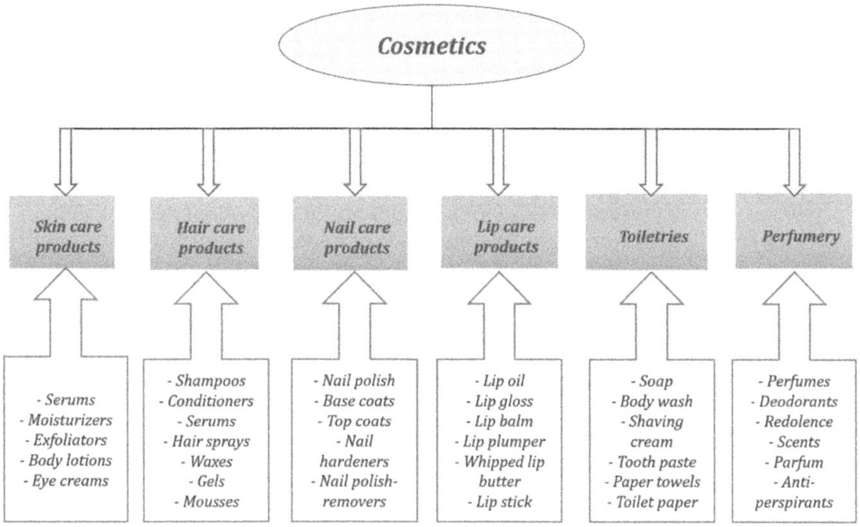


**Figure 1.2** The different cosmetic products used in the cosmetics sector.

## 1.2 History of Natural Product-based Cosmeceuticals

The word 'cosmeceuticals' was believed to be used for the first time in the 20th century. It was coined by Raymond Reed, a founding member of the United States Society of Cosmetic Chemists, in 1961. Albert M. Kligman, a professor at the University of Pennsylvania Medical School, discovered that topical retinoic acid (or tretinoin) can be used for both acne and wrinkle treatment, and was thus called the father of cosmeceuticals. However, cosmeceuticals only became popular on the market in 1996.[8] Table 1.1 depicts the history of natural cosmetics.

## 1.3 Cosmeceutical Properties

Natural cosmetics act as anti-oxidants, emulsifying agents, anti-aging, anti-inflammatory, anti-microbial, anti-tumor, anti-wrinkle and UV protective agents, exfoliants, emollients, and surfactants. For instance, ferulic acid is a derivative of plant phenolic content and is commonly found in grapes, apples, *etc.* It acts as an anti-oxidant and is also used as a skin whitening agent. Anti-oxidants also fight against skin aging by reducing the amount of free radicals.[15] Glycyrrhizic acid is a compound isolated from licorice that is used as an anti-inflammatory agent.[16] Coco beans contain triglyceride compounds such as linoleic acid and palmitic acid, which are used in reducing wrinkles. Several products, such as shea butter, soyabean oil and sunflower oil, act in reducing aging effects. Fatty acids, long chain

**Table 1.1** History of natural cosmetics.

| Era | Natural products used |
|---|---|
| **Natural cosmetics in the ancient world** | |
| 10000 BCE | Scented oils and ointments were used by Egyptian men and women to clean and soften their skin |
| 4000 BCE | Galena mesdemet (made of copper and lead ore) and malachite (a bright green paste of copper minerals) were applied by Egyptian women to their faces for color. Kohl (made up of burned almonds) was applied on the eyes. Women carried makeup kits at parties[9] |
| 3000 BCE | Fingernails were stained with gum Arabic, gelatin, beeswax, and egg by Chinese women[10] |
| | Grecian women painted their faces with white lead and applied crushed mulberries as rouge |
| 1500 BCE | Chinese and Japanese citizens used rice powder to make their faces white |
| 1000 BCE | Grecians whitened their complexion with chalk or lead face powder |
| **Natural cosmetics in the early common era** | |
| 100 | In Rome, people put barley flour and butter on their pimples and sheep fat and blood on their fingernails for polish[11] |
| 300–400 | Henna was used as hair dye and Mehndi in India[12] |
| **Natural cosmetics in the middle ages** | |
| 1200 | Perfumes are first imported to Europe from the Middle East |
| 1300 | In Elizabethan England, dyed red hair came into fashion. Women wore egg whites over their faces to create the appearance of a paler complexion[13] |
| **Renaissance cosmetics** | |
| 1400–1500 | Italy and France emerged as the main centers of cosmetics manufacturing in Europe. Arsenic was used as a face powder instead of lead |
| 1500–1600 | European women lightened their faces using white lead paint. Queen Elizabeth I of England used white lead, with which she created a look known as "the Mask of Youth" |
| **Natural cosmetics in the 19th and 20th century** | |
| 1800 | Zinc oxide was used as facial powder. Ceruse, which was made from white lead, was later discovered to be toxic and blamed for health problems including facial tremors[14] |
| 1900 | In Edwardian Society, pressure increased on middle-aged women to appear youthful while acting as hostesses. As a result, cosmetics use increased. Beauty salons were popularized |

esters, lanolin and beeswax are used in skin moisturizing formulations because of their emollient properties.[17] Egg yolk, mustard and honey act as natural emulsifiers, and play an important role in combining immiscible fluids into an emulsion.[18] Humectants have long-lasting moisturizing effects, and act by binding water. Some of the natural humectants are aloe, honey and glycerol. Exfoliants are found in milk, fruit and sugar cane. They have the ability to remove the outer layer of dead skin cells, and stimulate new cells to grow.[19] Potassium cocoate is derived from coconut oil and is used as a surfactant. Thus, it is used for cleansing and foaming, and has anti-microbial effects.[20]

## 1.4 Economic Overview of Green Cosmetics

There is increasing demand for the use of environment-based natural beauty products. With the rise of the chemical content in cosmetic products and awareness among consumers, companies have shifted themselves towards manufacturing natural cosmetics; for instance, L'Oreal has launched a plant-based hair dye named Botenea. There are various countries, such as in the Asia-Pacific region, Europe and North America, which have set some standards for regulating the natural cosmetics market, including Ecocert, the Soil Association, Cosmebio, ICEA, BDIH, OASIS, NPA, and NSF. There has been growth of the natural cosmetics industry in the past few years due to the rise in demand for synthetic chemical-free skin, and this trend is expected to continue during the forecast period. The revenue for the natural and organic market stood at 11.92 billion U.S. dollars in the year 2021, and there will be a rise to 18.07 billion U.S. dollars by the year 2027 according to the data (Figure 1.3) provided by Statista Consumer Market Outlook.[21]

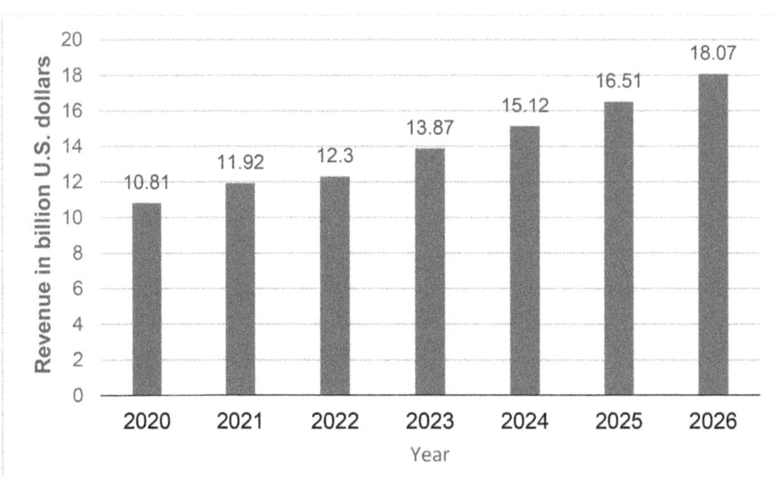


**Figure 1.3** Statistical data on the revenue of natural and organic cosmetics from year 2020–2026 in billion U.S. dollars. Reproduced from ref. 21 with permission from Statista.

# 1.5 Natural Cosmetic-based Formulations

## 1.5.1 Skin Care

Nowadays, advancements and developments in the cosmetic industries are emerging at a quite rapid rate. Products that cause no harm and are safe to use can be speculatively applied on the skin surface, *i.e.* locally or topically. Agricultural extracts are broadly used in cosmeceuticals because of their role in the maintenance of the well-being, appearance and purity of skin, hair and nails. The most accepted botanical products used by skin specialists include soy, teas (green, white and black), chamomile, caffeine and coffeeberry.[22]

Many compounds used in hand sanitizers attenuate the transference of contagions and thereby prevent the spread of pathogens. One such organic and humidifying hand sanitizer was manufactured based on the antibacterial effect of alcohol in addition to *Abelmoschus esculentus* (okra) polysaccharide, which helps to soothe, soften and increase the moisture levels in the skin. After a lot of preparatory assessments and quality examinations, certain outputs were revealed that suggest: (i) good moisturization potential; (ii) hydrated skin; (iii) non-irritability, (iv) better cleanliness of the hands; and (v) synergistic activity along with triclosan to avoid dry and rough skin.[23]

Utilizing natural polymers in cosmetic formulations has greatly expanded as a result of the cosmetics industry's desire to move towards sustainable biobased raw ingredients, such as natural polysaccharides. Emulsions are the most common formulations for skincare because of their feasibility and adaptability in the pharmaceutical industry. Baptista *et al.* studied emulsions of olive oil and α-tocopherol using an emulsifying agent such as FucoPol, an exopolysaccharide with a high molecular weight that contains carbohydrates, glucuronic acid, *etc.*[24] Herein, FucoPol was found to be better than other emulsifiers, available commercially. The synergistic anti-oxidant activity of FucoPol and α-tocopherol, along with the capability of FucoPol to heal wounds, has made it a promising biomaterial for the formulation of cosmetics.

Emulsions prepared using high frequency ultrasound technology (HFUT) without the need for any additive ingredients or emulsifying surfactants have differential properties. In comparison with reference emulsions, HFUT showed twice lower average droplet size and thrice greater viscosity, and furthermore it exhibited better applicability for cosmetic preparations.[25] The *in situ* gelation was evaluated for cosmetic emulsions prepared using two alginates, having a richer content of mannuronic acid or glucuronic acid compared with a biopolymer reference, such as xanthan gum. The results were consistent with those of traditional polymers, and the emulsion that contained the gelled form of the alginate had significantly higher viscoelastic characteristics and dynamic viscosity, and great consistency.[26]

Natural tri-glycerides obtained from derivatives of L-ascorbic acid that are lipophilic have been used as ingredients for cosmetics. Both plant and animal fats were investigated as trans-esterification substrates for ascorbic acid by lipase and their anti-oxidant capabilities in skin care products were assessed. The research showed that the formulation included ascorbyl esters, linseed oil, and coconut oil with a high potential to scavenge 2,2-diphenylpicrylhydrazyl (DPPH) radicals. Furthermore, the saturated medium side alkyl chain esters showed quicker diffusivity than the long chain ones.[27]

For a very long time, organic triglycerides have been included in cosmetic formulations. A natural ingredient in cosmeceuticals, bio-based algae oil, has a long shelf life, excellent thermal and oxidative stability, and high moisturizing capacity.[28] Skin has many barriers before absorption, of which the stratum corneum is the main one. Natural and organic products have been used to overcome this hurdle. A lipid called phytosphingosine (PS) is found in the stratum corneum naturally, both in its free form and as part of a major proportion of ceramides. The formulation containing phytosphingosine showed excellent antibacterial, antiseptic and anti-inflammatory activity for the treatment of acne.[29]

Natural product-based cosmeceuticals have grown in popularity as the demand for vitamins and plant-based goods has increased in a subset of the cosmetics industry. They are used in cosmetics and health-related products because of their anti-oxidant, anti-microbial, anti-aging, UV protection, cytoprotective, anti-carcinogenic, and anti-inflammatory characteristics. Their drawbacks, namely their poor solubility, instability, decreased skin penetration, and brief skin retention period, severely limit their applications in topical formulations. The utility of nanotechnology as an advanced tool may help to encapsulate them in order to overcome their drawbacks and deliver them to the skin in a controlled manner.[30] Similarly, Souza and co-workers worked using beeswax-based nanoparticles and incorporated them into gel-based formulations. These formulations have been shown to increase the stratum corneum water content and decrease trans epidermal water loss. The innovative natural bees wax-based nanoparticles demonstrated potential for skin restoration and opened the door to their possible commercial usage as a unique natural ingredient in dermatology and cosmetics for impaired skin barriers.[31] It is possible that hydroxycinnamic acids,[32] extracts or bioactive ingredients of mushrooms[33] and *Acmella oleracea*,[34] may serve as the basis for cosmeceutical formulations to deal with anti-ageing, anti-wrinkles, inflammation, and skin pigmentation. To attain the same bioactive goals, the extracts were further mixed into a basic cosmetic cream after analyzing the phenolic acids and ergosterol.

Skin aging is a normal, physiological, biochemical, and time-sensitive process that is caused by a complex interaction between internal and external variables. These long-term modifications to the skin's layers, morphology, metabolism, and appearance are detrimental. Gallic and ferulic acids, as well as herbal formulations containing *Thymus vulgaris*, *Panax ginseng*, *Triticum aestivum*, or *Andrographis paniculata*, are some of the most convincing natural ingredients for the formulation of novel dermo-cosmetics with anti-skin aging effects.[35] To slow down the progression of skin ageing, many natural organic compounds are incorporated into formulations, including moisturizers, herbal components, minerals, growth factors, exfoliants, vitamins, and proteins. Utilizing bioactive compounds that are derived from natural resources is a recent trend in anti-aging research. The anti-oxidant polyphenol is found in many consumable products and scavenges free radicals and gets rid of metabolic waste products. So, vitamin C and annatto seed powder acting as an anti-oxidant and natural colorant are added to reacetylated chitosan films and their use as an anti-aging skin mask was proposed.[36] Chitosan has been used for a variety of purposes, such as sun-protection, photo-aging, as a depigmenting agent in skin lighteners, in skin exfoliation, and in preventing the loss of fragrant ingredients.[37] Vitamin C and peptides, along with additives such as hyaluronic acid and mineralizing water, work as anti-aging agents, and have been widely used in dermatological products. These formulations are prepared at low pH and have been proven to have effectiveness in anti-wrinkles, anti-aging and skin regeneration.[38]

The fermentation of the black yeast fungus *Aureobasidium pullulans* produces pullulan, a naturally occurring edible polymer utilized in numerous cosmeceuticals. It is a water-soluble, edible, non-mutagenic, and non-toxic polymer. Pullulan and dioxybenzone are conjugated, which lessens the negative effects of lower photoabsorption intensity, with a considerable capacity for epidermal retention. Pullulan films are non-sticky, and they can be applied to the skin with ease because they are in a semi-liquid gel form. The hydrogel composition gives the skin on your face solidity, reduces lines, and restores suppleness.[39]

In contrast to land resources, marine resources have unique biochemical properties, and the demand for cosmeceuticals from these sources has increased tremendously in the past few years. The most significant repository of biodiversity is found in marine habitats, which also include biologically active compounds with untapped potential for use in medicines, nutraceuticals, and cosmeceuticals. Agar and carrageenan, which function as gelling and thickening agents to raise the viscosity of cosmetic compositions, are valuable bulk ingredients for the cosmetic industry. Various small compounds, such as ectoine, trichodin A, mytiloxanthin, kojic acid, sargafuran, chrysophanol, scytonemin, astaxanthin, mycosporine-like amino acids (MAAs) and PUFAs, have been used in the cosmetics industry due to their various characteristics, such as anti-wrinkle, skin hydration, anti-aging, anti-acne, anti-microbial, promoting skin hydration and reducing pigmentation.[40]

The search for alternative excipients that maintain functionality and cosmetic appeal has intensified as a result of growing customer demand for topical solutions that are safer, more organic, and durable. The primary variables include viscoelastic and sensory properties, the composition of the surfactant, and the presence of thickeners. A. Ali *et al.* utilized a human panel to analyze the tactile and sensory qualities of topical lotions. They compared starch and surfactant-based Pickering cream and the results showed that the starch-based formulation ranked lower for greasy, tacky, and squishy properties.[41]

Lignin, an aromatic component of plants rich in carbon content and having UV chromophoric properties, has been used as a photo-protective agent through the incorporation of submicrometric particles into cosmetological products. The results revealed that adding submicrometric lignin particles to a lotion would result in excellent anti-oxidant properties and UV protection properties, providing an opportunity for the usage of by-products in the lignocellulose biorefinery industry.[42] Dermatocosmetic products are used to remove cosmetic flaws and cause no interference in the skin rejuvenation process. By speeding up keratinization and exfoliation, abrasive ingredients can be found in many cleansing products to improve skin conditions. Kozlowska and co-workers developed sodium alginate microparticles and compared them with their combination with starch using peeling technology. The spherical, uniform shape of the produced microparticles reduces the possibility of skin irritation during peel application, thus creating possibilities for new applications in the cosmetics sector.[43]

Melanogenesis, a sequence of enzymatic chemical reactions, involves the manufacture of melanin, a pigment responsible for the pigmentation to the skin, eyes, and hair. The production of melanin is essential for enhancing the body's defense against the damaging effects of ultraviolet radiation, particularly in the genetic material of melanocytes. Its production in excessive amounts might result in dark patches and skin darkening. Researchers have examined natural sources of melanogenic inhibitors, such as *Leathesia difformis*, *Morus alba*, *Euphorbia supina*, *Syzygium polyanthum*, *Orostachys japonicus*, *Libidibia ferrea*, *Heracleum moellendorffii*, *Pouteria torta*, *Coix lacryma-jobi-adlay*, *Inula britannica*, sappanone A, cinnamaldehyde, *Artemisia argyi*, *Vitex negundo*, and *Gaillardia aristata*. The phenolic content present in these sources was reported to be anti-oxidant and anti-melanogenic.[44]

### 1.5.2 Hair Care

The addition of small concentrations, such as 0.1% for normal hair or 0.5% for damaged hair, of cationic polymers to shampoos can cause an increase in their efficacy and also improves the smoothening and creaminess properties of the hair. In a study reported in 2000, Sander and co-workers reported that guar gum cannot be used as a cationic polymer as it cannot be desorbed, which results in over-conditioning of the hair and causes build-up.[45]

Hair oil contains 70–80% of base oil, *e.g.* rice bran oil, light liquid paraffin oil or refined sesame oil, and 15–25% of additives, *e.g.* herb extracts, menthol, camphor, butylated hydroxytoluene, cinnamon, or vetiver oil. Hair oils are expected to provide strength and nourishment to the hair by reducing hair loss and inducing its rapid growth. The extent of penetration of hair is measured by laser scanning confocal microscopy (LSCM), a technique used to observe 3-D samples in their natural environment using fluorescent Nile red dye without destroying the sample integrity. In a study reported by Dandekar *et al.*, the penetration of mineral oil was faster than that of vegetable oil.[46]

Usually, polymers are used in hair care products as they make the consistency of hair thick and provide protection. Nevertheless, these materials are non-biodegradable and have some disadvantages. Polymers of a synthetic nature gather on the scalp and cause an exasperating coat through which any advantageous substances cannot permeate. Prolonged use of these materials causes permanent damage and consequently results in breakdown of the hair. Collagen extracted from fish waste was found to show positive action on hair according to the study conducted by Igielska-Kalwat and co-workers with the removal of extravagant sebum production and the maintenance of proper pH of the hair.[47]

Encapsulated nanotubes have been used for the treatment of hair and formulation of dyes, where the encapsulation created through the nanotubes provides a prolonged drug effect. The prominent penetration of drugs in the cuticle pores delivers the drug directly onto the surface of hair. Nanotube technology has shown good results on both human and other mammalian hair. The use of biodegradable and natural substances has resulted in better therapeutic and cosmetic formulations for human as well as veterinary use.[48]

The long-term accumulation of excessive arsenic in body tissues may lead to chronic arsenic poisoning. Tube wells are the leading sources of this arsenic rich water. The use of spirulina extract with a combination of anti-oxidant zinc produced notable positive results for the treatment of chronic arsenic poisoning. Randomized control studies conducted in Bangladesh evidenced that the use of a spirulina extract and zinc combination effectively reduced the concentration of arsenic from hair scalp and skin tissues by increasing its excretion through urine. By using this combination, symptoms of hyperpigmentation and skin lesions were significantly improved.[49]

In the hair-care sector, baldness and early greying of hair are the two main challenges that individuals endure. Drug repositioning is a drug discovery technique in which a current medication is used as a treatment for a different illness. Many drugs, like minoxidil, discovered to treat hypertension were found to provoke hair growth due to their vasodilation activity. Similarly, drugs like tofacitinib and ruxolitinib also induce hair growth. Bimatoprost is another drug that restores hair growth and melatonin is a hormone used for the treatment of patterned hair loss. Another problem faced by people is early greying of hair. Latanoprost, which was earlier used for the treatment of glaucoma, was found to stimulate hair pigmentation. Also, high doses of *p*-amino benzoic acid cause darkening of the hair.[50]

Anthocyanins are hydrophilic compounds present in fruits, vegetables and grains, which impart red color under acidic conditions and purplish-blue under alkaline conditions. Their anti-oxidizing and anti-inflammatory properties indicate their use in the pharmaceutical and cosmetics industries. Anthocyanin-based compounds are used in hair products as they show positive effects on the melanin content of hair. For instance, organic dyes excavated from natural origins, like blackcurrant fruit skin, impart intense blue color to the hair when paired with anionic base quinonoid in acidic conditions. The use of natural dyes serves as a replacement and mitigates the negative effects brought on by the use of synthetic dyes.[51]

Surfactants are used as emulsifying and cleansing agents in the cosmetics industry. Glycolipids are one of the natural surfactants that can be used in hair cosmetics. Four different rhamnolipids and an alkyl polyglucoside are promising glycolipid surfactants. Based on the study conducted by Laura Fernández-Peña and co-workers, when the beneficial effects of a combination of synthetic surfactants and polymers were compared with natural glycolipids and proteins, it was found that surfactants with short hydrocarbon chains show good efficacy.[52]

### 1.5.3 Toiletries

Natural conservatives or preservatives having anti-bacterial activity have been widely used as compared to synthetic ones owing to their safety and non-toxicity. Studies conducted to prove that the extract of *Rubus rosaefolius* Smith is a better organic conservative primarily included compositions in the form of an emulsion or gel.[53] A number of bacteria and fungi were tested by preserving them with the mentioned preservative in either emulsion or gel form and it has shown satisfactory efficacy as an organic conservative to be applied on the skin surface against Gram positive and negative bacteria, rather than fungi.

According to the European pharmacopoeia, essential oils are defined as the formulated compounds of refined organic components derived from scented plants through various methods. Volatile oils and their particular ingredients are extremely aqua-phobic and act as either the main component or as a conservative in different cosmeceuticals and toiletries. Some examples of volatile oils are *Apium graveolens*, *Salvia officinalis*, *Artemisia judaica*, *etc*. The components of such volatile oils play a major part in the destruction of bacteria and fungi, and thus have acquired a wide place in the manufacturing of cosmeceuticals; however, they do not always have a definite impact, as they do have side effects like hypersensitive reactions. The important ingredients of volatile oils are isoprenoids, cinnamic acids and terpentines. Volatile oils carrying *Calendula officinalis* and *Osimum basilicum* are exceedingly used to prepare sun-tan lotions as they soak up most of the ultraviolet light, avoid dermatoheliosis, and protect from tanning, sun creasing and other skin problems. However, applying them for a long duration might result in allergies and skin infections, and could even be noxious to the liver. The volatile oils having high incidences of these side effects are peppermint oil, tea tree oil, eucalyptus oil, and lavender oil. Thus, additional examinations are required to get a better outcome of these products.[54]

The crystalline dextrins, such as cyclic glycans, 'cyclodextrins' and their derivates, are used as a medium to intensify the solvability of weakly soluble substances, increase the penetration of drugs through the mucosal/skin membrane and upgrade the bio accumulation of drugs. They are often used in nutritional agents and toiletries. Because of their aquaphilic nature and high molecular weight, they do not permeate through the oleophilic membranes and hence make the membrane flawless. The most used organic Schardinger dextrins are alpha, beta and gamma cyclodextrins that are ring shaped, having an aquaphilic exterior region and being oleophilic in the interior. Their physical and chemical characteristics are the same as those of their straight chain by-products, but because of their cyclo-structure they are only a bit affected by the enzymes. Also, they are far better building and dispersive agents. They elevate the consumption of drugs coming under bio-pharmaceutics classification system (BCS) class II that are weakly dissolved in water but have a good penetrating ability.[55]

### 1.5.4 Nail Care

The nail plate is a tough, extended and bent framework used to protect the phalanx bones at the tips of the fingers and toes that are more prone to fractures as they are fragile and small. On the formation of newer nails, lanulae force the already grown nails to the higher and exterior side of the skin and on maturation of these new nail cells, they appear transparent. Various forms of nanotechnology are now being widely used for nails: (i) nanoencapsulation strategies including nanoliposomes, vesicles that conduct and circulate fluids having a double layer of lipid on the surface and formed by the discharge of phospholipids, generally regarded as safe (GRAS) products to carry necessary ingredients into the nail foldings; (ii) niosomes, unionized surface active agents being better carriers for weakly consumed ingredients, but not considered under GRAS due to dermatotoxicity; (iii) solid lipid nanoparticles, which protect the ingredients from disintegration for a longer period of time; and (iv) nanostructured lipid carriers, which are more spacious and allow the collection of extra ingredients over the nail folding.[56]

Anthocyanins, derived from the Greek word 'anthos' meaning flower and 'kyneous' meaning dark blue tint, represent a major group of chemical substances obtained from plants, fruits and vegetables that are biologically active but not nutritive. They are particular dyes that have a phenolic nature and are soluble in water. They are able to provide a massive collection of various colors from red to purplish blue and hence play the part of native colorants having demand in the food, pharmaceutical and cosmeceutical industries. The cosmetic formulations possessing anthocyanins are efficient for the extraction of melanin and reduce the formation of reactive oxygen species of skin cells that produce keratin and fibroblasts that develop connective tissues. Many cosmeceutical companies such as Nivea®, Oriflame and L'Oreal follow the strategy of using derivatives of anthocyanins in their products.[51]

### 1.5.5 Lip Care

The formulation of lip care products primarily involves different oils and waxes for the prophylaxis and treatment of conditions such as dryness and roughness of the lips. The main reason for the lips getting dry and rough involves the shortened moisture holding capacity of the stratum corneum and feeble barrier function. Thus, to assess the degree of moisturization, lip skin penetration, lip hydration, dryness of the lips, stratum corneum barrier function, *etc.*, a tri-method approach of three different lipids, such as berry fruit wax, coconut oil and olive oil, was carried out using the technique of confocal Raman spectroscopy (CRS), corneometry (skin capacitance) and trans-epidermal water loss (TEWL). The berry fruit wax, coconut oil and olive oil are thus considered as the active demulcents in this formulation. The proportion of olive oil was found to be 5–10 µm deep through the lip skin surface for 6 h after the application of the product, which on comparison with coconut oil and berry fruit wax was 5–10 µm for 2 h after product application. TEWL was curtailed and skin hydration was boosted for 2 h after product application. An improvement was seen during the ocular interpretation of soreness, roughness and tearing of the lip skin surface for 2 h after product application. From the above parameters, a definite conclusion was made that olive oil had the maximum efficacy, since its effect was observed after up to 6 h of product application. Hence, the tri-method study is authentic and compatible for the administration of all the three lipids in the stratum corneum of the lips.[57]

Nowadays, an appeal for elements of natural origin and organic ingredients has been constantly surging. So, there has been extreme demand for the use of naturally acquired dispersing agents in the formulation of the cosmetics dealing with lip care, *i.e.* lipsticks. Recently, castor oil, octyldodecanol and meadowfoam seed oil have been studied as dispersing agents for lipsticks made up of Red 7 Lake powder and it was revealed that castor oil had the highest viscidity and retention rate among the others. Meanwhile, the lipstick containing meadowfoam seed oil was found to have moderate solid nature and gentle creaminess, suggesting its suitability as a dispersing agent.[58]

The properties of lipsticks, such as their composition, consistency, rigidness and melting point, can be modified by changing the amount and quality of elements used in the formulation. An experiment was carried out by D-optimal mixture design, which is one of the most efficient and impressive means of resolving the problem of escalation associated with lipsticks. During the study, this method was used to escalate the liquefaction point of organic lipstick considering components like pitaya seed oil, virgin coconut oil, beeswax, candelilla wax and carnauba wax. D-optimal mixture design is generally used to analyze the number of individual variables together with a focus on the physical properties of the formulation. On performing the D-optimal analysis, the proportion of pitaya seed oil and virgin coconut oil had a reduced effect on the liquefaction point, whereas increasing the proportion of beeswax boosted the liquefaction point. An appropriate lipstick formulation holds a liquefaction point in the range of 40–56 °C, and on experimentation it was detected to be 46 °C, which matched the theoretical value of 46.5 °C.[59]

### 1.5.6 Perfumes

In general, perspiration is released in the armpit and it is without any type of smell. The usual underarm smell is released from non-aromatic particles that are present in the exocrine glands and ooze out due to active Corynebacteria (mainly) and *Staphylococcus* species. Studies by Schmid and co-workers involved the segregation of the axillary malodor releasing enzyme (AMRE) and its prior structure 3-methyl, 2-hexanoic acid, both of which help to activate the smelling parameters from the discharge of non-aromatic armpits.[60] These segregates were developed and cultivated under suitable conditions and were then altered with the salt of carbamic acid and further split by zinc-dependent aminocyclase into scented ethanol along with glutamine that attenuates the passing out of the foul smell and additionally covers it with a floral one. Thus, this study has two purposes: (i) assessment of the process that controls the foul smell of cosmetic products, by segregation of corynebacteria; and (ii) alteration in the formation of prior foulness with aromatic particles.

## 1.6 Limitations of Natural Cosmeceuticals

Natural products aim to avoid any potentially harmful ingredients in products applied to the skin, but some are toxic and allergenic in nature. For instance, terpenes that are used predominantly in fragrances may cause liver toxicity in humans.[61] Limonene and linalool, used as cleaning agents, detergents and deodorants, cause allergies and irritation in their oxidized state.[62] Tea tree oils used in the preparation of cosmetic products cause dermatitis on the eyelids, hands and face when using soap, cream, shampoo, body lotion, and shaving cream containing tea tree oil.[63] Cinnamic acid used in perfuming, masking agents and skin conditioning causes allergies when used in excess amounts.[64] Caffeine is used in the preparation of various creams and lotions. A concentration of caffeine above 15 mg $L^{-1}$ causes neurodegenerative disorders, irritability and nervousness, and this can even lead to antagonism of aminobutyric acid, which can result in seizures.[65] Berberine, a component in *Coptis japonica* used as an anti-oxidant and anti-wrinkle agent, causes skin conditions like swelling, sagging, *etc.*[66] Kaempferol, used in skin care products having anti-collagenase and anti-tyrosinase activities, shows mutagenic as well as carcinogenic effects because of the conversion of Kaempferol to quercetin.[67] Propolis is a bee product that possesses anti-oxidant, antimicrobial and regenerative properties, but propolis ingestion may result in allergies, stomatitis, labial edema, and oral pain.[68]

## 1.7 Conclusion

Cosmeceuticals are biologically active components that exhibit cosmetic as well as pharmaceutical benefits. Natural components exhibit various cosmeceutical properties, such as anti-oxidant, anti-aging, anti-wrinkle, anti-acne, skin lightening, hair smoothening, emulsifying, surfactant *etc.*, and are tremendously increasing in market demand. In this chapter, the importance of natural products for the production of natural cosmeceuticals has been discussed. Although natural products for cosmetic uses have been widely explored, there is still a lot to explore in nature in order to obtain more components that may be formulated in cosmetic formulations. During their consideration, allergic responses should be studied in detail to prevent allergic responses. Amendments to the regulations are needed in order to fulfil the costumer's demand for safe and effective cosmeceuticals.

## References


1. C. Barros and R. B. G. Barros, *J. Cosmetol. Trichology*, 2020, **6**, 1.
2. M. V. Setshego, A. O. Aremu, O. Mooki and W. Otang-Mbeng, *BMC Complementary Med. Ther.*, 2020, **20**, 81.
3. L. S. Joshi and H. A. Pawar, *Nat. Prod. Chem. Res.*, 2015, **3**, 170.
4. M. Aćimović, V. Tešević, K. Smiljanić, M. Cvetković, J. Stanković, B. Kiprovski and V. Sikora, *Adv. Technol.*, 2020, **9**, 54.
5. A. Pandey, G. K. Jatana and S. Sonthalia, *Cosmeceuticals*, StatPearls, 2022.
6. A. Abd Aziz, Z. M. Taher, R. Muda and R. Aziz, *Recent Trends Malaysia Med. Plants Res.*, 2017, p. 126.
7. https://www.byrdie.com/natural-vs-organic-beauty-products.
8. B. K. Nanjwade, *World J. Pharm. Pharm. Sci.*, 2017, 643–691.
9. D. A. Scott, *Stud. Conserv.*, 2016, **61**, 185.
10. B. Han, J. Chong, Z. Sun, X. Jiang, Q. Xiao, J. Zech, P. Roberts, H. Rao and Y. Yang, *Archaeometry*, 2021, **63**, 1042.
11. K. Olson, *Classical World*, 2009, **102**, 291.
12. K. Patkar, *Indian J. Plast. Surg.*, 2008, **41**, 134.
13. https://www.cosmeticsinfo.org/get-the-facts/a-history-of-cosmetics-from-ancient-times/.
14. J. Drobnik and K. Wełna, *Pol. J. Cosmetol.*, 2017, **20**, 349.
15. I. Kusumawati and G. Indrayanto, *Stud. Nat. Prod. Chem.*, 2013, **40**, 485.
16. P. Ciganović, K. Jakimiuk, M. Tomczyk and M. Z. Končić, *Antioxidants*, 2019, **8**, 445.
17. J. K. Liu, *Nat. Prod. Bioprospect.*, 2022, **12**, 40.
18. D. J. McClements and C. E. Gumus, *Adv. Colloid Interface Sci.*, 2016, **234**, 3.
19. N. Packianathan and R. Kandasamy, *Funct. Plant Sci. Biotechnol.*, 2011, **5**, 94.
20. S. Pengon, N. Chinatangkul, C. Limmatvapirat and S. Limmatvapirat, *Asian J. Pharm. Sci.*, 2018, **13**, 409.
21. https://www.statista.com/forecasts/1264932/worldwide-revenue-natural-organic-cosmetics-market.
22. A. F. Stallings and M. P. Lupo, *J. Clin. Aesthet. Dermatol.*, 2009, **2**, 36.
23. M. Kanlayavattanakul, C. Rodchuea and N. Lourith, *Int. J. Cosmet. Sci.*, 2012, **34**, 280.

24. S. Baptista, J. R. Pereira, C. V. Gil, C. A. V. Torres, M. A. M. Reis and F. Freitas, *Polymers*, 2022, **14**, 2349.
25. H. Kichou, Y. Dancik, C. Eklouh-Molinier, N. Huang, M. Soucé, L. Gressin, G. Gillet, I. Chourpa, E. Munnier and F. Bonnier, *Int. J. Cosmet. Sci.*, 2022, **44**, 262.
26. D. Terescenco, L. H. Benali, F. Canivet, M. B. le Gelebart, N. Hucher, E. Gore and C. Picard, *Int. J. Cosmet. Sci.*, 2021, **43**, 573.
27. M. Ćorović, A. Milivojević, M. Simović, K. Banjanac, R. Pjanović and D. Bezbradica, *Sustainable Chem. Pharm.*, 2020, **15**, 100231.
28. H. B. Nejad, L. Blasco, B. Moran, J. Cebrian, J. Woodger, E. Gonzalez, C. Pritts and J. Milligan, *Int. J. Cosmet. Sci.*, 2020, **42**, 237.
29. T. Pavicic, U. Wollenweber, M. Farwick and H. C. Korting, *Int. J. Cosmet. Sci.*, 2007, **29**, 181.
30. A. C. Santos, D. Rodrigues, J. A. D. Sequeira, I. Pereira, A. Simões, D. Costa, D. Peixoto, G. Costa and F. Veiga, *Int. J. Pharm.*, 2019, **572**, 118787.
31. C. Souza, L. A. P. de Freitas and P. M. B. G. M. Campos, *AAPS PharmSciTech*, 2017, **18**, 2505.
32. O. Taofiq, A. M. González-Paramás, M. F. Barreiro and I. C. F. R. Ferreira, *Molecules*, 2017, **22**, 281.
33. I. Ahmad, S. Irfan, A. A. Dera, G. S. Zaman, H. C. Chandramoorthy, M. A. Mir and P. Rajagopalan, *Biologia*, 2020, **75**, 1739.
34. S. M. Savic, N. D. Cekic, S. R. Savic, T. M. Ilic and S. D. Savic, *Int. J. Cosmet. Sci.*, 2021, **43**, 530.
35. E. F. Costa, W. V. Magalhães and L. C. Di Stasi, *Molecules*, 2022, **27**, 7518.
36. C. R. Afonso, R. S. Hirano, A. L. Gaspar, E. G. L. Chagas, R. A. Carvalho, F. V. Silva, G. R. Leonardi, P. S. Lopes, C. F. Silva and C. M. P. Yoshida, *Int. J. Biol. Macromol.*, 2019, **132**, 1262.
37. A. Muxika, A. Etxabide, J. Uranga, P. Guerrero and K. De La Caba, *Int. J. Biol. Macromol.*, 2017, **105**, 1358.
38. S. Escobar, A. Valois, M. Nielsen, B. Closs and D. Kerob, *Int. J. Cosmet. Sci.*, 2021, **43**, 131.
39. R. Raychaudhuri, S. Naik, A. B. Shreya, N. Kandpal, A. Pandey, G. Kalthur and S. Mutalik, *Int. J. Biol. Macromol.*, 2020, **161**, 1189.
40. A. Alves, E. Sousa, A. Kijjoa and M. Pinto, *Molecules*, 2020, **25**, 2536.
41. A. Ali, L. Skedung, S. Burleigh, E. Lavant, L. Ringstad, C. D. Anderson, M. Wahlgren and J. Engblom, *Int. J. Pharm.*, 2022, **613**, 121370.
42. S.-X. Li, M.-F. Li, J. Bian, X.-F. Wu, F. Peng and M.-G. Ma, *Int. J. Biol. Macromol.*, 2019, **132**, 836.
43. J. Kozlowska, W. Prus and N. Stachowiak, *Int. J. Biol. Macromol.*, 2019, **129**, 952.
44. C. F. G. Neto, P. do Nascimento, V. C. da Silveira, A. B. N. de Mattos and C. D. Bertol, *Int. J. Cosmet. Sci.*, 2022, **44**, 143.
45. P. Hössel, R. Dieing, R. Nörenberg, A. Pfau and R. Sander, *Int. J. Cosmet. Sci.*, 2000, **22**, 1.
46. A. Srivastav, P. Dandekar and R. Jain, *J. Cosmet., Dermatol.*, 2019, **18**, 1947.

47. J. Igielska-Kalwat, E. Kilian-Pięta and S. Połoczańska-Godek, *Polymers*, 2022, **14**, 749.
48. A. C. Santos, A. Panchal, N. Rahman, M. Pereira-Silva, I. Pereira, F. Veiga and Y. Lvov, *Nanomaterials*, 2019, **9**, 903.
49. M. Misbahuddin, *Clin. Toxicol.*, 2006, **44**, 135.
50. G. Sotiropoulou, E. Zingkou and G. Pampalakis, *Exp. Dermatol.*, 2021, **30**, 628.
51. J. S. Câmara, M. Locatelli, J. A. M. Pereira, H. Oliveira, M. Arlorio, I. Fernandes, R. Perestrelo, V. Freitas and M. Bordiga, *Nutrients*, 2022, **14**, 5133.
52. L. Fernández-Peña, E. Guzmán, F. Leonforte, A. Serrano-Pueyo, K. Regulski, L. Tournier-Couturier, F. Ortega, R. G. Rubio and G. S. Luengo, *Colloids Surf., B*, 2020, **185**, 110578.
53. E. A. Ostrosky, E. M. C. Marcondes, S. de O. Nishikawa, P. S. Lopes, G. H. C. Varca, T. de J. A. Pinto, T. V. O. Consiglieri, A. R. Baby, M. V. R. Velasco and T. M. Kaneko, *AAPS PharmSciTech*, 2011, **12**, 732.
54. E. Guzmán and A. Lucia, *Cosmetics*, 2021, **8**, 114.
55. P. Jansook, N. Ogawa and T. Loftsson, *Int. J. Pharm.*, 2018, **535**, 272.
56. D. Schoon, Nanotechnology in Nail Care, in *Nanotechnology in Dermatology*, ed. A. Nasir, A. Friedman and S. Wang, Springer, 2012, pp. 19–30.
57. S. Bielfeldt, J. Blaak, S. Laing, M. Schleißinger, C. Theiss, K. Wilhelm and P. Staib, *Int. J. Cosmet. Sci.*, 2019, **41**, 281.
58. B. Maktabi, M. W. Liberatore and G. Baki, *Int. J. Cosmet. Sci.*, 2021, **43**, 484.
59. N. Kamairudin, S. S. A. Gani, H. R. F. Masoumi and P. Hashim, *Molecules*, 2014, **19**, 16672.
60. A. Natsch, H. Gfeller, P. Gygax and J. Schmid, *Int. J. Cosmet. Sci.*, 2005, **27**, 115.
61. T. Zárybnický, I. Boušová, M. Ambrož and L. Skálová, *Arch. Toxicol.*, 2018, **92**, 1.
62. H. Audrain, C. Kenward, C. R. Lovell, C. Green, A. D. Ormerod, J. Sansom, M. M. U. Chowdhury, S. M. Cooper, G. A. Johnston, M. Wilkinson, C. King, N. Stone, H. L. Horne, C. R. Holden, S. Wakelin and D. A. Buckley, *Br. J. Dermatol.*, 2014, **171**, 292.
63. A. C. de Groot and E. Schmidt, *Contact Dermatitis*, 2016, **75**, 129.
64. A. Gunia-Krzyżak, K. Słoczyńska, J. Popiół, P. Koczurkiewicz, H. Marona and E. Pękala, *Int. J. Cosmet. Sci.*, 2018, **40**, 356.
65. M. Kolahdouzan and M. J. Hamadeh, *CNS Neurosci. Ther.*, 2017, **23**, 272.
66. A. F. G. Cicero and A. Baggioni, *Adv. Exp. Med. Biol.*, 2016, **928**, 27.
67. D. Kashyap, A. Sharma, H. S. Tuli, K. Sak, S. Punia and T. K. Mukherjee, *J. Funct. Foods*, 2017, **30**, 203.
68. V. D. Wagh, *Adv. Pharmacol. Sci.*, 2013, **2013**, 11.

CHAPTER 2

# Natural Products as Cosmeceuticals in Different Countries: A Regulatory Perspective

AVANTIKA AGRAWAL[a] AND SUMIT S. CHOURASIYA[b*]

[a]Natura-Inland, Nature Cure Research Centre, Bilaspur, C.G, 495001, India;
[b]IOL Chemicals and Pharmaceuticals Ltd, Barnala, Punjab, 148101, India
*E-mail: sumitchourasiya@iolcp.com

## 2.1 Introduction

The regulatory guidelines of any product designed/formulated for human use are intended to assure the quality of the product for its safe and effective usage. In order to monitor and inspect the marketed cosmetic products, cosmetic regulations are made to cover similar standards, safety and potency, along with certain major legislative wants, like nomenclature, labeling, *etc.* However, the safety requirements for any marketed beautifying products across the globe are influenced by the differences in guidelines charted by their regulatory authorities.

Cosmeceuticals, the advanced form of cosmetics, have certainly captivated the personal hygiene industries around the world. Cosmetics, known to be derived from the Greek "kosmtikos", are products or preparations intended for external application mainly to decorate/beautify, to clean, to improve odor or to protect and keep in a good looking state.[1] For a long time in different

Bioprospecting of Natural Sources for Cosmeceuticals
Edited by Deepika Kathuria, Ajay Sharma, Meenakshi Verma and Gulzar Ahmad Nayik

Published by the Royal Society of Chemistry, www.rsc.org

cultures, many of these products have been claimed to contain extracts from natural sources. Usually, natural products are secondary metabolites without any known direct role in growth or development, yet many of the marketed products have ingredients of natural origin. This is clear from the fact that the roots of the application/use of cosmetics date back to the time of the Romans, Greeks and Egyptians, where people in those days used to paint their body with clay or arsenic (natural products), or make tattoos of soot and bone charcoal (which are the base of modern-day tattoo products). At the present time, chemical ingredients extracted from natural sources like amino acids, flavones, tannins, vitamins, *etc.* are used in cosmetic products and are also known to effect the biological processes in the dermis and participate in certain skin ailments.[2–5]

Similar findings brought into light "cosmeceuticals", a new breed of cosmetics intended to provide effects further than just beautification but still staying behind the qualities of being a drug. Though there is confusion as to the exact definition and scope of this term, still around 40% of the world's dermatologist prescriptions include cosmeceuticals. Thus, cosmeceuticals can be any cosmetic with therapeutic ability to positively affect the site of application in addition to beautification. The term cosmeceutical has been immersed into dermatologist prescriptions, literature, academic symposia, lectures and discussions across the globe, yet it still remains formally un-recognized in the US-FDA and EMA, possibly due to differences in the definitions stated by the Federal Food, Drug, and Cosmetic Act (FD&C Act) for 'cosmetics' and 'drugs'.[6–8]

As stated earlier, different countries have different regulatory requirements. For example, in the U.S. there are stringent guidelines for drug products, whereas cosmeceuticals/cosmetics are not required to meet stringent safety and efficacy norms for marketed and registered over-the-counter (OTC) medicines.[9] In Japan, certain skin care products are described as 'quasi-drugs' (instead of being called cosmeceuticals) when they are a mix of cosmetics and drugs.[10,11] The Japanese authority allows cosmetics with mild and safe active pharmaceutical agents. Japan and Europe classify cosmeceuticals as sub-cosmetics, while in the US they are kept under the subclass of drugs.[12] It has also been stated that the same active ingredient can be a cosmetic in one proportion, while a cosmeceutical constituent at another concentration. For example, at less than 2% concentration, kojic acid is an ingredient in cosmetic creams, while when added at 2 to 5% concentration the formulation becomes cosmeceutical. The same is true for salicylic acid, ascorbic acid, co-enzyme Q10 and many others.[13]

Preferably, unlike for drugs, for cosmeceuticals with natural products, the registering procedure should not be as stringent. However, as per Good Clinical Practice (GCP), it is essential to perform sufficient clinical studies to check the desired activity for the treatment of any minor skin ailments and also to assure optimal safety without any known side effects.[14] For example, there was a well-known legal argument related to the labeling of minoxidil, prescribed dermatologically for treating male pattern baldness, which although available OTC is considered a drug as it possesses pharmacological activity.[15] Cosmeceuticals have no doubt had a significant impact on global GDP, thereby improving the community relations of humans nationwide. However, there is a big catch with the increasing desire among people globally to look younger and healthier.

According to the Strategic Market Research survey (shown in Figure 2.1), the natural product-based cosmeceuticals market is expected to see a vigorous increment of compound annual growth rate (CAGR) of 8.6% with the value forecasted increasing from \$45.56 billion in 2021 to \$95.75 billion by 2030.[16] Also, for the Indian market, Indian Cosmetic Sector Analysis suggested a steep growth from 24.53 billion USD in 2022 to 33.33 billion USD in 2027 at a CGAR of 6.3% over a period of five years.[17] In the coming five years, the Asia-Pacific area is anticipated to have the fastest growth of natural and organic cosmetics, according to a recent analysis from Mordor Intelligence.[18] The growing demand for organic-based, safer and effective cosmetic products has motivated manufacturers to innovate and produce new safe and effective products containing natural active ingredients. Also, the increased awareness related to the risk from synthetic chemicals over plant or other natural resource-derived products among the end users through reviews from social media and the internet has added to this.

As it represents an exceedingly innovative, expeditious and complicated sector, cosmeceutical production ought to be regulated. This will not only ensure the safety and quality of beauty products, but ward off detrimental influences to the users' health.[12] However, the regulatory frameworks are some distance from being harmonized as they vary among markets/international locations, which imposes a significant challenge for global enterprises to promote identical products in all markets. This lack of

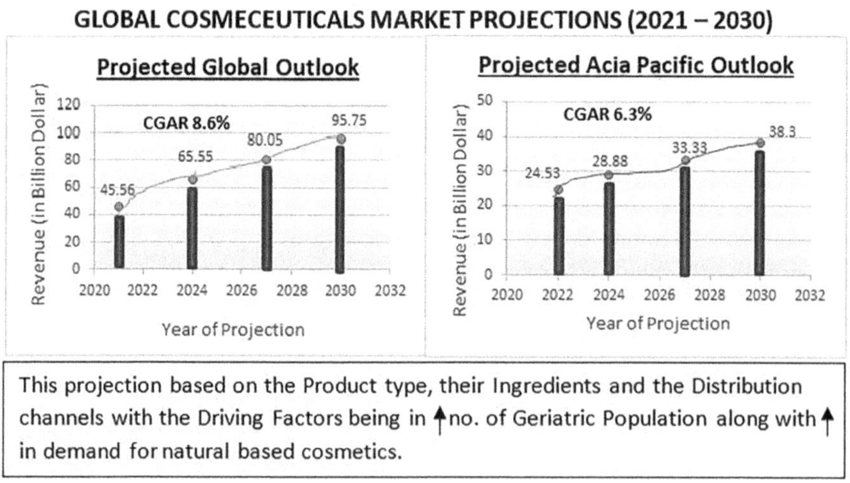


**Figure 2.1** Market research survey trends for 2021 to 2030.

harmonization not only hinders the regulatory bodies in ascertaining the standards of product compliance, but such disparities can also have an impact on international trade. In order to foster innovation, boost the market, and remove trade obstacles, it is crucial to discover solutions that can result in the harmonization of the regulatory frameworks for cosmetics. Various global groups are working together to attain this goal. Mentioning a few, the International Cooperation on Cosmetics Regulation (ICCR),[19] the Organization for Economic Co-operation and Development (OECD)[20] and the International Organization for Standardization (ISO)[21] include regulatory authorities from various countries that meet annually to discuss the safety, efficacy, testing guidelines *etc.* for cosmeceutical innovations. This book chapter provides an overview of the laws governing cosmetic items in different nations from the USA to the People's Republic of China, the European Union, Canada, Japan, and India, which constitute some of the world's major markets.

## 2.2 Regulations in Different Countries

From a regulatory perspective, the word "cosmeceuticals" is un-recognized and such items are seen as cosmetics both within Europe and the United States. As a result, in Europe, the European Cosmetic Products Regulation (EC) No. 1223/2009 must be complied with before a cosmetic product is sold in these markets.[22] Like cosmetics, cosmeceuticals which are also applied on the body surface are basically a cross between cosmetics and pharmaceuticals (a kind of hybrid). They frequently contain many natural and organic ingredients, like various oils, botanical extracts, herbs, antioxidants, peptides, organic anti-inflammatories, pigment-lightening agents *etc.* in addition to vitamins. A product enters the cosmeceutical category if its stated purpose is to "treat or avoid illness or otherwise influence the structure of the cutis" yet it is promoted/branded as a cosmetic. Figure 2.2 briefly outlines the steps in the design and production of any cosmeceutical.

Some examples of cosmeceuticals are given below:

- An eye cream with unique nature-based components that may rejuvenate cells or increase collagen synthesis.
- Essential oil products for sleep improvement.
- A shampoo that fights dandruff.
- Skin hydrating gels that also ease joint or muscular discomfort.

### 2.2.1 United States of America

Cosmetic products containing active ingredients that do fall under both the cosmetics and drug categories, and thus assert therapeutic benefits, need to receive New Drug Application (NDA) clearance or adhere to the proper monograph for an over-the-counter (OTC) drug in the US. Additionally, there are

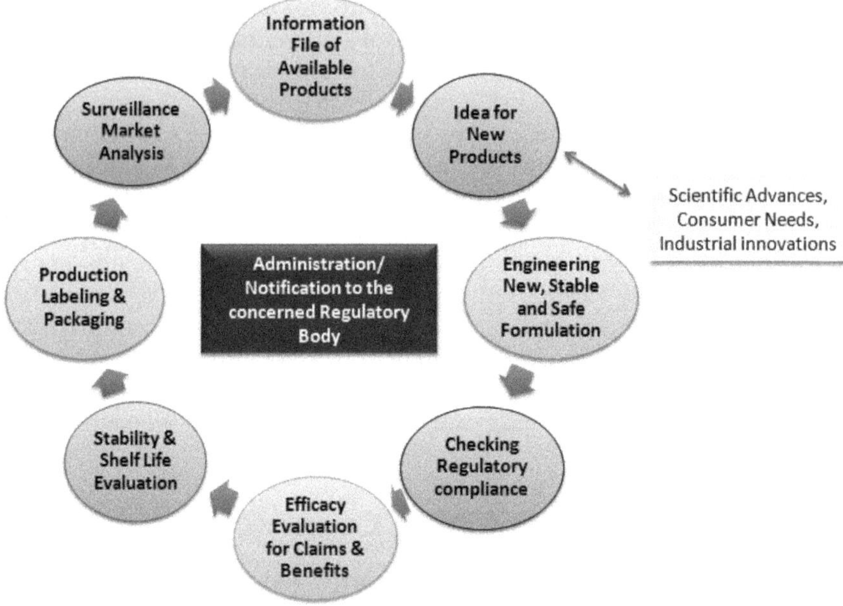


**Figure 2.2** Stages in the production of any cosmeceutical formulation from innovation to marketing.[8,23]

defined Good Manufacturing Practice (GMP) regulations as per the FDA for cosmetics.

#### *2.2.1.1 Food, Drug, and Cosmetic Act (FD&C Act)*

As discussed earlier, although the term "cosmeceutical" is not acknowledged by the Federal Food, Drug, and Cosmetic Act (FD&C Act), the cosmetic industry uses it to describe cosmetic products that offer added therapeutic or drug-like advantages.[24] Unlike pharmaceuticals, other than color additives, the law does not mandate any FDA clearance for cosmetic goods and ingredients prior to being put on the market, although there are rules and regulations applied to the cosmetics sold in interstate commerce.[25] The Federal Food, Drug, and Cosmetic Act (FD&C) and the Fair Packaging and Labeling Act (FPLA) are the two key regulations governing cosmetics sold in the US. These statutes provide the FDA with the power to control cosmetics. These two statutes came into existence in 1938 and 1966, and since then they have stayed principally unchanged except for a few amendments.[26] Congress is responsible for passing federal legislation in the United States. Congress permits some government firms, like the FDA, to develop rules in order to ensure that the laws are implemented on a daily basis. Congress must amend the statute before the FDA's legal power over cosmetics is changed.

According to section 201(i) of the FD&C Act, cosmetics are defined as "articles intended to be rubbed, poured, sprinkled, or sprayed on, introduced into, or otherwise applied to the human body for washing, beautifying, enhancing attractiveness, or modifying the look." Skin moisturizers, fragrances, lipsticks, nail polishes, eye and face cosmetics, cleaning shampoos, permanent waves, hair colors, and deodorants are among the items included in this description. It also includes any chemical created to be used as a cosmetic product component.

However, the article may be a drug (as defined by the FD&C Act, 201(g)) or, in some cases, a medical device (as defined by the FD&C Act, 201(h)), if it is planned to have any therapeutic purpose, like the treatment or prevention of any disease, or even to affect the structure or function or appearance of the body. Other "personal care items" can be subject to regulation as consumer goods or dietary supplements.[26] Therefore, the Regulatory laws for both the articles differ in their process of registration, approval, labeling, and GMP, with drug-like products needing approval from the FDA's Center for Drug Evaluation and Research (CDER) and given a Drug Identification Number (DIN) before reaching the market.

In the case of cosmeceuticals, when the product has two planned/intended purposes and falls under both the cosmetics and drug classifications, it must adhere to both drug and cosmetic regulations. A shampoo, for instance, is a cosmetic since its main purpose is to wash and clean the hair. But if it cures dandruff also as its main purpose in an antidandruff therapy, then it is a medicine as well. As a result, an antidandruff shampoo is both a medication and a cosmetic. Likewise, some deodorants that also act as antiperspirants, fluoride-containing toothpastes, sun protection cosmetics, moisturizers *etc.* are a few more examples of cosmetic/drug combinations. The same holds true for "essential oils" as well. Let's say a scent/perfume is a cosmetic that promotes attractiveness. However, when it is advertised with any assertions for "aromatherapy", like being a sleep enhancer or happy digestion *etc.*, then based on its intended purpose, it would fall under the definition of a medicine. Likewise, when a massage oil is meant to lubricate the skin it is a cosmetic, but if it claims to relieve muscular discomfort, it is a medicine.[27,28]

As per the US FDA, the intended use of any product can be recognized in many ways. For example:

- From the statements made in advertisements either online/on social media or in print media, or in any marketing resources. Even though a product is advertised as a cosmetic, certain claims might cause it to be classified as a medication based on the projected purpose, like to cure, alleviate, or otherwise influence the structure, function or appearance of the human body. Examples include products promising to renew skin cells, heal varicose veins, lessen cellulite, prevent alopecia, affect melanin (pigment) synthesis in the skin, *etc.*
- From the buyers' opinion, as to why they prefer one product over another, based on what they believe the product will do and its reputation.

- From any constituent known publicly and industrially to have potential therapeutic benefits, that give the product a drug-like status, *e.g.* fluoride in toothpaste.

### 2.2.1.2 New Laws

In addition to these regulations, on 24 June 2011, the U.S. Safe Cosmetic Act (H.R. 2359) was presented. Also in 2019, in the House of Representatives, the Safe Cosmetics and Personal Care Products Act (H.R. 4296) was proposed.[29] Since the FDA did not previously mandate safety testing of even the substances included in cosmetics and personal care products, this modification was seen as a significant advancement. The new federal law also mandates supply chain openness and industrial exchange of safety data, and makes cosmetic firms liable for the safety of the components in their products. Additionally, the FDA promotes the use of non-animal testing methods and recommends the most appropriate method to the manufacturers. Also, if a final product poses a substantial risk of injury, the FDA may force a recall.

The Safe Cosmetics and Personal Care Products Act (H.R. 4296) requires businesses to register their ingredients used, the production facilities, and the final products with the FDA in addition to fulfilling the GMP requirements. The Act also protects susceptible groups, *i.e.* children, pregnant women, infants, and other highly unprotected populations. Under the Act, reports of major undesirable effects are also made accessible to the public.

Certain scented components, though frequently harmful, and 'hidden' in cosmetics, were earlier permitted by federal law. However, complete disclosure of such components is mandated by HR 4296. The bill states that cosmetic labeling should mention any fragrance allergen at a concentration of 0.01% or above (in rinse-off cosmetic products) or 0.001% (in leave-on cosmetic products) present in a cosmetic and designate it as a fragrance allergen if it is listed in Annex III of the European Union Cosmetics Regulation (No. 1223/2009).[30]

In order to promote and control the labeling compliances and public health issues, the FDA has collaborated with the Federal Trade Commission (FTC), and has designed a website to inform and raise awareness among the general population. In accordance with the FDA's definition of 'natural' for food products, the FTC also emphasizes that items claiming to be "all-natural" or "100% natural" should not contain any artificial substances or chemicals.[31]

However, a division of the Council of Better Business Bureaus (CBBB), a non-governmental national advertising organization, The National Advertising Division (NAD) allows the use of the term 'natural' on a product even if it contains only a small portion of actually natural ingredient.

Due to this lack of uniformity, chaos is common among manufacturers regarding the use of 'natural'. To deal with this situation, in 2021, in the 117th Congress, Sean P. Maloney introduced the Natural Cosmetics Act (the Bill – H.R. 5872) to have clear definitions for "natural" and "naturally-derived ingredients" with respect to cosmetics and personal care products.[32]

According to the Bill, "natural" refers to a product made up of at least 70% natural ingredients (unprocessed ingredients) and a "naturally-derived ingredient" denotes that the initial raw material was either derived directly from a plant feedstock or was of mineral, plant, animal, or microbial origin.

#### *2.2.1.3 USDA (United States Department of Agriculture)*

The term "organic" or "natural", used to indicate components in cosmetics, body care, or personal care products, is **not defined or regulated** by the United States Food and Drug Administration (FDA) in the cosmetics industry. Through the National Organic Program (NOP) rule, 7 CFR Part 205, the USDA controls the term "organic" as it pertains to agricultural goods.[33]

Any personal care item or cosmetic or body care item qualifies for the NOP certification if it contains or is prepared from agricultural products and can stand by the organic production, processing, handling, and labeling criteria from USDA/NOP. Additionally, the USDA-accredited organic certifying agent must also certify the enterprises that generate the organic agricultural substances, the handlers of these agricultural substances, and the producer of the finished products.

It should also be noted that the cosmetic products claiming organic labeling should fulfil the standards by both the USDA regulations and the FDA regulations.[34] After having qualified, these organic products become eligible for labeling under the following 4 organic categories of all other agricultural products, based on their organic content and other factors:[35]

1. *100% organic*: when all the ingredients used in the product are strictly produced organically, with the exception of salt and water. In labeling, the USDA Organic Seal **may be** shown on products' principal display panels (PDP), and the certifying agent's name and address **must be** displayed on information panels (IP).
2. *Organic*: At least 95% of the components in the product must be organically grown (excluding salt and water). Other components in the product must be non-agricultural substances approved for use on the National List or agricultural substances not grown organically but are otherwise permitted for use on the National List. The USDA Organic Seal **may be** shown on products' principal display panels (PDP), and the certifying agent's name and address **must be** displayed on information panels (IP).
3. *Made with organic ingredients*: Such products must include at least 70% organic components, and the main display panel (PDP) of the product label may list up to three of those ingredients. Body lotion, for instance, may be labelled as "body lotion consisting of rosemary, organic lavender, and chamomile" or "body lotion created with organic herbs" if it contains at least 70% organic ingredients (excluding salt and water). The USDA Organic Seal **may not be** shown on the products' principal

display panels (PDP), and the certifying agent's name and address **must be** displayed on information panels (IP).

4. *Less than 70% organic ingredients*: The word "organic" is **not** allowed to appear anywhere on a product's main display panel (PDP). On the information panel's ingredients statement, they could nonetheless list the precise USDA-certified components as being obtained organically (excluding salt and water). The USDA Organic Seal and a certifying agent's name and address **may not be** displayed on the products.

Any cosmetic or personal care item that does not adhere to the aforementioned production, labeling, processing, handling, and certification criteria is prohibited from making any claims about its organic USDA-certified status or compliance with USDA organic requirements.

However, for cosmetics, personal care, or body care items derived from non-agricultural materials, the USDA neither has any control over their manufacturing and labeling nor asserts that they adhere to USDA organic criteria. Additionally, certification from private standards like eco-labels, earth friendly, *etc.* are also accepted in US markets and such labels are presently not regulated by the USDA's NOP.

### 2.2.2 Canada

Health Canada, the regulatory body in Canada that governs the production, distribution and marketing of cosmetics, does not use the word "cosmeceutical" (to indicate cosmetics with pharmaceutical-like advantages). Thus, these kinds of products are included either as cosmetics or as drugs (based on the assertions made and/or the product's composition) and are governed by the Consumer Packaging and Labeling Act (CPLA) and its regulations as well as the Food and Drugs Act's cosmetic regulations on composition, labeling, safety, and advertising.[36] The product labeling, safety of chemicals used and the final products, and obligatory notification of all cosmetic items are the three most important aspects of the Canadian cosmetic regulatory system.

A "cosmetic" under the cosmetic regulation of Health Canada is "any substance or mixture of substances, containing deodorants and fragrances, produced or marketed for the purpose of use in cleansing, enhancing or modifying the skin, complexion, teeth, or hair". These are governed by the Cosmetic Regulation Act (1977; C.R.C., c. 869) and Food and Drugs Act (1985; sections 16, 17, 18, 27 and 30; R.S.C., 1985, c. F-27), almost unchanged except for a few amendments over the years.[36]

In Canada, cosmeceutical products are classified as over-the-counter (OTC) or natural-health products (NHP), the latter of which is regarded as a subset of "drugs",[37] and these are classified as products at the cosmetic–drug interface (PCDI). For example, sunscreens with sunscreen claims, formulated for use on the face or skin, as cosmetics or skincare. These can be categorized, depending on their ingredients, as natural-health products (NHPs) or non-prescription medications as per the Natural and Non-prescription Health Products Directorate (NNHPD).[38,39]

In Canada, the safety of cosmetic products is under the manufacturer's sole responsibility. However, within ten days of the product's initial sale, the manufacturers must submit an online Cosmetic Notification Form (CNF) (as per section 30, 31 of CR) for each product individually, though this notification does not mean that the product is legally approved for sale or complies with laws. This Cosmetic Notification Form (CNF) provides facts about the manufacturer's address and phone number, the cosmetic, its use and appearance, and also its components with concentrations.[40] Additionally, under Health Canada's Consumer Product Safety Program (CPSP), all cosmetic items for manufacturing, distribution, and usage are subjected to "risk evaluations" to make sure they don't cause unintended harm while serving their intended/labelled purpose.[41]

Unlike the EU, Health Canada has certain less extensive lists for control over the usage of ingredients in cosmetics. The compounds that are prohibited or restricted for use in cosmetics are included in Canada's "Cosmetic Ingredient Hotlist", a scientific document that is routinely reviewed and updated, under the sections of FDA and CR.[42] Another requirement of the Food and Drugs Act and the Cosmetic Regulations statute is flawless labeling (clearly mentioning any natural or botanical ingredients). In addition, the Canadian legislation requires that the components be listed in both English and French if they are not named in accordance with the International Nomenclature of Cosmetic Ingredients (INCI).

In addition to NNHPD, Canadian legislation has Canadian Organic Standards (COS) for naturals and organics. However, the labeling of personal care items or cosmetics is NOT regulated by the Canadian Organic Standards and these are overseen by Health Canada, so at present there are still no standards or regulations regarding the use of the terms "natural," "organic," *etc.*

However, EcoCert (organic certification organization from France) has certified a lot of Canadian cosmetic firms as organic.[43] As per EcoCert, products can be certified in one of two categories:

- *EcoCert Organic Cosmetic*: containing at least 90% plant-derived constituents and at least 10% of ingredients from organic farming (excluding salt and water).
- *EcoCert Natural Cosmetic*: containing at least 50% plant-derived constituents and at least 5% of ingredients from organic farming (excluding salt and water).

In addition, there are private firms as well in Canada that grant "Organic Certification" for certain skin care products, though not under any Canadian administrative body including Health Canada, like ACS Canada.[44] These firms often get their testing done in ILAC-ISO 17025 accredited laboratories.

### 2.2.3 Japan

In Japan, cosmetic products are regulated under the Pharmaceutical and Medical Devices Law (PMDL), which came into effect in 2014, replacing the old Pharmaceutical Affair Law (PAL; Article 42, Paragraph 2, Law No. 145) of 1960. The Law is administered by the Ministry of Health, Labor, and Welfare (MHLW).[10] In Japan the term "cosmeceuticals" isn't used; instead products with mixed actions, *i.e.* cosmetic and therapeutic, are classified as "quasi-drugs" (as discussed earlier in this chapter). These are the two distinct categories into which most of the beauty products fall. In Japanese legislation, cosmetics are defined as "Articles having a moderate effect on the human body, designed for application by sprinkling, rubbing, or other techniques, in order to beautify, clean, modify the look, and improve the attractiveness, or to preserve the skin or hair in excellent condition". These are further divided into 6 sub-categories, namely cosmetics: perfume and eau de cologne, makeup, skin care, hair care, special purpose cosmetics, and cosmetic soaps.[45]

A cosmetic manufacturing license and a cosmetic marketing license must first be obtained in Japan in order to register a cosmetic product, each with separate requirements. However, in order to preserve the quality and safety of their products, marketing license holders are required to adhere mainly to two standards: The Good Vigilance Practice (GVP) standard and the Good Quality Practice (GQP) standard. The producers, after receiving the necessary permissions, must submit a cosmetic marketing notification to the same region that issued the cosmetic marketing license. The product can finally be released onto the market if all of the aforementioned standards have been met. Also, safety issues, if any, must be reported to the Minister of Health, Labor and Welfare within 30 days.[45]

As per Japan's law, the labeling of cosmetics is required to be in Japanese. Therefore, in order to comply with the PMDL's mandate that all component names be listed on the labeling, the Japan Cosmetic Industry Association (JCIA) has created a Japanese translation of the "List of Cosmetic Ingredient Label Names." In the case of quasi-drugs, the name of the active ingredient must be mentioned on the label as per the MHLW.[45]

For quasi-drugs, a label must list the names of particular substances that the MHLW has prescribed. The legislation has created lists for restricted and forbidden compounds, as well as lists for helpful preservatives, UV filters, and colorants, in order to regulate the type of ingredients used in cosmetics. For example, for the assessment of quasi-drug whitening products and anti-wrinkle products, the Japanese Cosmetic Science Society (JCSS) has issued guiding principles.[46,47]

Japan is Asia's second-largest market for natural cosmetics. The use of natural ingredients and eco-friendly packaging is a trend that many local Japanese firms have adopted. However, the absence of standardized criteria presents a significant obstacle for organic/natural cosmetics. Due to limitations in the official certification criteria, consumers and producers in Japan use a variety of and sometimes contradictory definitions for organic beauty products. Also the standard is not valid for the whole of Japan, as the country's present official organic certification, "Yuki JAS", only applies to food and beverages. As a result of the absence of an official organic certificate in the Japanese cosmetics industry, the market is captured by certified imported organic trademarks like Melvita, Weleda and Neal's Yard Remedies. Apparently, in chase, smaller firms have also started seeking international certificates, like USDA organic, NaTrue, Ecocert *etc.*[48]

The lack of formal certification has also resulted in a second principle—the general positioning of 'naturals' in the consciousness of the target audience. This covers trademarks that advertise themselves as natural but the amount of natural or botanical ingredients is not fixed in their formulations. Because of this general positioning of 'naturals', despite the vagueness in definition and lack of criteria, this section is well-liked by Japanese customers. The increased acceptance can also be understood due to the lesser importance given to the claims related to natural/organic by Japanese buyers. There are many local and foreign participants in this bigger market for natural cosmetics, ranging from those that specialize in natural goods to a limited number of "natural" product lines being added by already established brands.

Examples include the addition of "Luminique Botanical Pure", a shampoo and treatment, by Unilever Japan to its Lux line, and Kose Corp.'s significant expansion of its "Nature & Co" skin and hair care brand. Manufacturers of mass cosmetic products use a wide variety of natural components, ranging from locally available Japanese naturals like indigenous herbs, rice, yuzu citrus, *etc.*, to internationally available botanicals like lavender, rose, tea tree, *etc.*[48]

### 2.2.4 China

The State Council of China officially released the long-awaited "Cosmetics Supervision and Administration Regulation (CSAR)" in June 2020, on which work started formally from 2013. CSAR came into effect on 1 January 2021.[49] CSAR superseded the older "Regulations on Hygiene Supervision of Cosmetics (CHSR)", which was released back in 1989. The National Medical Product Administration (NMPA), the agency looking after medicines, medical devices, and cosmetics (formerly known as the China Food and Drug Administration [CFDA]), is developing a comprehensive regulatory structure to support CSAR.[50]

At present, the State Administration for Market Regulation (SAMR), National Medical Products Administration (NMPA), and General Administration of Customs (GAC) are the three primary regulating expert bodies in the cosmetic industry. Continuing this development, from these laws, several subsidiary laws dealing with the registration and notification procedure, GMP, adverse effect monitoring, *etc.*, have been released.[51]

Though cosmeceuticals is not an officially recognized term, cosmetics in China are separated into 'special cosmetics' and 'general cosmetics' categories, under the new CSAR. The former include hair curling products, agents preventing hair loss, hair dyes, whitening agents, sunscreen, and "cosmetics with unique effectiveness claims." However, a product may be classified in more than one group, like a hair product can be a 'general cosmetic' if it only improves hair quality and thereby prevents breakage. It can be a 'special cosmetic' if it prevents hair loss by affecting scalp health and it can be a 'medication' if it promotes hair growth by getting involved in human physiological activities.[49] These special cosmetics, as per the new rule, need NMPA approval before their production, while general cosmetics are allowed to be marketed just with a notification. However, before NMPA registration or notification, it is mandatory (since 1 January 2022) for the registrant or notifier to complete a self-assessment safety review of its products themselves or through a qualified agency and disclose the results of the product safety assessment during the registration and notification.

Furthermore, NMPA issued "Rules and Catalog for Categorization of Cosmetics" defining a coding-based classification system for cosmetics comprising the five areas in which they may belong. These five areas are[52] the efficacy claims (covering 28 different claims), site of application (covering 10 different body parts), administration form (covering 14 various available dosage forms), target consumers based on age groups (3 different age groups) and finally the manner of usage (*i.e.* rinse-off or leave-on). Each sub-category is given a two-digit number or letter, re-arranged to form a code. For example, a moisturizing facial lotion with sunscreen protection desired for a leave-on effect and meant for the general population will be coded as 0409-05-02-01-02.

Like in Japan, China has also listed the harmful and helpful ingredients to be used in cosmetics. Additionally, China has clearly demarcated the "new" and "existing" cosmetic ingredients. The ingredients that are not present in the "Inventory of Existing Cosmetic Ingredients in China" (IECIC 2021) are regarded as "new". They can be natural or synthetic compounds, being utilized for the first time in Chinese cosmetics.[50] The IECIC 2021 includes the standard Chinese name and INCI (International Nomenclature Cosmetic Ingredient) name/English name of cosmetic ingredients in reference to the documents issued by the Personal Care Products Council (PCPC). Businesses may instantly check the listing of a cosmetic ingredient in the most recent edition of the Inventory of Existing Cosmetic Ingredients in China (IECIC), using CIRS's free Chinese cosmetic ingredient regulatory database (China CosIng).[53,54]

Chinese legislation also lacks a clear definition for "organic," "natural," or "clean" cosmetics and there are no clear laws to govern this category of cosmetics because of the difficulty in establishing homogenous standards. As a result, China's regulatory framework has been unsuccessful in keeping check on the growing market and compliance concerns with the production, labeling, selling, and advertising of organic and natural products.[55] Broadly in China, there are four regulatory bodies governing organic/natural products, namely the Organic Product Certificate Catalogue (2018), National Standard for Organic Products (the present active one is GB/T 19630-2019), Organic

Product Certificate Management Measures (2014), and Organic Product Certification Implementation Rules (2014).[56] However, organic cosmetics are not listed either in the recent National Standard for Organic Products or Organic Product Certificate Catalogue, which include lists of products from animals, plants and microbes. Also, the organic certification was excluded by the People's Republic of China's Certification and Accreditation Administration (CNCA) authority on 1 July 2012.[57] Since then, no regulatory body has been established to grant certification to organic cosmetics. Also since 2015, the Chinese government prohibited the use of the terms "organic," "zero-added" or "100% natural" in product labeling claims in 'Chinese language'. Consequently, many firms have obtained international certifications from Ecocert or German natural organic certification BDIH (Bundesverband Deutscher Industrie-und Handelsunternehmen, a Federal Association of German Industry and Trade) (see Figure 2.3).

## 2.2.5 European Union

European Medicines Agency (EMA) standards clearly distinguish between medications and cosmetics in the region, and a product cannot belong to both categories if it is meant to be used externally and has a cosmetic purpose. The old Cosmetics Directives were replaced on 30 November 2009, by the new Cosmetic Products Regulation, EU Regulation 1223/2009. As a result of the European Union's new Cosmetics Regulation, Europe claims to have a strong, globally recognized system that strengthens product safety while taking into account the most recent technical advancements. The majority of the rules in this new regulation took effect on 11 July 2013.

**Figure 2.3** Some of the international organic certification systems adopted by Chinese manufacturers.

The principal set of regulations governing completed cosmetic goods that are put on the EU market is Regulation (EC) No. 1223/2009 on cosmetic products. By strengthening the safety of cosmetic products, and streamlining the regulatory framework for all players in the sector, it enhances consumer safety. This law makes processes much simpler so that internal marketing of cosmetics has become feasible.[58]

The older Directive 76/768/EC, which was issued in July 1976 and has undergone multiple significant revisions, was the one replaced by the new legislation. The new legislation offers a strong, globally recognized system that strengthens product safety while also taking into account the most recent technical advancements, along with the potential usage of nanomaterials. Regulations prohibiting animal testing from this derivative still remain in place.

The major improvements brought about by the cosmetic legislation are:

- Reinforced safety criteria for cosmetic items. Before releasing any product onto the market, manufacturers must adhere to certain guidelines while creating a product safety report.
- The concept of a "responsible person" was introduced. The EU only allows "responsible persons" who are either legal entities or natural persons to advertise cosmetic items. The new cosmetics rule makes it possible to identify the accountable party accurately and spells out their responsibilities.
- A centralized reporting system for all cosmetics sold in the EU. Through the EU cosmetic goods notification site (*i.e.* CPNP), manufacturers will have to inform about their items, though only one time, before launching.
- The requirement to report serious undesirable effects (SUEs) was introduced. It will be the responsibility of the responsible party to alert national authorities of any major adverse impacts. Additionally, information from consumers, medical experts, and other sources will be gathered by the government. They will have the responsibility to provide this information to other EU nations as well (SUE reporting guidelines).[59]
- New guidelines for the usage of nanomaterials in cosmetics. It is necessary to obtain unambiguous authorization for various preservatives, colorants and UV filters, including those made of nanomaterials. If the Commission has concerns, products containing nanomaterials that are not otherwise prohibited by the cosmetics rules will be subjected to thorough safety evaluation at the EU level. In labeling, the term "nano" must be used in brackets after the chemical name in the ingredient list to identify nanomaterials.[60]

In the European Union (EU), medications and cosmeceuticals are governed by the European Medicines Agency (EMA), which has its headquarters in London, England. The EU's Traditional Herbal Medicinal Products Directive 2004/24/EC, or THMPD, went into effect on 30 April 2011, and it governs the use of herbal products in medicines. The rule was created as a sub-directive for the act on Human Medicinal Products Directive 2001/83/EC, and it asserts that each EU Member State must have a distinct set of information for herbal substances or herbal preparations. This might be used to assess marketing submissions for herbal medicines from businesses and includes pharmaceuticals that incorporate herbal ingredients.[61]

According to the EMA, for any herbal formulation to be sold, it must fall into one of the following three categories:

a. Based on appropriate safety data and reasonable efficacy, a substance can be categorized under traditional medical use provisions (referred to as "traditional use"). It is then given a traditional use registration (simplified registration procedure) by a Member State.
b. A product may fall within the definition of "well-established usage" for therapeutic purposes, provided that there is scientific literature demonstrating that the active ingredients have been in widespread use as medicines inside the Union for at least 10 years, with acknowledged efficacy and an acceptable degree of safety. As a consequence, a Member State or the European Medicines Agency often grants the medicine marketing authorization.
c. After assessment of a marketing authorization application entailing safety and effectiveness data from the company's own development (a "stand-alone" application) or a blend of the company's research and bibliographic data (a "mixed application"), a product may be approved. If all conditions are satisfied, the product is then given marketing authorization by a Member State or by the Agency *via* the centralized method.

Thus, in conclusion, although a product's safety must be demonstrated, proving efficacy is not always necessary, and simply the conventional indications in listed circumstances must be acceptable. However, the quality of the herbal medical products must always be confirmed, regardless of the legal methods to reach the market.[62]

The following definitions are included in the Directive for herbal medicines, herbal preparations, and herbal substances:[61]

- *Herbal medical products*: When the active ingredient in any pharmaceutical is entirely one or more herbal substances, or herbal preparations, or a combination of these as active ingredients.
- *Herbal substances*: These are mostly entire, broken up, or cut plants, plant parts, algae, fungus, and lichen in their natural state, generally dried out but occasionally fresh. It also includes the exudates not submitted to any particular treatment. According to the binomial system, herbal compounds are defined accurately in terms of the plant component employed and the botanical name (genus, species, variety and author).

- *Herbal preparations*: The products made from herbal substances after they have undergone a number of processes, including expression, distillation, fermentation, extraction, fractionation, concentration or purification. Fragmented or ground herbal products, essential oils, extracts, expressed juices, tinctures, and processed exudates are some examples.

Thus, in the EU, the burden of proof for the safety of a cosmetic product being marketed rests on its maker or importer. At the national level, laws are enforced, and each EU nation has commanding organizations accountable for ensuring compliance.[63]

In order to provide technical guidance for the safety evaluation of cosmetics, the Scientific Committee on Consumer Safety (SCCS) produced "SCCS Notes of Guidance for the Testing of Cosmetic Ingredients and Their Safety Evaluation" in the EU, with the 10th version released in 2018. Animal testing for cosmetic chemicals and products has been prohibited in the EU and in several other nations or areas due to the 3R principle trend. These days, studies on alternative methodologies are a frequent source of worry, and SCCS has highlighted replacement techniques in more recent iterations of the guidelines. Additionally, a brand-new technological method termed next-generation risk assessment (NGRA) has surfaced in recent years. The International Cooperation of Cosmetics Regulation (ICCR) is now working on NGRA, which intends to include new approach techniques into an integrated strategy for risk assessment of cosmetic compounds.[50] Leading businesses are working on formulas that can easily be evaluated on 3D bioprinted living human tissue, doing away with the necessity for time- and money-consuming complex animal and human experiments.[8]

### 2.2.5.1 *Under EU Legislation Some of the Certifications That Are Accepted for the Use of Natural Ingredients in Cosmetics Markets*[64]

**2.2.5.1.1 Different Natural and Organic Cosmetics Standards.** As per 'Ecovia Intelligence', in Europe, due to the growing demand for organic and natural cosmetics, many cosmetic businesses (makers and formulators) are creating certified products and/or using recognized components in their formulations. In 2019, the European natural cosmetics industry was worth €3.9 billion, accounting for 5% of the overall European cosmetics market.[65]

With the increasing importance of being certified in terms of natural and organic standards for raw materials used, which adds to legitimacy, there are around 25 different natural and organic cosmetics standards in Europe, applied at the national level, like Nature & Progrès, CCPB, Organic Farmers & Growers, Demeter, COSMOS and NaTrue.[66] These specifications include certification programs for both completed goods and raw materials.

Among all the certifications available, NaTrue and COSMOS are the widely used ones, which are regarded as indicators of excellence. These labels are frequently used by ingredient purchasers and cosmetic makers on their packaging and marketing materials.

**2.2.5.1.2 The Fair Trade Standard.** Initially, the fair trade standard was an agricultural standard. It is, nevertheless, widely utilized for cosmetic goods. Fairtrade International maintains the most well-known certification programs for fair trade goods. Due to the fact that the Fairtrade standard was primarily made for north–south trading and it addresses the social component of sustainable production, it is well-liked among suppliers in developing nations. Fair for Life/For Life Certification, Ecocert Fair Trade, and FairWild are some other certification programs.

Fair trade certification is considered as ethical certification for meeting the safety and efficacy demand among the suppliers of natural components in developing nations. Examples of exporters of products with fair trade or FairWild certification are Guru Nanak Oil Mills (U) Ltd in Uganda exporting certified shea butter, and EcoProducts from South Africa exporting certified baobab oil to European markets, including FairTale and The Savannah Fruits Company among others.

### *2.2.5.2 Quality Management*

In addition to the increasing demand for standards of the natural/organic raw materials used in cosmetics among European consumers, there is also demand for standards of production quality. Henceforth, adopting certain quality management standards will increase the brand's reputation, creating trust and value among the customers. Also, this will help in maintaining batch-to-batch uniformity in the quality of natural raw materials procured from developing nations, meeting compliance with legal requirements. Some of the standards that help validate the quality of cosmetic production using natural ingredients are:

- Good Agricultural and Collection Practices (GACP) – EMA/HMPC/398706/2021 (under consideration) and EMEA/HMPC/246816/2005 (currently applicable).[67,68]
- Good Manufacturing Practice (GMP) developed by the European Federation for Cosmetics Ingredients,[69] which came into effect from 1 January 2019.
- International Fragrance Association (IFRA) standards – for regulating the use of essential oils from natural sources, in perfumes.[70]
- ISO 9001:2015,[71] ISO 16128-1:2016(en),[72] and ISO 16128-2:2017/Amd 1:2022(en)[73] from the International Organization for Standardization (ISO).
- Food Safety Net Services (FSNS), part of the Food Safety System Certification (FSSC), providing cosmetic and supplement testing through its ISO/IEC-17025 endorsed laboratories across North America.[74]

In the future, quality requirements will become more significant. Due to the stringent EU code of practice and rising consumer demand for high-quality goods, buyers are searching for raw material suppliers that can guarantee uniform quality levels (see Table 2.1). Therefore, quality management standards and certification should be adopted by suppliers of natural cosmetic components since this is likely to boost their eligibility for European markets.

## 2.2.6 India

The Indian cosmeceutical and personal care market has also shown significant growth in recent years and is expected to have a CGAR of 6.3% by 2027. The Indian cosmetic governing bodies are The Drugs and Cosmetics Act 1940 and Rules 1945 (DCAR), and Bureau of Indian Standards (BIS), with the latter governing the Labeling Declarations of Schedule S items of The Drugs and Cosmetics Rules of 1945. The Drug and Cosmetic Act 1940 and Rules 1945, under section 3(aaa), has defined a cosmetic as "any article intended to be rubbed, poured, sprinkled or sprayed on, or introduced into, or otherwise applied to, the human body or any part thereof for cleansing, beautifying, promoting attractiveness, or altering the appearance, and includes any article intended for use as a component of a cosmetic".[75] Additionally, specifications for skin creams and lipsticks are also provided by BIS in Indian Standards (IS) 6608:2004 and 9875:1990. Furthermore, manufacturers are required to test heavy metals and arsenic in the raw materials for these products in accordance with Indian Standard 6608:2004. As per the rules, once passed, it's not mandatory to re-test the final products.[76] The Central Drugs Standard Control Organization (CDSCO), which is part of the Indian Drug Controller General's office (DCGI), is the licensing body responsible for the registration of all cosmetics produced and/or imported in India. The Indian Drug and Cosmetics Act of 1940, which regulates the production, sale, import, export, and clinical research of drugs and cosmetics in India, is administered by the DCGI's office.[77] This CDSCO is the National Regulatory Authority (NRA) of India, governed under the Directorate General of Health Services, Ministry of Health & Family Welfare, Government of India.

In India, the respective state governments appoint the State Licensing Authorities that govern the manufacturing of cosmetics by their system of scrutiny and licensing, while the Central Licensing Authority (CLA) appointed by the central government governs the import of cosmetics by a system of registration, all in accordance with the provisions of the Drugs and Cosmetics Act and Rules made thereunder (see Figure 2.4). The CLA is the Drugs Controller General India (DCGI), who grants the Import Registration Certificate and regulates the import vide Gazette Notification G.S.R 763(E), in accordance with the Drugs and Cosmetics Act, 1940. If the cosmetic meets the mentioned criteria, the manufacturer must register its details, *i.e.* site of production, package size, and variants (if any), before importing it.

As per the new Cosmetics Rules, 2020, released by the central government under the Drugs and Cosmetics Act, 1940, the CLA is appointed under rule

**Table 2.1** List of some of the important certifications asked for by European buyers of natural cosmetic ingredients.

| Certification | Logo | Category | Cost of acquisition | Market(s) accepted | Further information (reference) |
|---|---|---|---|---|---|
| COSMOS | | For organic and natural ingredients (independently) | Depends on the company market share, types of natural/organic ingredients and certification type | In 71 countries including the UK, France, and Germany | 89 |
| NaTrue | | For organic and natural ingredients (independently) | Price is fixed – EUR 220 per certified product for 2 years. If more than 100 products, the label fee is EUR 170 per product for 2 years | UK, France, Germany, Italy | 90 |
| Fairtrade International | | For social components of ethical and sustainable production | Depends on the type, quality, quantity, and form of product to be certified and also on the scope of the producer and the country/region applied from | UK, France, Germany, Italy | 91 |
| Fair For Life | | For Social responsibilities and ethical production | Depends on the type of certification required, the country/region applied from, if the applicant has any other certification and also on the size and complexity of the supply chain involved | UK, France, Germany, Italy | 92 |
| FairWild | | Mainly for ingredients harvested from wild sources and also for social responsibilities | Depends on a number of things like the complexity involved to get the ingredients (in pure form), location *etc.* | UK, France, Germany, Italy | 93 |
| UEBT | | For biodiversity and ethical sustainable conservation | Depends on the company market share | UK, France, Germany, Italy | 94 |

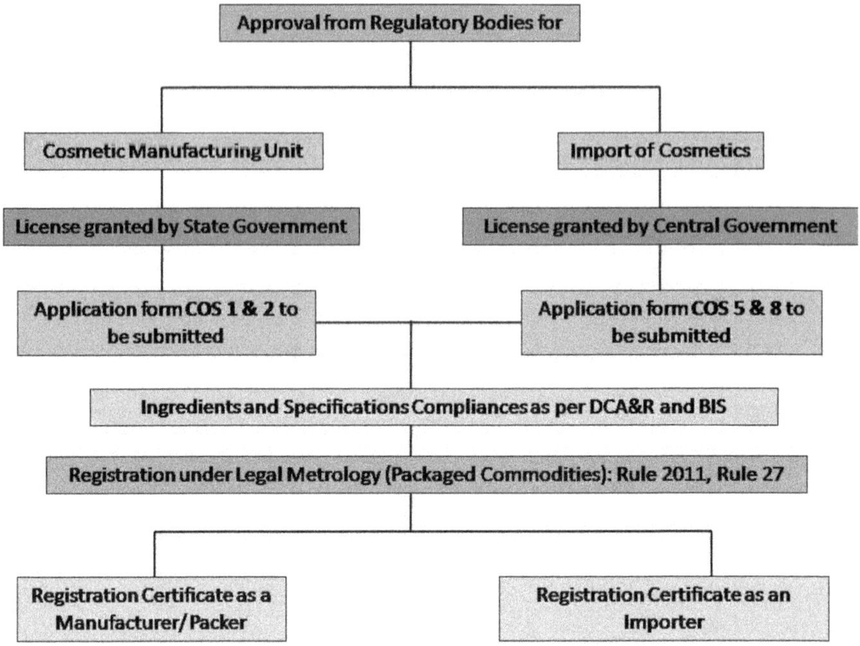


**Figure 2.4** Brief illustration of the regulation of cosmetics and beauty products in India (for import and manufacture).[78]

3(f) of Cosmetic Rules 2020 12(1), or by a person to whom such authority may be delegated under rule 5.

According to the Ninth Schedule of Rule 39 of the Cosmetic Rules, 2020, in order to be imported or manufactured, a cosmetic product should confirm all the specifications outlined in the schedule, as well as any other restrictions under the rule, including those related to the quality and safety standards. However, if the cosmetic is not included in the Ninth Schedule, it must nonetheless adhere to all relevant regulations, specifications, and standards in the country of origin.

No cosmetic may assert, pretend to assert, or transmit to the intended user any concept that is untrue or deceptive (according to Cosmetic Rules, 2020, Rule 36).[75,79]

In India, CDSCO has created a portal for online registration of cosmetics (being imported or manufactured). Once the application is submitted, a designated officer reviews it for various compliances. The respective State Licensing Authority approves the application, if it is deemed complete in every way, and the acceptance letter becomes accessible through the SUGAM dashboard.

Use of cosmetics comprising dyes, colors, and pigments other than those listed by the Bureau of Indian Standards (IS: 4707 Part 1 as modified) and

Schedule Q is restricted as per Rule 134 of the Drugs and Cosmetics Rules.[80,81] The use of lead and arsenic for coloring is banned by Rule 145 and imports of such cosmetics are prohibited under Rule 135. Cosmetics containing mercury compounds are prohibited from being manufactured and imported, under Rule 145 D and 135 A.

Indian cosmetics legislation also does not have a clear demarcation for the use of natural/organic ingredients in cosmetics. However, from 2019 the Agricultural and Processed Food Products Export Development Authority (APEDA), Ministry of Commerce and Industry, Government of India has drafted a number of standards under the National Program for Organic Production (NPOP) for organic cosmetics and personal care products. As per the standards of the first version of the draft, ingredients from plants, animals, or microorganisms must be the main component of organic cosmetics and personal care products, and these must be produced, processed, and handled in accordance with the requirements outlined in these regulations. Unless specifically permitted as an exception, manmade elements (from chemical processes or petrochemicals) should not be made use of.[82,83]

An updated annual organic system plan (OSP) must be submitted by the manufacturer, processor, or handler of organic cosmetics and personal care products every year. As per these regulations under NPOP, ingredients permitted for use include: water, minerals (processed as per Annexure 1 and 2), plant products (certified organic as per Appendix 1, 5, 7, 8 or created as per Annexure 1), microbial and marine products (processed as per Annexure 1) and animal or livestock products (certified organic as per Appendix 2, 3).[82]

APEDA has also set standards for the processing and manufacture of organic cosmetics and personal care products:

*Labeling claims as per the rule*: APEDA has set the following standards for the labeling of cosmetics and personal care products that claim to be organic or made from organic/natural ingredients:

(i) Salt, water, and mined or processed minerals should not be labelled as organic ingredients as they are regarded as neutral substances.

(ii) Products prepared using only one ingredient of natural origin as per the rule will be labelled as organic.

(iii) The following conditions must be met when, under the mentioned standards, all components are not from organic origin or do not appear on the list of permitted ingredients:

   a. "ORGANIC" – when the product contains at least 95% of ingredients certified organic or from permitted origin.
   b. "MADE WITH ORGANIC 4 INGREDIENTS *x*%" – when in the product less than 95% but more than 70% of ingredients are certified organic or from permitted origin. Here '*x*' specifies the proportion of organic ingredients in the finished product.

c. The product "cannot be labelled as organic" if it contains less than 70% of organically certified ingredients. However, the list of ingredients can have the name of any organic ingredient in the product.

Additionally, APEDA has also established standards for the post-production cleaning and disinfection of the production environment and equipment as well as for hygienic storage and transport of organic products (specified in Annexure 5 of the rule).[83]

Apart from the APEDA established standards, the new Cosmetic Rules, 2020, has defined standards for the use of synthetic organic colors and natural organic colors in cosmetics. Accordingly, any cosmetic product must not contain:[79]:

a. Arsenic in the form of arsenic trioxide at a concentration more than 2 parts per million.
b. Lead in the form of lead at a concentration more than 20 parts per million.
c. Heavy metals (except lead) in the form of total metal elements at a concentration more than 100 parts per million.

*NOTE*: In the Drug and Cosmetic Act 1940, and rules of 1945, under clause 127 (2) it is mandatory to mention the name of the permitted color used in the preparation. However, this holds only for drugs and is not applicable/mandatory for cosmetics. As of now, The Drug and Cosmetic Act has not laid down any defined nomenclature for the use of coloring agents in cosmetics. For example, Garnier Wrinkle Lift anti-aging cream and Pond's White Beauty cream both use color index 14 700 on their labels, which is not a mentioned standard and does not reveal any specific information, such as the name, chemical nature, toxicity profile *etc.* Thus there is a need for harmonizing the color nomenclature in Indian cosmetic legislation.[84]

Very recently in July 2022, the Union Health Ministry of the Indian Government published a draft of The Drugs, Medical Devices and Cosmetics Bill, 2022, which is considered to be a drastic change over the pre-independence Drugs and Cosmetics Act, 1940, legislation. The new bill aims to address the advancements in technology and the changing consumer needs with time.[85,86] With this the Ministry has also formed a new committee called the Drugs, Medical Devices, and Cosmetics Consultative Committee (DMDCCC), headed by the Drugs Controller General of India, which will ensure homogeneity in the bill implementation all over the nation.[87]

### 2.2.7 ISO 16128-1:2016 and ISO 16128-2:2017/Amd 1:2022(en)

ISO, The International Organization for Standardization, is a worldwide federation of national standards organizations. It is an independent, non-governmental organization that creates international standards through its ISO Technical Committees (TC). When creating ISO/IEC Directives, ISO works closely with the International Electrotechnical Commission (IEC). The definitions and standards for natural, natural origin, organic, and organic origin cosmetic ingredients and products are outlined in ISO 16128. In 2016, part one of ISO 16128 was released. In order to promote safe innovations, ISO 16128 aimed to "encourage the use of a greater range of natural and organic components in the development of a diverse array of cosmetic products" (https://www.iso.org/standard/62503.html). The second portion (ISO 16128-2), was released in 2017 (https://www.iso.org/standard/65197.html). It stated criteria for the classification of 4 categories of natural/organic ingredients mentioned in ISO 16128-1.[88]

As per the ISO definition, a natural cosmetic product should contain at least 90% ingredients of natural origin. For items that are certified as organic, at least 95% of the natural ingredients used in it must be produced from organic farming. Additionally, rinse-off cosmetics must have more than 10% organic components, while leave-on products must have at least 20% of their total constituents come from organic farming (as opposed to non-emulsified products, which must have at least 80% mineral or mineral origin ingredients).

However, unlike fair trade, ISO 16128 does not cover considerations related to the product claims and labeling, human and environmental safety, socioeconomic factors, properties of the packaging materials, or any other regulations that apply to cosmetics. As a result, several businesses and organizations claim that ISO 16128 might not be sufficient for complete standardization of any cosmetic product. Additionally, the ISO standards are not accessible to common customers since they must be purchased and are not offered for free. Therefore, ISO standards that offer a globally consistent criterion may be used in conjunction with other certifications like COSMOS and NaTrue, that can be mentioned clearly on labels, if the products are found to be compliant with them.

## 2.3 Conclusion

Cosmeceuticals are specially designed cosmetics containing biologically active ingredients to impart pharmaceutical benefits. Though not an authorized term by any of the legislation, it can't be denied that at the present time, cosmeceuticals constitute more than 40% of the global cosmetic market. As discussed in this chapter, the regulatory frameworks for cosmetics appear different in various nations around the globe (summarized in Table 2.2), but still have a certain level of similarity to allow inter-nation transitions. To meet the changing needs of consumers and increasing demand for safe and effective cosmeceuticals, a revision to regulatory guidelines is warranted.

**Table 2.2** Summary of all the regulatory aspects in different countries as discussed in this book chapter.

| Regulatory topics | USA | EU | Canada | Japan | China | India |
|---|---|---|---|---|---|---|
| Cosmetic regulatory body | The Federal Food, Drug, and Cosmetic Act (FD&C) and the Fair Packaging and Labeling Act (FPLA), enacted in 1938 and 1966, The U.S. Safe Cosmetic Act (H.R. 2359) (in 2011) and The Safe Cosmetics and Personal Care Products Act (H.R. 4296) (in 2019) | European Medicines Agency (EMA) Directives [(EC) No. 1223/2009] | Cosmetic Regulation Act (1977) and Food and Drugs Act (1985) administered by Health Canada | Pharmaceutical and Medical Devices Law (PMDL) administered by the Ministry of Health, Labor, and Welfare (MHLW) | Cosmetics Supervision and Administration Regulation (CSAR) under The National Medical Product Administration (NMPA) | Cosmetics Rules, 2020; The Drugs and Cosmetics Act 1940 and Rules 1945 (DCAR), and Bureau of Indian Standards (BIS) under The Central Drugs Standard Control Organization (CDSCO) |
| Regulation of natural/organic ingredients | Natural Cosmetics Act (the Bill – H.R. 5872) (in 2021); the United States Food and Drug Administration (FDA) through National Organic Program (NOP) | The EU's Traditional Herbal Medicinal Products Directive 2004/24/EC | Natural and Non-prescription Health Products Directorate (NNHPD) and Canadian Organic Standards (COS) | No law as such except for some international certifications | No law as such except for some international certifications | National Program for Organic Production (NPOP) under the Agricultural and Processed Food Products Export Development Authority (APEDA) |

| | | | | | | |
|---|---|---|---|---|---|---|
| Pre-marketing product safety | Manufacturers ensure it in accordance with FDA, needing CDER approval | Manufacturer maintains product safety report in PIF | Manufacturer's sole responsibility under CPSP | Manufacturer must adhere to GVP and GQP std | Manufacturers themselves or through a trusted professional agency | Manufacturer needs to maintain safety records |
| Pre-marketing approval | Not required | Not required by Cosmetic Directive | Need to submit CNF | Must submit a marketing notification to regional legislation | Only for special cosmetics by the National Institutes for Food and Drug Control (NIFDC) | Required under state government licensing |
| Ingredient safety concern | The Cosmetic Ingredient Review (CIR) expert panel conducts independent safety reviews, with the results published in the *International Journal of Toxicology* and on the CIR website | As per notes of Scientific Committee on Consumer Safety (SCCS) guidance | Cosmetic Ingredient Hotlist under the sections of FDA and CR | As per MHLW and JCSS | Inventory of Existing Cosmetic Ingredients in China (IECIC) and STSC std | As per The Drugs and Cosmetics Act 1940 and Rules 1945 (DCAR), and Bureau of Indian Standards (BIS) |
| Display of safety information | On principal display panel | On outer and inner packages | On principal display panel | On outer and inner packages | On outer and inner packages | On inner label only |

*(continued)*

**Table 2.2** (*Continued*)

| Regulatory topics | USA | EU | Canada | Japan | China | India |
|---|---|---|---|---|---|---|
| Cosmeceuticals marketing as cosmetics | United States has required extra regulatory hurdles because they are classified as drugs. Some of the substances include sunscreens, anti-caries toothpaste and lip balms | EU allows the marketing of cosmetic products with certain medicinal effects | Governed by the Consumer Packaging and Labeling Act (CPLA) and the Food and Drugs Act's cosmetic regulations; marketed as OTC/NHP | As per MHLW and JCSS | Needs NMPA approval under new CSAR rules | Regulated by CDSCO |
| Color additives | Color additives are not classified as over-the-counter (OTC) drug actives, they are also subject to more regulatory scrutiny in the U.S.; 21 CFR 73, 74, 81, 82 | As per Annex IV of Regulation (EC) No. 1223/2009 | Regulated as per Cosmetic regulation under Food and Drugs Act and Cosmetic Ingredient Hotlist | By MHLW under Pharmaceutical and Medical Devices Law (PMDL) | As per Safety and Technical Standards for Cosmetics (STSC) | Applicable as per BIS and Schedule Q |
| Animal testing | Animal testing bans in eight states (California, Hawaii, Illinois, Maine, Maryland, Nevada, New Jersey and Virginia). For the rest of the country animal testing is not mandatory | Ban on animal testing and marketing for ingredients and products | No bans | No bans but in the process of phasing out. Animal testing is mandatory for quasi-drugs | Testing is not mandatory, but there are preconditions/exceptions | Ban on animal testing for ingredients and products |

| | | | | | | |
|---|---|---|---|---|---|---|
| Labeling | Should comply with FD&C and FP & L; FDA 21 CFR 701 & 740 | Based on latest Council Directive No. 1223/2009; Article 6 of Cosmetics Directive 76/768/EEC | As per INCI, both in English and French | As per List of Cosmetic Ingredient Label Names and MHLW for quasi-drugs | As per Management Measures for Cosmetic Labeling drafted by NMPA | Should comply with part XV of D&C rules 1945; BIS and PCRO |
| Claim of natural/organic | As per USA/NOP and National Cosmetic Act | Under allowed European certifications | From International certifications | From International certifications | From international certifications | As per APEDA rules |
| Expiry date | Not required | Date of minimum durability if it is less than 30 months. Period after opening if durability is more than 30 months | Not required | Must mention "best before" or "expiration" | Must be mentioned as use period/expiry date | Indicated as "use before date" |
| Post marketing reporting system | By Voluntary Cosmetic Registration Program (VCRP) | By the Platform of European Market Surveillance Authorities for Cosmetics (PEMSAC) | By the Consumer Product Safety Program (CPSP) within Health Canada | GVP (Good Vigilance Practices) standards specified by the MHLW Ordinances | Through CSAR | N/A |

## List of Abbreviations

| | |
|---|---|
| FDA | Food and Drug Administration |
| EU | European Union |
| US FDA | United States Food and Drug Administration |
| EMA | European Medicines Agency |
| FD&C Act | Food, Drug, and Cosmetic Act |
| OTC | Over-The-Counter |
| GCP | Good Clinical Practice |
| GDP | Gross Domestic Product |
| CAGR | Compound Annual Growth Rate |
| ICCR | International Cooperation on Cosmetics Regulation |
| OECD | Organization for Economic Co-operation and Development |
| ISO | International Organization for Standardization |
| EC | European Cosmetic Products Regulation |
| NDA | New Medicine Application |
| GMP | Good Manufacturing Practice |
| FPLA | Fair Packaging and Labeling Act |
| CDER | Centre for Drug Evaluation and Research |
| DIN | Drug Identification Number |
| FTC | Federal Trade Commission |
| CBBB | Council of Better Business Bureaus |
| NAD | National Advertising Division |
| NOP | National Organic Program |
| USDA | United States Department of Agriculture |
| CFR | Code of Federal Regulation |
| PDP | Principal Display Panel |
| IP | Information Panel |
| CPLA | Consumer Packaging and Labeling Act |
| NHP | Natural Health Products |
| PCDI | Product at the Cosmetic–Drug Interface |
| NNHPD | Natural and Non-prescription Health Products Directorate |
| CNF | Cosmetic Notification Form |
| CPSP | Canada's Consumer Product Safety Program |
| CR | Cosmetic Regulations |
| INCI | International Nomenclature of Cosmetic Ingredients |
| COS | Canadian Organic Standards |
| ACS | Advanced Cosmetic Sciences |
| ILAC | International Laboratory Accreditation Cooperation |
| PMDL | Pharmaceutical and Medical Devices Law |
| PAL | Pharmaceutical Affair Law |
| MHLW | Ministry of Health, Labor, and Welfare |
| GVP | Good Vigilance Practice |
| GQP | Good Quality Practice |
| JCIA | Japan Cosmetic Industry Association |
| JCSS | Japanese Cosmetic Science Society |

| | |
|---|---|
| CSAR | Cosmetics Supervision and Administration Regulation |
| CHSR | Regulations on Hygiene Supervision of Cosmetics |
| CFDA | China Food and Drug Administration |
| SAMR | State Administration for Market Regulation |
| NMPA | National Medical Products Administration |
| GAC | General Administration of Customs |
| IECIC | Inventory of Existing Cosmetic Ingredients in China |
| PCPC | Personal Care Products Council |
| CIRS | Centre Information Retrieval System |
| CNCA | People's Republic of China's Certification and Accreditation Administration |
| BDIH | Bundesverband Deutscher Industrie-und Handelsunternehmen |
| CPNP | Cosmetic Products Notification Portal |
| SUE | Serious Undesirable Effect |
| THMPD | Traditional Herbal Medicinal Products Directive |
| SCCS | Scientific Committee on Consumer Safety |
| NGRA | Next-Generation Risk Assessment |
| ICCR | International Cooperation of Cosmetics Regulation |
| CCPB | Certification Body for cosmetics and personal care |
| COSMOS | COSMetic Organic and Natural Standard |
| NaTrue | International Natural and Organic Cosmetics Association |
| GACP | Good Agricultural and Collection Practices |
| FSNS | Food Safety Net Services |
| FSSC | Food Safety System Certification |
| DCAR | Drugs and Cosmetics Act 1940 and Rules |
| BIS | Bureau of Indian Standards |
| CDSCO | Central Drugs Standard Control Organization |
| DCGI | Indian Drug Controller General's office |
| NRA | National Regulatory Authority |
| CLA | Central Licensing Authority |
| APEDA | Agricultural and Processed Food Products Export Development Authority |
| NPOP | National Program for Organic Production |
| OSP | Organic System Plan |
| DMDCCC | Drugs, Medical Devices, and Cosmetics Consultative Committee |
| TC | Technical Committees |
| IEC | International Electrotechnical Commission |

## References


1. R. Arora, G. Aggarwal, G. A. Dhingra and M. Nagpal, *Asian J. Pharm. Clin. Res.*, 2019, **12**, 7.
2. B. Fonseca-Santos, M. Antonio Corrêa and M. Chorilli, *Braz. J. Pharm. Sci.*, 2015, **51**, 17.
3. I. Binic, V. Lazarevic, M. Ljubenovic, J. Mojsa and D. Sokolovic, *J. Evidence-Based Complementary Altern. Med.*, 2013, **2013**, 1.

4. H. De Wet, S. Nciki and S. F. Van Vuuren, *J. Ethnobiol. Ethnomed.*, 2013, **9**, 1.
5. J. C. Lee, M. F. Hou, H. W. Huang, F. R. Chang, C. C. Yeh, J. Y. Tang and H. W. Chang, *Cancer Cell Int.*, 2013, **13**, 1.
6. P. Morganti, G. Morganti, A. Gagliardini and A. Lohani, *Cosmet*, 2021, **8**, 65.
7. I. Aranaz, N. Acosta, C. Civera, B. Elorza, J. Mingo, C. Castro, M. L. Gandía and A. H. Caballero, *Polymers*, 2018, **10**, 213.
8. A. Pandey, G. K. Jatana and S. Santhalia, *Cosmeceuticals*, StatPearls Publishing, 2019.
9. S. J. De Salva, *Regul. Toxicol. Pharmacol.*, 1985, **5**, 101.
10. Japan Ministry of Health L and W of J, Act on Securing Quality, Efficacy and Safety of Pharmaceuticals, Medical Devices, etc. (Act No. 145 of August 10, Showa 35), 2014.
11. Signicent LLP, Quasi-Drugs Report: The Regulated Cosmeceuticals with Growing Market, 2022.
12. M. Ferreira, A. Matos, A. Couras, J. Marto and H. Ribeiro, *Cosmetics*, 2022, **9**, 72.
13. G. Karakaya, A. Türe, A. Ercan, S. Öncül and M. D. Aytemir, *Bioorg. Chem.*, 2019, **88**, 102950.
14. R. S. Stern, *N. Engl. J. Med.*, 1994, **331**, 1348.
15. B. J. Vermeer, B. A. Gilchrest, V. Vermeer and S. L. Friedel, *Arch. Dermatol.*, 1996, **132**, 337.
16. https://www.strategicmarketresearch.com/market-report/cosmeceuticals-market.
17. Research and Markets, India Beauty and Personal Care Market is Projected to Register a CAGR of 6.32%, in Terms of Value, Between 2022 & 2027.
18. Mordor Intelligence, Organic Skincare Products Market Analysis - Industry Report - Trends, Size & Share, 2022.
19. International Cooperation on Cosmetics Regulation, About ICCR, standards and structure, accessed on 14 January 2023.
20. The Organization for Economic Cooperation and Development, About the OECD, accessed on 14 January 2023.
21. International Organization for Standardization (ISO) Technical Commitee, ISO/TC 217 - Cosmetics, accessed on 14 January 2023.
22. F. Pistollato, F. Madia, R. Corvi, S. Munn, E. Grignard, A. Paini, W. Andrew, A. Bal-Price, P. Prieto, S. Casati, E. Berggren, S. K. Bopp and V. Zuang, Current EU regulatory requirements for the assessment of chemicals and cosmetic products: challenges and opportunities for introducing new approach methodologies, *Arch. Toxicol.*, 2021, **95**, 1867.
23. Cosmetics Europe - The Personal Care Association, Innovation and Future Trends in the Cosmetics Industry, accessed on 06 January 2023.
24. C. M. Choi and D. S. Berson, Cosmeceuticals, *Semin. Cutaneous Med. Surg.*, 2006, **25**, 163.
25. FDA, Cosmetics Guidance & Regulation, 2022.

26. U.S. Food & Drug Administration, FDA Authority Over Cosmetics: How Cosmetics Are Not FDA-Approved, but Are FDA-Regulated | FDA, 2022.
27. FDA, Is It a Cosmetic, a Drug, or Both? (Or Is It Soap?), 2022.
28. FDA, Cosmetics Safety Q&A: Personal Care Products, 2022.
29. J. D. Schakowsky, H.R.4296 - Safe Cosmetics and Personal Care Products Act of 2019, in 116th Congress, 1st Session. House - Energy and Commerce; Education and Labor, 2019.
30. M. Pinto, United States (U.S.) Cosmetics Regulation – CRITICAL CATALYST, 2021.
31. R. Appel and T. Esposito, How is "natural" regulated in the US? Ingredients/ Regul., 2020.
32. S. P. Maloney, H.R.5872 - 117th Congress (2021-2022), Natural Cosmetics Act. House - Energy and Commerce, 2021.
33. USDA, Labeling Organic Products, Agricultural Marketing Service, accessed on 07 January 2023.
34. U.S. Food & Drug Administration, "Organic" Cosmetics, 2022.
35. L. Mayordomo, Understanding organic and natural certifications in cosmetics, blog.weareprovital.com, accessed on 07 January 2023.
36. Regulatory information for cosmetics - Canada.ca, 2022.
37. Health Canada, Guidance document: Classification of Products at the Cosmetic-Drug Interface, 2008.
38. Health Canada, About Natural Health Product Regulation in Canada - Canada.ca, 2022.
39. Health Canada, Draft: Guidance document - sunscreen monograph - Canada.ca, 2012.
40. Notification of Cosmetics - Canada.ca, 2017.
41. Evaluation of the Consumer Product Safety Program - Canada.ca, 2022.
42. Cosmetic Ingredient Hotlist: Prohibited and Restricted Ingredients - Canada.ca, 2022.
43. Canadian Organic Standards - Canadian Food Inspection Agency, 2022.
44. ACS Canada, CanadaCerts | Natural Cosmetic Certification, accessed on 08 January 2023.
45. Trial Expert C, Cosmetics Registration Process In Japan | Credevo Articles, 2022.
46. Task Force Committee for Evaluation of Whitening Function, Guidelines for Evaluation of Quasi-Drug Whitening Products for New Efficacy Claims Task Force Committee for Evaluation of Whitening Function, *J. Jpn. Cosmet. Sci. Soc.*, 2007, **31**, 432.
47. Task Force Committee for Evaluation of Anti-Aging Function, Guidelines for Evaluation of Anti-Wrinkle Products, *J. Jpn. Cosmet. Sci. Soc.*, 2007, **31**, 411.
48. Y. Mitani, Tapping into Organic/Natural Beauty Opportunities in Japan, Euromonitor.com, 2017.
49. State Council G of C, Regulations on the Supervision and Administration of Cosmetics (National Decree No. 727), 2020.
50. Z. Su, F. Y. Luo, X. R. Pei, F. L. Zhang, S. X. Xing and G. L. Wang, *Cosmetics*, 2020, 7, 98.

51. W. Xu, Asia Pacific Regulatory Updates: GMPs, Safety, Ingredient Restrictions, Claims and More, *Cosmet. Toiletries*, 2022, **137**, 30–35.
52. NMPA Announcement on Issuing the Classification Rules and Classification Catalogue of Cosmetics, nmpa.gov.cn, 2021.
53. CIRS, China CosIng - Chinese Cosmetic Ingredient Regulatory Database, *Online Database*, 2021.
54. Inventories of Cosmetic Ingredients in China are Fully Updated - Regulatory News - Personal and Home Care Products - CIRS Group, *Regul News*, 2021.
55. L. Li, Decoding the Natural and Organic Cosmetic Regulations in China, South Korea, and Japan, 2021.
56. Organic Products Production, procession, labelling and management system requirements, 2020, openstd.samr.gov.cn.
57. CNCA Announcement on Publishing the New Version of < Organic Product Certification Catalogue>" - Certification and Accreditation Administration of China. CNCA Announc No 22. 2019.
58. Legislation, https://single-market-economy.ec.europa.eu/sectors/cosmetics/legislation_en, accessed on 05 January 2023.
59. Market surveillance, https://single-market-economy.ec.europa.eu/sectors/cosmetics/market-surveillance_en, accessed on 05 January 2023.
60. Nanomaterials, https://single-market-economy.ec.europa.eu/sectors/cosmetics/cosmetic-products-specific-topics/nanomaterials_en, accessed on 05 January 2023.
61. Herbal medicinal products, European Medicines Agency, https://www.ema.europa.eu/en/human-regulatory/herbal-medicinal-products, accessed on 05 January 2023.
62. J. Krause, G. Tobin and J. Krause, Discovery, Development, and Regulation of Natural Products, *Using Old Solut to New Probl - Nat Drug Discov 21st Century*, InTech, 2013.
63. P. Ravi, D. S. Reddy, B. Ramu and B. Rajkamal, *World J. Pharm. Pharm. Sci.*, 2016, **5**, 1.
64. What requirements must natural ingredients for cosmetics comply with to be allowed on the European market? CBI, 2022.
65. What is the demand for natural ingredients for cosmetics on the European market? CBI, 2022.
66. T. Grum, Natural and organic cosmetic certifications and standards for EU beauty industry, 2022.
67. *Good Agricultural and Collection Practice for Starting Materials of Herbal Origin - Scientific Guideline*, European Medicines Agency, 2022.
68. World Health Organization (WHO), *WHO Guidelines on Good Agricultural and Collection Practices (GACP) for Medicinal Plants*, 2003.
69. European Federation for Cosmetic Ingredients, Personal Care Products Council, *Effci GMP for Cosmetic Ingredients Including the Certification Scheme for GMP for Cosmetic Ingredients*, 2017.
70. Introduction - enjoying fragrance with confidence, ifrafragrance.org, accessed on 14 January 2023.

71. ISO/TC 176/SC 2 Quality systems, ISO 9001:2015 - Quality management systems - Requirements, Last updated in 2021.
72. ISO Technical Committee, *ISO 16128-1:2016(en), Guidelines on technical definitions and criteria for natural and organic cosmetic ingredients and products — Part 1: Definitions for ingredients*, 2016.
73. ISO Technical Committee 217, *ISO 16128-2:2017/Amd 1:2022(en), Cosmetics — Guidelines on technical definitions and criteria for natural and organic cosmetic ingredients — Part 2: Criteria for ingredients and products — AMENDMENT 1*, 2022.
74. Food Safety Net Services - FSNS, https://www.fssc.com/development/training-organizations/courses/food-safety-net-services, accessed on 14 January 2023.
75. Government of India, *The Drugs and Cosmetics Act 1940 and Rules 1945*, Ministry of Health and Family Welfare, 2016.
76. K. Kavya and H. S. Chawra, *Res. J. Top. Cosmet. Sci.*, 2022, **13**, 14.
77. ASSOCHAM INDIA, Indian Cosmeceutical, Cosmetics & Personal Care Market 2022, in *COSMé India*, 2017.
78. N. Bindra and K. Bindra, *Cosmetics Regulatory Services India*, NKG Advis Bus Consult Serv Pvt Ltd, 2021.
79. Central Drugs Standard Control Organization, Directorate General of Health Services, *Cosmetics Rules, 2020*, Ministry of Health & Family Welfare, Government of India, 2022.
80. Bureau of Indian Standards, *Classification for Cosmetic Raw Materials and Adjuncts Part 1 Colourants (Fourth Revision)*, The National Standards Body of India, 2020.
81. S. R. Jha, *Indian Standard of Cosmetics Raw Materials (Is 4707 Part – II)*, S J EXIM Serv, 2019.
82. Draft Traceability Standards for Organic Cosmetics and Personal Care Products, Agricultural and Processed Food Products Export Development Authority; Ministry of Commerce and Industry, Govt. of India, 2019.
83. Important announcements by Agricultural and Processed Food Products Export Development Authority, APEDA.gov.in, Minist. Commer. Ind. Gov. India, 2022.
84. S. Kumar and R. N. Gupta, Regulation for safety and quality of cosmetics vis-a-vis colourants in India compared with other nations, *Pharm. Lett.*, 2012, **4**, 181.
85. M. Rewari and V. Sharma, New Drugs, Medical Devices And Cosmetics Draft Bill, 2022 - India, *Life Sci Biotechnol Nanotechnol*, by mondaq.com, 2022.
86. Ministry of Health and Family Welfare, *Draft of New Drugs, Medical Devices and Cosmetics Bill*, Department of Health and Family Welfare; Government of India, 2022.
87. M. Ilancheran, *India's New Drugs Medical Devices and Cosmetics Draft Bill 2022*, Guest Column Clin, 2022.

88. L. Teles, *ISO 16128 – Guidelines For 'Natural' And 'Organic' – Critical Catalyst*, 2021.
89. COSMOS, https://www.cosmos-standard.org/en/certification, accessed on 05 January 2023.
90. NaTrue, https://natruc.org/our-standard/certification-process-step-step, accessed on 05 January 2023.
91. FAIRTRADE, https://www.fairtrade.net/act/become-a-fairtrade-licensee, accessed on 05 January 2023.
92. FAIRFORLIFE, https://www.fairforlife.org/pmws/indexDOM.php?client_id=fairforlife&page_id=become&lang_iso639=en, accessed on 05 January 2023.
93. FAIRWILD, https://www.fairwild.org/fairwild-certification, accessed on 05 January 2023.
94. UEBT, https://uebt.org/certification-bodies, accessed on 05 January 2023.

CHAPTER 3

# Methods of Isolation, Characterization and Authentication of Essential Oils for Cosmetics

NIBEDITA NATH[a], SUBHENDU CHAKROBORTY[b*] AND KAUSHIK PAL[c]

[a]Department of Chemistry, D.S Degree College, Laida, Sambalpur, Odisha, 768214, India; [b]Department of Basic Sciences, IITM, IES University, Bhopal, Madhya Pradesh, 462044, India; [c]Department of Physics, University Centre for Research and Development (UCRD), Chandigarh University, Gharuan, Mohali, Punjab, 140413, India
*E-mail: subhendu.cy@gmail.com

## 3.1 Introduction

It's in our nature to appreciate aesthetics. As a result, the history of cosmetics and other forms of personal care products can be traced back to ancient times, when people tried everything from lemon juice and honey to clay and milk to mud and even arsenic in the name of beauty. Today, we use materials such as these to make cosmetics and other personal care products. At present, the personal care and cosmetics industry is a significant one that provides a significant contribution to the overall GDP of the world.[1] According to a survey published by Grand View Research, the total value of the worldwide market for organic goods is anticipated to amount to $42.1 billion by the year 2030.

Bioprospecting of Natural Sources for Cosmeceuticals
Edited by Deepika Kathuria, Ajay Sharma, Meenakshi Verma and Gulzar Ahmad Nayik

Published by the Royal Society of Chemistry, www.rsc.org

This pattern demonstrates that people are becoming increasingly worried about the effects that their actions have on the environment and their health.[2]

Natural ingredients for skin care are becoming increasingly popular due to their defensive and protective roles against the production of free radicals and decreased production of oxidative enzymes. These oxidative enzymes include collagenase and elastase, which are typically responsible for the breakdown of the extracellular matrix of the skin, as well as tyrosinase, the enzyme that catalyzes the most crucial step in the biosynthesis of melanin.[3] Natural ingredients for skin care are also gaining popularity as a consequence of this, and there is rising interest in the discovery of natural sources that are capable of blocking these enzymes and the use of these natural sources as prospective cosmetic components in the creation of cosmetics for topical application.[4]

Cosmetics are "items meant to be poured, sprayed, sprinkled, rubbed, or otherwise applied to the human body for washing, beautifying, boosting attractiveness, or altering the look without hurting structure or function," according to the Food, Drug, and Cosmetic Act.[5] Objects that are poured, sprayed, sprinkled, or rubbed against the skin, or that attempt to penetrate the skin, are also considered cosmetics. "Cosmeceuticals" are products that combine elements of both cosmetics and medications.[6,7] Cosmeceuticals, as these items are known, include active ingredients that might alter bodily structure or function.[8] Cosmetics may be chemically made or biologically derived. The ideal ingredients for cosmeceuticals should be those that can be digested by the skin and are safe, effective, new, stable, economical to produce, or all of the above.[9]

## 3.2 Essential Oils

Essential oils (EOs), often referred to as essences, aetheroleum, etheric oils or volatile oils, are the organic compounds that combine to generate these substances.[10,11] Terpenes, terpenoids, and phenylpropenes are only a few of the bioactive chemical components that make up EOs. More than 17 000 species of aromatic plants, many of which are members of angiosperm families like Lamiaceae, Zingiberaceae, and Asteraceae, may generate them.[12] Essential oils, commonly referred to as "essences," are odorous and flammable compounds present in plants that are extracted either through co-distillation with solvents or by steam distillation.[13]

Essential oils can originate from various parts of the plant, such as the flowers and inflorescences (including ylang–ylang, rose, lavender, and chamomile), leaves (including rosemary, peppermint, laurel, lemongrass, and basil), seeds (including cumin, anise, and cardamom), fruits (including nutmeg and black pepper), peel (including bergamot, orange, tangerine, and lemon), bark, roots, and berries (including juniper and all spice).[14–17] Different plant sources for EOs are described in Table 3.1.[18]

Essential oils from fruits and vegetables, like oranges, have generated the most money in the cosmeceutical industry in particular because they have a

**Table 3.1** Sources of EOs from different plant families.

| Type of plant | | Plant family | Essential oil-bearing plants |
|---|---|---|---|
| Gymnosperms | | *Pinaceae* | Fir and pine |
| | | *Cupressaceae* | Cedar leaf, cedarwood, juniper |
| Angiosperms | Monocotyledons | *Zingiberaceae* | Ginger and cardamom |
| | | *Poaceae* | Vetiver and aromatic grass |
| | Dicotyledons | *Apiaceae* | Coriander and fennel |
| | | *Geraniaceae* | Geranium |
| | | *Asteraceae* | Chamomile, wormwood |
| | | *Lamiaceae* | Lavender, patchouli, mint, oregano |
| | | *Myristicaceae* | Mace and nutmeg |
| | | *Lauraceae* | Cinnamon, camphor, *Sassafras* |
| | | *Rosaceae* | Rose |
| | | *Santalaceae* | Sandalwood |
| | | *Oleaceae* | Jasmine |

variety of health benefits, including improving treating scars, firmness, skin elasticity, pimples, and stretch marks. Due to their antibacterial qualities, lemon and orange EOs are excellent ingredients for hair care and skin products. The antibacterial qualities of such oils have also created new opportunities for commercial expansion.[19]

## 3.3 Chemistry of Essential Oils

Essential oils may contain a combination of aliphatic (non-aromatic) and aromatic molecules, each of which contribute to the aroma that is experienced.[20] Volatile oils are liquid and colorless at room temperature. Exceptions include cinnamon, cloves, *Sassafras*, and vetiver, which have a unique fragrance and low density. They also have a refractive index.[21] Diverse structures create EOs with varying numbers and quality. Essential oils are found in the cytoplasm of inner secretory cells, secretory hairs, secretory pockets, and epidermal cells. These are the types of plant cell secretions that produce EOs. These oils are intricate blends that could include more than 300 different substances.[22] They are made up of volatile organic substances that typically have low molecular weights (under 300). These volatile molecules include types of aldehydes, amines, ethers or oxides, amides, alcohol, phenols, ketones, esters, heterocycles, and primary terpenes. A wide range of aromatic notes, including citrus (limonene), herbal (selinene), floral (linalool), fruity ((*E*)-nerolidol), *etc.*, are offered by alcohols, aldehydes, and ketones.

Terpenoids and phenylpropanoids are two main chemical groups that make up plant EOs. Terpenoids have a wide range of oxygenated derivatives, including peroxides, phenols, ketones, alcohols, esters, aldehydes and ethers, as well as diverse carbon skeletons (see Figure 3.1). Essential oils contain different types of terpenes, some of which are functionalized derivatives of ketones or derivatives of aldehydes (citronellal, sinensal), alcohols

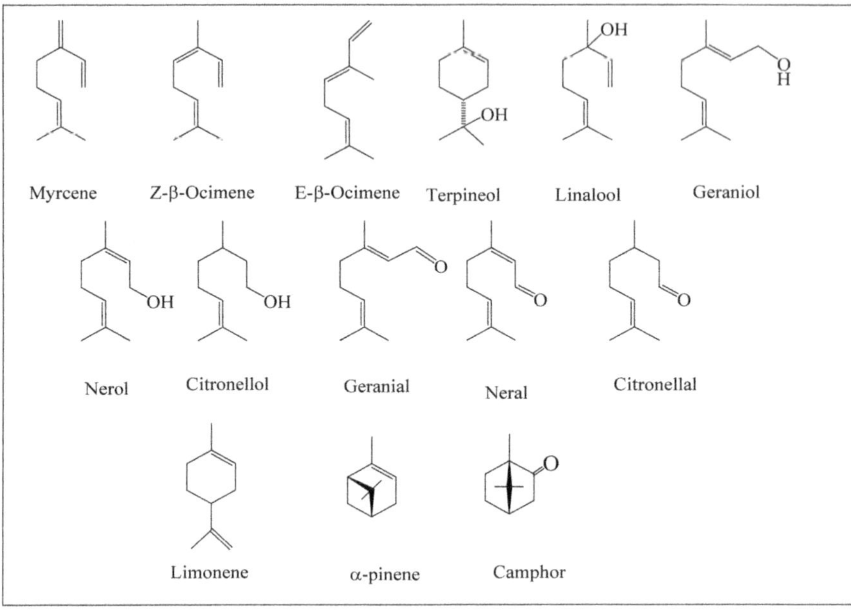


**Figure 3.1** Structures of important terpenes.

(geraniol, bisabolol), phenols (menthone, *p*-vetivone) and esters (terpinyl acetate, cedryl acetate). Essential oils are also used for their aromatherapy properties (thymol).[23]

Phenylpropanoids possess one too many $C_6$–$C_3$ fragments, with the $C_6$ unit often consisting of a benzene ring. It is not possible to provide a standard categorization for these chemicals.[16] The plant families Apiaceae, Lamiaceae, Myrtaceae, and Rutaceae are the most common ones to be found among the plants that contain phenylpropanoids as a primary source.[24] Essential oils are known to include a number of different phenylpropanoids, including anethole, apiol, cinnamaldehyde, eugenol, myristicin, and safrole (see Figure 3.2).

## 3.4 Methods of Isolation

Essential oils are extracted from several aromatic plants. The local population's reliance on these plants as a medicine has helped propel them to the forefront of public consciousness. All parts of the plant, from the buds and flowers to the seeds and leaves and even the stem, can be used to extract EOs. Epidermal cells, together with the channels, secretary cells, and cavities, are storage sites for EOs.[25] Essential oils are what give plants their distinctive aromas. Before the EOs are extracted, the plant components are entirely dried, left fresh, or partially dehydrated. The pace at which EOs may be taken from

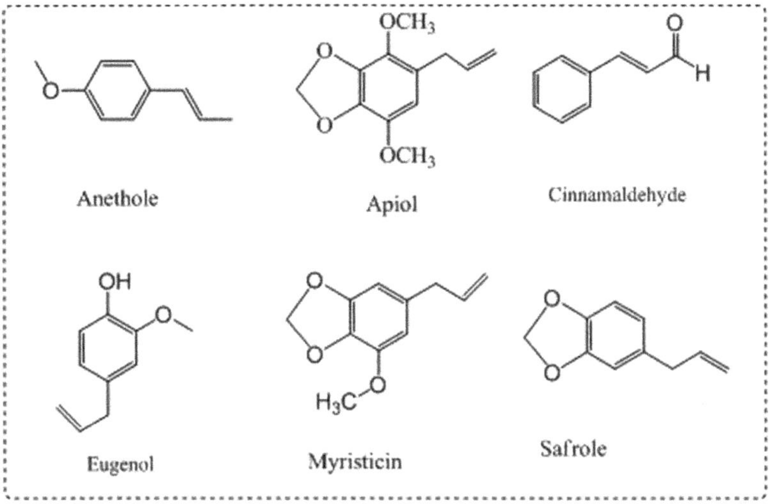


**Figure 3.2** Structure of some phenylpropanoids.

a plant is determined by diffusion through the plant's tissues, which involves the surface from where the oil has been removed.

The consistency of the EO is an important factor to consider when trying to extract it. Both the steam distillation technique and the hydro distillation procedure are considered to be among the most significant strategies for the extraction of EOs. These approaches are the most suitable and efficient ones that may be used for the extraction process.[26] Other methods of extraction can be utilized; however, these approaches are not going to be particularly useful for this operation. These include the use of microwaves or liquid carbon dioxide, as well as high-pressure or low-pressure distillation using either steam or hot water.[27]

The steam distillation (SD) method is the one that is utilized the most frequently in the process of extracting EOs. The extraction of EOs using this process gives a yield of 93%, while the remaining 7% can be obtained *via* the use of alternative techniques.[28] In its most basic form, the procedure starts with the heating of the plant material by means of steam that is created by a steam generator. Heat is the major factor that determines how successfully the structures of the plant materials deteriorate, burst, and release aromatic components or EOs.[29] Steam distillation is the method that is most usually used to produce profitable extraction. Essential oils have been extracted from *L. cubeba* fruits, leaves, and roots with high yields and their use in clinical applications has been predicted.[30] Masango created a novel steam distillation extraction technique in order to enhance the isolated EO yields while simultaneously reducing the amount of wastewater produced during the extraction process. This technique makes use of a bed that is packed with plant samples very tightly and positions it directly over the steam source. It is strictly prohibited for boiling water to come into contact with the organic components of the plants, and only steam is permitted to travel through them. Because of this, the process requires less steam, and as a consequence, it is possible to minimize the amount of water that is used during the distillation step.[28]

Avicenna was the first person to establish extraction using a vessel known as an alembic, which is the oldest and simplest method of oil extraction. Avicenna also discovered the hydrodistillation (HD) method. Rose was the first plant extract to be utilized and cleaned up with the help of this process. After the plant materials have been submerged in water within the alembic, the next step is to bring the entire mixture to the boil. This is the first step in the process. A heating source, the alembic, a condenser that transforms the vapor that is released from the vessel into a liquid, and a decanter that receives the condensate and isolates the EOs from the water are the components that make up the piece of equipment.[31] This extraction technique is used for extracting plant materials such as flowers or wood, and also used for extracting plant materials having a high boiling point. This approach permits the extraction of EOs to take place at a regulated temperature without causing oil overheating. This is made possible by the fact that the oils are surrounded by water. The most significant advantage of utilizing this extraction method is that it allows for the separation of plant components at temperatures lower than 100 °C.[32]

Orio *et al.* coupled conventional HD with microwave-generated HD in the processing of mint leaves (MGH). The MGH principle states that increasing the microwave polarity will improve EO extraction since it will influence the water solubility in the sample.[33] Rivera *et al.* devised a new extraction technique for the purpose of extracting caraway (*Carum-carvi* L.) EOs using HD by direct induction of heat generated by a magnetic field (DIHMF). This approach was utilized to extract caraway.[34–36] In their research on *Zataria multiflora* Bioss EOs, Farahnaky *et al.* developed a unique method called Ohmic-assisted hydrodistillation (OAHD). This helped them cut down on the extraction time as well as other factors while achieving effective results. By coupling electrodes to a power supply that utilizes alternating current, the unique method known as was developed. This method makes it possible to heat an object *via* the ohmic heating process.[35] Table 3.2 shows the extraction of EOs from various plant species using the HD technique.

**Table 3.2** EO extraction using the HD technique from various plants.

| Technique | Time (h) | Species | Reference |
|---|---|---|---|
| Hydrodistillation | 4 | Leaves of *Annona Foetida* | 37 |
| Clevenger-type drying | 3 | *Hypericum* taxa | 38 |
| salt: Sodium sulphate | 5 | *Litsea cubeba* | 39 |
| and anhydrous mag- | 2.5 | *Eucalyptus* species | 40 |
| nesium sulphate | 2.5 | *Chamomile* species | 41 |
| | 3 | *Citrus* species | 42 |

The standard methods have a number of drawbacks, one of which is that they encourage the EOs to go through chemical changes such as oxidation, isomerization, and hydrolysis. Because of these drawbacks, more changes have been made to the extraction processes. These procedures entail high temperatures, which not only lower the overall quality of the EOs but also make the extraction process take significantly longer. During the process of extracting EOs, it is absolutely necessary to preserve the chemical contents of the oils as well as their original proportions. Traditional extraction techniques, such as steam distillation and solvent extraction, require more time and a large amount of organic solvents to complete the extraction process.[42] In addition, the disadvantages of these techniques, such as volatile component losses, low oil extraction efficiency, unsaturated compound degradation, and so on, must be taken into consideration by new extraction techniques.[32] New extraction techniques must take into consideration parameters such as solvent used, energy consumption, extraction time, and carbon dioxide emissions.

In this context, an innovative process known as supercritical fluid extraction (SFE) combines 96.5% carbon dioxide with other factors.[43,44] As of late, there has been a lot of attention paid to the use of supercritical carbon dioxide (S-$CO_2$) as a medium for extraction, and this trend is expected to continue. Researchers are becoming increasingly interested in supercritical fluids as a result of the fact that their solvent power and selectivity may be adjusted. In this method, carbon dioxide was utilized as a supercritical fluid (S-$CO_2$). Because of its positive effects on the environment, the chemistry, and the economy, S-$CO_2$ is an attractive candidate for use as a solvent. An alternative to existing methods and technologies that have limits in terms of extraction efficiency, time, and different environmental challenges is the use of solvent fractionation (SF) to extract EOs.[45]

Modernization, extraction parameters, and material preparation are being taken into consideration for development of SFE. Freon™ 22, nitrous oxide, and hexane were the solvents of choice for Wang *et al.*'s process of extracting polar chemicals.[46] Because of their essential features, the use of polar molecules is restricted. Wai *et al.* conducted research on the effect that several modifiers, including water, acetonitrile, ethanol, acetone, and methanol, had on SFE. Because of its active polarity, methanol is the polar co-solvent modifier that is employed most frequently. It is miscible with $CO_2$ to a concentration of 20% and is less hazardous than ethanol. When using fresh plants for solvent-free extraction of EOs, mechanical challenges arise due to the high moisture content of the plants. A mixture of sodium sulphate and silica gel is added to the plant material in order to prevent such problems from occurring.[47–50] This helps the plant material to keep its moisture and improves the effectiveness of the SFE process.

In a broad sense, the idea of SFE refers to the process of reusing and recycling the same fluid *via* several cycles of compression and decompression. The supercritical state of $CO_2$ may be achieved by first subjecting it to extremely high levels of compression, and then heating it. After that, the raw plant material is processed in order to load volatile materials and plant extracts. After that, the combination of $CO_2$ and plant extracts is sent to two separators, where it undergoes gradual decompression so that the $CO_2$ may be separated from the extracted plant material. As a result of the rapid transformation that takes place when $CO_2$ is exposed to the normal pressure and temperature of the atmosphere, the $CO_2$ is released from the second separator and recirculated back into the storage tank without leaving any trace of solvent residue in the final product.

Conventional extraction techniques, such as solvent and hydrodiffusion, face challenges such as the loss of variety of evaporative components, inadequate isolation coherence, and hazardous solvent residues at the end product stage. These challenges were the impetus for beginning research on solvent-free microwave technology. Solvent-free microwave extraction (SFME) may be utilized in a variety of contexts.[28] Using this technique, EOs derived from spices, aromatic herbs, and dry seeds may be quickly isolated. SFME also offers a variety of other benefits. Researchers have reported the advantages of this method, which can be summed up as follows: obtaining EOs with a shorter extraction time, an environmentally friendly process, and high yields and selectivity.[28] In SFME, first the plant samples undergo heating using a microwave, followed by dry distillation at atmospheric pressure without any solvent. When it comes to the environment, SFME is a great choice. Oregano was extracted by Bayramoglu *et al.* using the SFME technique at four distinct microwave power levels: 249 W, 498 W, 373 W, and 622 W. Essential oil yields were obtained for each microwave power level. At power levels of 249 W, 498 W, 373 W, and 622 W, respectively, the highest possible yields of oregano EO were produced. These yields were 0.049 mL $g^{-1}$, 0.053 mL $g^{-1}$, 0.052 mL $g^{-1}$, and 0.054 mL $g^{-1}$, respectively.[51]

The extraction of puerarin from the plant *Radix puerariae* using a microwave takes only one minute. Phutdhawong and his colleagues conducted research on *Cinnamomum iners Reinw.* Linalool-containing EOs were successfully extracted, and a significant amount of oil was obtained. When comparing extraction by traditional HD with extraction by microwave-aided HD, GC chromatograms were utilized, and the results revealed that extraction by microwave-assisted HD resulted in a better yield of the components when compared to conventional HD.[52–54] Table 3.3 shows various parameters for extracting EOs from various plant species using the microwave extraction method.

For the purpose of extraction, sonication is regarded to be quite trustworthy, and the efficiency with which it operates improves the amount of material extracted.[61] The production and kinetics are entirely controlled by the ultrasonic frequency. However, ultrasound's features in terms of extraction yield and kinetics vary conditionally based on the source plant that is being separated. A study showed that EOs could be successfully extracted from pyrethrum flowers with a high yield, which contrasts with the impact of kinetics and woad seed oil production. When the temperature

**Table 3.3** Extraction of EOs using a microwave-assisted extraction method.

| Technique | Parameters | Species | Reference |
|---|---|---|---|
| Microwave | 30 min, 800 W | *Cinnamomum iners Reinw. Ex* | 55 |
| | 15 min, 600 W | *Eucalyptus × hybrida* Maiden | 56 |
| | 1 h, 800 W at 2450 MHz | *Laurus nobilis* L. | 57 |
| | 45 min, 800 W | *Salvia triloba* L. and *Laurus nobilis* L. | 58 |
| | 10 min | *Nigella sativa* L. seed EO | 59 |
| | 10 min, 170 W | EOs | 60 |
| | 150 min, 800 W | *Geranium sylvaticum* | 61 |
| | 150 min, 800 W | *Cymbopogon citratus* | 62 |
| | 20 min | Mint leaf EOs | 34 |

**Table 3.4** Extraction of EOs using an ultrasound-assisted extraction method.

| Technique | Parameters | Source of EOs | Reference |
|---|---|---|---|
| Ultrasound | 40 °C, 20–30 min | *Salvia* sp. | 67 |
| | 25 °C, 10 min | *Oliveria decumbens* Vent | 68 |
| | 40–60 °C, 45 min | Pomegranate seeds | 69 |
| | 70 °C, 28 min | Cumin seeds | 70 |
| | 32.5 °C, 10.5 min | *Elettaria cardamomum* Maton | 71 |
| | 50 Hz, 5 min | *Curcuma* rhizomes and *Curcumae Radix* | 72 |

reaches 66 °C, the solid–liquid extraction of pyrethrum flowers, also known as solid–hexane extraction, produces a greater yield than when the temperature is lower. Therefore, ultrasonography has a substantial impact on the temperature range that is optimal, which is 40 to 66 °C.[62] UAE is an innovative and cost-effective method with simple and easy to handle equipment. The initial step in solvent extraction is ultrasonication.[63,64] Similar to the popular and effective Soxhlet extraction, which can be used to isolate many different types of EOs, UAE works in much the same manner. However, the quantity and dynamics of the plant matrix influence the characteristics of the extraction performed using this approach. The UAE method was created by researchers as a result of higher extraction yields of 15 to 30 percent greater quantities of ginseng EOs.[65,66] The criteria that must be satisfied in order to extract EOs using UAE from a wide variety of plant species are outlined in Table 3.4.

The components that are contained inside the extracts that are formed by the aforementioned procedures are a complex combination of naturally occurring compounds that range in polarity. For the production of pure bioactive chemicals, separation and purification are both necessary steps. Their separation continues to be a key problem in the characterization and identification of pure bioactive natural compounds. In recent years, there has been progress made in the purification and isolation of natural products.[71]

Paper chromatography, HPTLC, gas chromatography, HPLC, TLC, OPLC, and column chromatography are all methods that have been utilized in order to isolate and purify a variety of bioactive natural products. Column chromatography and thin-layer chromatography (also known as TLC) are still utilized often[72] because they are simple, cost-effective, and readily available with a variety of stationary phases.

The analytical method known as gas chromatography (GC) is used to separate substances in accordance with their vaporization rates. GC can provide both qualitative and quantitative information on the individual chemicals that make up a sample. While the liquid phase remains steady and unmoving, the gas phase is now in motion. The distribution of the chemical species in the gas phase is what determines the pace of migration of the species. For instance, a species that entirely disperses itself into the gas phase would migrate at the same pace as the moving gas, but a species that totally disperses itself into the stationary phase will not move at all. The migration rate of species that are distributed partially in both phases will be intermediate.[73] Micalizzi *et al.* used the GC-MS method to assess the enantiomeric distribution of chiral compounds in an EO. This method was also used to determine the authenticity of samples. Species that are distributed partially in both phases will migrate at an intermediate rate.[74]

For the purpose of isolating natural products, a method known as high performance liquid chromatography (HPLC) provides a flexible approach that is also very reliable. The HPLC methodology is an analytical method that is used to separate and identify the organic and inorganic solutes present in a range of samples. These samples might come from the pharmaceutical, food, environmental, or industrial sectors. Among the several analytical methods now available, this one is quickly becoming the favorite for fingerprinting research aimed at ensuring the purity of medicinal plants. Selecting a detector is the first step in using HPLC for substance identification. The relative intensities of the stationary phase and movable phase are the primary determinants of the degree of separation. Contemporary HPLC typically employs a non-polar solid phase (C18) and a polar liquid phase (water or a water-based solvent combination).[75]

Innovative in the field of parallel chromatography, OPLC brings together the best features of both traditional TLC and high-performance thin-layer chromatography (HPTLC). OPLC's versatility as an analytical and preparative tool makes it a great choice for use in research and quality control laboratories. It's a very effective method of liquid chromatography that combines the advantages of thin-layer chromatography (TLC), high-performance liquid chromatography (HPLC) and flash chromatography (FC). Finally, we have a technique that can compete with, if not surpass, HPLC and FC. In OPLC, like in other chromatographic methods, a pump is used to push a mobile phase (usually a liquid) through a stationary phase (typically a solid such as silica). This method is utilized by OPLC in order to separate substances.[73]

## 3.5 Methods of Characterization

Spectroscopic data from several methods, including mass spectrometry, nuclear magnetic resonance (NMR), UV-visible, and infrared (IR), are analyzed to determine the structure of EOs. These spectra can be used to help deduce the chemical structures of EOs.[76]

UV-visible spectroscopy is used to definitively identify classes of compounds in pure samples as well as in biological mixtures. Because aromatic compounds are powerful chromophores in the UV region, UV-visible spectroscopy is the method of choice for quantitative analysis. One can determine the identity of naturally-occurring chemicals using UV-visible spectroscopy.[77] In addition, it was found that spectroscopic UV-visible methods were less selective and gave information on the composition of the overall polyphenol concentration. Infrared spectroscopy that makes use of the Fourier transform is a good approach for finding functional groups in plant extracts. Comparatively speaking, this method is less costly and takes less time. It is helpful in both the identification of molecules and the determination of their structures.[78] It uses analytical equipment with a high resolution that can decode structural compounds and determine the components of chemical substances. The Fourier transform infrared spectroscopy (FTIR) method allows for the rapid and non-destructive fingerprinting of herbal powders or extracts.

Through the use of NMR spectroscopy, one can get insight into the material's physicochemical and biological properties. In place of the more common practice of employing one-dimensional NMR techniques, it is possible to determine the intricate structure of the molecules by employing two-dimensional NMR techniques. A technique that may be described as solid-state NMR spectroscopy can be used to analyze the molecular structure of solids. Radio labelled $^{13}$C NMR is used to determine the different types of carbon atoms present in the chemicals. $^{1}$H-NMR is used to determine the kinds of hydrogen that are present in the chemical as well as the connections between the hydrogen atoms. This may be done in addition to determining the connections between the hydrogen atoms. With the use of mass spectrometry (MS), it is possible to recognize and quantify known substances, and this technique is also beneficial in gaining an understanding of the structural and chemical features of molecules. Utilizing the MS spectra, one is able to arrive at an estimate of the sample's molecular weight. This approach, which simultaneously specifies the molecular weight and a diagnostic segment, is used for structural elucidation of organic compounds, sequencing of peptides or oligonucleotides, and monitoring the presence of previously defined compounds in complicated mixtures with great specificity. It determines a diagnostic molecule fragment and molecular weight.[75]

## 3.6 Methods of Authentication

One of the main issues with manufacturing EOs on a large scale is the relatively low yield per mass of raw material, which appears to be the cause of the typically high costs that the oils command on the global market.[79] This is one of the apparent causes of EO adulteration, together with the rising demand for them. Adulteration may potentially make a substance more dangerous, especially when it comes to skin reactions. Concern also exists regarding the co-existence of pollutants and adulterants.[80] Therefore, authentication is extremely important for both users and chemical companies.[81]

Essential oils have been studied for a very long time using both physical and chemical constants. These data provide insight into the quality of the oils and are essential for determining whether or not they have been adulterated. Examples of physical constants are the optical rotation, specific gravity, and refractive index. When attempting to describe solubility, it might be helpful to talk about how effectively it dissolves in ethyl alcohol. On the other hand, the chemical constants that are used to assess EOs include the concentrations of alcohol, aldehyde, and ketone, in addition to the acid value and ester value after acetylation, and a few more variables. The consideration of these aspects makes it possible to assess the quality of EOs and to display the characteristics of such oils, as shown in Figure 3.2.[82] For this, a universal definition of an EO is needed. According to the "Association Française de Normalisation" (AFNOR) and the European Pharmacopoeia (Ph. Eur.), EOs are items produced from pure, recognized plant raw materials and acquired by hydrodistillation, steam distillation, mechanical methods, or dry distillation for some woods. The Ph. Eur.[81] created this definition.

A number of quality standards may be defined by the standardization processes. Non-compliance with standardization testing doesn't always indicate tampering. Ageing, processing, or storage may cause racemization or polymerization, which affects critical attributes without adulteration.[83] Adulteration can involve the use of non-volatile chemicals, cheaper synthetic or natural molecules, volatile compounds, EOs from other natural sources, or vegetable oils.[81,84] On the other hand, it is also a very popular practice to adulterate EOs by completely or partially substituting the plant for a different species.[85,86] It is important to note that adulteration lowers the quality of the EOs and, in many situations, can have negative effects on safety. As a result, authentication is crucial for assuring both consumer safety and the caliber of the methods used to produce EOs, which has an impact on regulatory and standards issues.[87]

It should be noted that adulterants are regularly added to EOs at relatively low concentrations (5–8% w/w), which makes it difficult to identify them using standard analytical techniques.[88] Two methods are frequently used to evaluate product adulteration: one is finding product fingerprints and the other is identifying a particular marker in the products.[81] As a result, it is possible to assess whether a product complies with regulatory entity standards, guaranteeing the EO's identity, quality, safety, and effectiveness.[89]

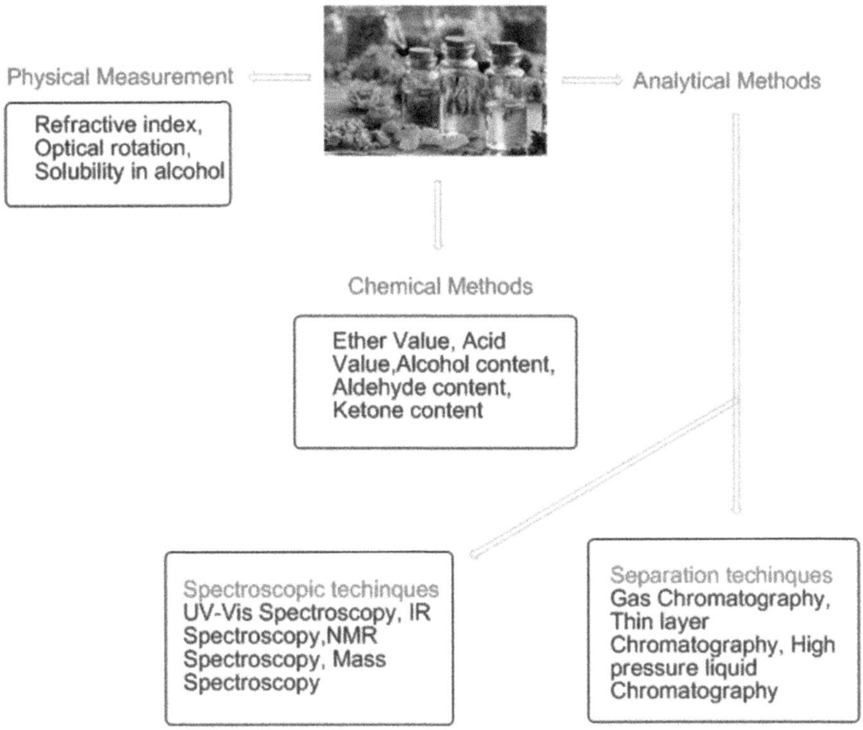


**Figure 3.3** Analysis of EOs.

The purity of EOs can be assessed on two separate scales. The first involves a panel of sensory analysts evaluating the product's organoleptic characteristics.[90] The foundation of the second is the analysis of a number of physicochemical constants, such as optical rotation, refractive index, and specific gravity. The ability of the EO to be dissolved in certain solvents, such ethanol, is another characteristic that may be used to determine whether or not it has been tampered with. The investigation of certain chemical characteristics, such as acidity and the concentrations of ester, ketone, aldehyde, or alcohol, on the other hand, may also reveal information that is extremely helpful when determining the degree of purity of a particular EO. Essential oils are characterized by these factors, which are of considerable assistance in determining the oils' overall quality (see Figure 3.3).[82]

## 3.7 Application of Essential Oils in Cosmetics

The growing market demand for herbal cosmetics has piqued the interest in EOs, widely used as enhancers or active ingredients in aromatherapy, cosmetics, and fragrances.[91–93] The use of EOs for health, beauty, and healthy lifestyle dates back to prehistoric times. Concerns about the potential dangers of synthetic ingredients to human health have grown in recent decades. As a result, the utilization of natural materials to improve human health and beauty has been a rising trend.

Clove, citronella, patchouli, and nutmeg are the four primary EOs found in Indonesia that have been shown to have antiaging properties due to the presence of functional groups in their chemical compositions. On rat skin that had been exposed to UVB, these oils had the effect of reducing the wrinkle score (exogenous factor). The skin irritation potential of patchouli oil was 0; the skin irritation potential of nutmeg oil was 0.02; the skin irritation potential of citronella oil was 0.09; and the skin irritation potential of clove oil was 0.39. The following is the sequence in which the antiaging benefits of these oils were discovered: in order of effectiveness, clove is superior to patchouli, nutmeg, and citronella oil as an endogenous factor.[94]

Both the EOs of *P. dioscoridis* and *E. bonariensis* have been shown to have anti-aging effects. The primary components of *P. dioscoridis* EO are muurolene, aromadendrene oxide-2, berkheyaradulen, dehydro-cyclolongifolene oxide, and maaliene. While the predominant components of *E. bonariensis* EO are *trans*-β-farnesene, *O*-ocimene, and maaliene, these compounds are found in very low concentrations. The EOs of *P. dioscoridis* and *E. bonariensis* in a 1 : 1 mixture demonstrated significant anti-aging activity, which was greater than that of the EOs of *P. dioscoridis* or *E. bonariensis* alone.[95]

Both geranium essential oil (GEO) and calendula essential oil (CEO) were put through tests to determine their antioxidant and sun protection factor (SPF) properties. Oils are elevated to the role of active component in a variety of cosmeceutical formulations because of the protective and preventive capabilities that they possess. The combination of these two features raises oils to this role. The GEO SPF value was determined to be 6.45, while the CEO SPF value was discovered to be 8.36. Both CEO and GEO have a greater SPF than that of GEO; therefore, these EOs can be utilized as sunscreens to prevent sunburn. GEO and CEO's antioxidant and SPF capabilities can deliver a synergistic photoprotective impact or boost the formulation's added value.[96]

Lipstick has been regarded as one of the most iconic beauty products for some time. Pigments are essential to the production of lipstick since they impart color not only to the product but also to the user's lips.[97] Castor oil and octyldodecanol were examined with meadowfoam seed oil as a lipstick pigment dispersion. Meadowfoam seed oil, octyldodecanol, and castor oil were used to spread and stabilize lipstick Red 7 Lake powder. Meadowfoam oil, octyldodecanol, and castor oil are dispersants. Castor oil absorbed more than meadowfoam seed oil or octyldodecanol; therefore, more was needed to moisten and stabilize the pigment. Octyldodecanol made smooth, pigmented lipsticks. They were softer than castor oil lipsticks, but harsher than octyldodecanol lipsticks. The lipsticks made from 40% meadowfoam seed oil were softer than castor oil lipsticks, as they had the lowest friction and payoff. To the untrained eye, each shade of lipstick appeared to be almost identical. It is possible to say the same thing about the color on printer paper.[98]

EOs are also used in hair care products. The hair growth potential of oleogels containing rosemary oil and cedarwood oil was evaluated using beeswax as the organogelator.[99]

## 3.8 Conclusions and Future Prospects

People are showing an increased interest in utilizing skin care products, and this interest is fast growing. Because of this factor, cosmetics businesses all over the world are expanding at an accelerating rate. Herbal formulations are currently receiving priority status in this establishment as a direct result of the increased requests placed by clients. The use of EOs in herbal skin care formulations has gained popularity in recent years due to their beneficial effects on the skin. These oils, for instance, are effective against a wide range of skin conditions, including acne, pimples, eczema, and psoriasis, and they promote the appearance of healthy, radiant skin that makes you seem years younger. In addition, since some EOs have been shown to cause skin allergies or other types of irritations, more study in this area is strongly encouraged. Fragrances and active components made from them find widespread applications in the cosmetics industry. Their bioactive properties (anti-aging, antibacterial, sun protection, and whitening) and their ability to impart a wide range of distinctive and enjoyable fragrances to cosmetic products make them highly sought-after and treasured. Essential oils are used to treat a broad variety of medical issues, including infections, depression and anxiety as well as being used as antifungal, antibacterial, and anticancer agents, and to improve wound healing. Essential oils are being utilized increasingly frequently in the medical industry, and they are typically administered to the areas of the body that are visible to the outside world in order to alleviate pain. They are used all over the world, and because of their improved utilization, the global market for EOs is expanding swiftly and gaining more significance on a daily basis.

Because it is possible for pathogens to become immune to certain EOs, it is imperative that more research is conducted on a wider range of EOs. EOs are widely used in edible coatings and films for preservation, in the cosmetics industry as perfumes and skin repairs, as drug enhancers in the pharmaceutical industry *etc*.

## References

1. I. Barbalova, Global Beauty and Personal Care: The Year in Review and Winning Strategies for The Future, InCosmetics/documents/IC11EuromonitorIntGlobalBeautyAndPersonalCare, 2011, http://www.incosmetics.com/RXUK/RXUK.
2. Grand View Research, Organic personal care market size worth $42.19 billion by 2030, 2022, https://www.grandviewresearch.com/press-release/global-organic-personal-care-market.

3. S. A. T. Tamsyn, H. Pauline and P. N. Declan, *BMC Complementary Altern. Med.*, 2009, **9**, 1–11.
4. M. L. Soto, E. Falqué and H. Domínguez, *Cosmetics*, 2015, **2**, 259–276.
5. U.S. Food and Drug Administration Centre for Food Safety and Applied Nutrition, CFSAN/Office of Cosmetics and Colors Fact Sheet: Is it a Cosmetic, a drug or both? (or it is Soap?), 2002, Retrieved from http://www.cfsan.fda.gov/dms/cos-218.html.
6. P. Elsner and H. I. Maibach, *Cosmeceuticals: Drugs vs Cosmetics*, Marcel Dekker, New York, 2000.
7. L. E. Millikan, Cosmetology, Cosmetics, Cosmeceuticals: Definitions and Regulations, *Clin. Dermatol.*, 2001, **19**(4), 371–374.
8. U.S. Food and Drug Administration Centre for Food Safety and Applied Nutrition, CFSAN/Office of Cosmetics and Colors Fact Sheet: Is it a Cosmetic, a drug or both? (or it is Soap?), 2002, Retrieved from http://www.cfsan.fda.gov/dms/cos-218.html.
9. A. C. De Groot and E. Schmidt, *Essential Oils: Contact Allergy and Chemical Composition*, CRC Press, Boca Raton, FL, USA, 2016, p. 17.
10. N. S. Sangwan, A. H. A. Farooqi and F. Shabih, *et al.*, *Plant Growth Regul.*, 2001, **34**, 3–21.
11. C. S. Sell, *The Chemistry of Fragrance. From Perfumer to Consumer*, The Royal Society of Chemistry, Cambridge, UK, 2nd edn, 2006, p. 329.
12. Q. H. Zeng, J. B. Zhao, J. J. Wang, X. W. Zhang and J. G. Jiang, *LWT J. Food Sci. Technol.*, 2016, **68**, 595.
13. E. Lubinic, *Les Huilesessentielles et Leurutilisation*, Edition Vigot, Paris, 2003, p. 270.
14. P. Tongnuanchan and S. Benjakul, *J. Food Sci.*, 2014, **79**, R1231–R1249.
15. K. H. C. Baser and G. Buchbauer, *Handbook of Essential Oils: Science, Technology, and Applications*, CRC Press, Boca Raton, FL, USA, 2015.
16. K. H. C. Baser and F. Demirci, Chemistry of essential oils, in *Flavours and Fragrances: Chemistry, Bioprocessing and Sustainability*, ed. R. G. Berger, Springer, New York, NY, USA, 2007, pp. 43–86.
17. S. Babita, P. Sellam, M. Jayoti and R. Puja, *Hort. Flora Res. Spectr.*, 2014, **3**, 7–13.
18. C. Franz and J. Novak, Sources of essential oils, in *Handbook of Essential Oils and Science, Technology and Applications*, ed. K. HüsnüCanBaser, and G. Buchbauer, CRC Press Taylor & Francis Group, Boca Raton, FL, USA, 2010, pp. 39–81.
19. K. Ahuja and S. Singh, *Essential Oils Market Size by Application (Orange oil, Lemon oil, Eucalyptus oil, Clove oil, Peppermintoil, Jasmine oil, Rosemary oil, Cornmint oil, Citronella oil, Geranium, Spearmint oil, Lavender oil, Tea tree oil and others) byApplication (Food & Beverage, Aromatherapy, Cosmetics & Toiletries, Pharmaceuticals, Cleaning & Home Care, Animal Feed,Fragrances and Others) Industry Analysis Report, Regional Outlook, Growth Potential, Competitive Market Share & Forecast, 2019–2026*, Global Market Insights, Inc., 2019.
20. N. Sadgrove and G. Jones, *Agriculture*, 2015, **5**, 48–102.

21. W. Dhifi, S. Bellili, S. Jazi, N. Bahloul and W. Mnif, A critical review, *Medicines*, 2016, **3**, 25.
22. C. S. Sell, *The Chemistry of Fragrance. From Perfumer to Consumer*, The Royal Society of Chemistry, Cambridge, UK, 2nd edn, 2006, p. 329.
23. A. Modzelewska, S. Sur, K. S. Kumar and S. R. Khan, *Curr. Med. Chem.: Anti-Cancer Agents*, 2005, **54**, 477–499.
24. F. Bakkali, S. Averbeck and D. Averbeck, *et al.*, *Food Chem. Toxicol.*, 2008, **46**, 446–475.
25. A. H. Gilani, A. U. Khan, Q. Jabeen, F. Subhan and R. Ghafar, *J. Ethnopharmacol.*, 2005, **100**(3), 347–352.
26. T. Kulisic, A. Radonic, V. Katalinic and M. Milos, *Food Chem.*, 2004, **85**(4), 633–640.
27. N. Bousbia, M. A. Vian, M. A. Ferhat, E. Petitcolas, B. Y. Meklati and F. Chemat, *Food Chem.*, 2009, **114**(1), 355–362.
28. P. Masango, *J. Cleaner Prod.*, 2005, **13**, 833–839.
29. K. G. D. Babu and V. K. Kaul, *Flavour Fragrance J.*, 2005, **20**, 222–231.
30. O. Zhao, J. W. Zhou and D. M. Ban, *J. Chin. Med. Mater.*, 2010, **33**, 1417–1419.
31. P. Tongnuanchan and S. Benjakul, *J. Food Sci.*, 2014, **79**, 1231–1249.
32. Z. A. A. Aziz, A. Ahmad, S. H. MohdSetapar, A. Karakucuk, M. M. Azim, D. Lokhat, M. Rafatullah, M. Ganash, M. A. Kamal and G. M. Ashraf, *Curr. Drug Metab.*, 2018, **19**, 1100–1110.
33. L. Orio, G. Cravotto, A. Binello, G. Pignata, S. Nicola and F. Chemat, *J. Sci. Food Agric.*, 2012, **92**, 3085–3090.
34. L. L. Rivera and G. Vilarem, *Flavour Fragrance J.*, 2007, **22**, 178–183.
35. M. Gavahian, A. Farahnaky, M. Majzoobi, K. Javidnia, M. J. Saharkhiz and G. Mesbahi, *Int. J. Food Sci. Technol.*, 2011, **46**, 2619–2627.
36. E. V. Costa, M. L. B. Pinheiro and J. R. A. Silva, *Quim. Nova*, 2009, **32**(1), 78–81.
37. F. Maggi, C. Cecchini, A. Cresci, M. Coman, B. Tirillini, G. Sagratini, F. Papa and S. Vittori, *Chem. Biodiversity*, 2010, **7**(2), 447–466.
38. L. Si, Y. Chen, X. Han, Z. Zhan, S. Tian, Q. Cui and Y. Wang, *Molecules*, 2012, **17**, 7057–7066.
39. F. Sefidkon, M. H. Assareh, Z. Abravesh and M. M. Barazandeh, *Iran. J. Pharm. Res.*, 2007, **6**(2), 135–140.
40. N. S. Radulovic, M. Z. Mladenovic, P. D. Blagojevic, Z. ZStojanović-Radic, T. Ilic-Tomic, L. Senerovic and J. N. Runic, *Food Chem. Toxicol.*, 2013, **6**, 554–565.
41. G. M. Kamal, F. Anwar, A. I. Hussain, N. Sarri and M. Y. Ashraf, *Int. Food Res. J.*, 2011, **18**(4), 1275–1282.
42. C. Deng, N. Yao, A. Wang and X. Zhang, *Anal. Chim. Acta*, 2005, **536**, 237–244.
43. G. Guiochon and A. Tarafder, *J. Chromatogr. A*, 2011, **12**(18), 1037–1114.
44. K. D. Klerck, D. Mangelings and Y. Vander Heyden, *J. Pharm. Biomed. Anal.*, 2012, 77–92.

45. S. M. Pourmortazavi, P. Baghaee and M. Mirhosseini, *Flavour Fragrance J.*, 2004, **5**, 417–420.
46. L. Wang and C. L. Weller, *Trends Food Sci. Technol.*, 2006, **17**, 300–312.
47. D. Hurren, *Filtr. Sep.*, 1999, **36**, 25–27.
48. Q. Lang and C. M. Wai, *Talanta*, 2001, **53**, 771–782.
49. M. Hamburger, D. Baumann and S. Adler, *Phytochem. Anal.*, 2004, **15**, 46–54.
50. S. Chemat, A. Lagha, H. AitAmar, P. V. Bartels and F. Chemat, *Flavour Fragrance J.*, 2004, **19**, 188–195.
51. B. Bayramoglu, S. Sahin and G. Sumnu, *J. Food Eng.*, 2008, **88**, 535–540.
52. Z. Guo, Q. Jin, G. Fan, Y. Duan, C. Qin and M. Wen, *Anal. Chim. Acta*, 2001, **436**, 41–47.
53. W. Phutdhawong, R. Kawaree, S. Sanjaiya, W. Sengpracha and D. Buddhasukh, *Molecules*, 2007, **12**, 868–877.
54. S. C. Verma, C. L. Jain, A. Kumari, M. M. Padhi and R. B. Devalla, *J. Sep. Sci.*, 2013, **36**, 1255–1262.
55. G. Flaminia, M. Tebanoa, P. L. Cionia, L. Ceccarinib, A. S. Riccic and I. Longoc, *J. Chromatogr. A*, 2007, **1143**(1–2), 36–40.
56. M. Kosar, Z. Tunalier, T. Ozek, M. Kurkcuoglu and K. H. C. Baser, *Z. Naturforsch., C*, 2005, **60**, 501–504.
57. F. Benkaci-Ali, A. Baaliouamer and B. Y. Meklati, *Chromatographia*, 2006, **64**, 227–231.
58. P. Miletic, R. Grujic and Z. M. Balaban, *Chem. Ind. Chem. Eng. Q.*, 2009, **15**(1), 37–39.
59. N. Kahriman, G. Tosun, H. Genc and N. Yayli, *Turk. J. Chem.*, 2010, **34**, 969–976.
60. A. H. Nour, R. Mathialagan and A. H. Nour, *Int. J. Chem. Environ. Eng.*, 2013, **4**(4), 221–226.
61. M. Romdhane, C. Gourdon and G. Casamatta, *Ultrasonics*, 1995, **33**, 221–227.
62. M. Romdhane and C. Gourdon, *Chem. Eng. J.*, 2002, **87**, 11–19.
63. H. Li, L. Pordesimo and J. Weiss, *Food Res. Int.*, 2004, **37**, 731–738.
64. L. Hui, O. Etsuzo and I. Masao, *Jpn. J. Appl. Phys.*, 1994, **33**(5B), 3085–3087.
65. D. T. Velickovica, D. M. Milenovica, M. S. Ristidb and V. B. Veljkovicc, *Biochem. Eng. J.*, 2008, **42**(1), 97–104.
66. H. Sereshti, Y. Izadmanesh and S. Samadi, *J. Chromatogr. A*, 2011, **1218**(29), 4593–4598.
67. H. Abbasi, K. Rezaei and L. Rashidi, *J. Am. Oil Chem. Soc.*, 2008, **85**, 83–89.
68. P. Hashemi, M. Shamizadeh, A. Badiei, A. R. Ghiasvand and K. Azizi, *Chromatographia*, 2009, **70**(7–8), 1147–1151.
69. H. Sereshti, A. Rohanifar, S. Bakhtiari and S. Samadi, *J. Chromatogr. A*, 2012, **1238**, 46–53.
70. D. Lua, Y. Caoa, L. Lia, Z. Y. Zhua, X. Donga, H. Zhangc, Y. F. Chaia and Z. Y. Loua, *J. Pharm. Anal.*, 2011, **1**(3), 203–207.

71. S. Sasidharan, Y. Chen, D. Saravanan, K. M. Sundram and L. Y. Latha, Extraction, isolation and characterization of bioactive compounds from plants'extracts, *Afr. J. Tradit. Complement. Altern. Med.*, 2010, **8**(1), 1–10.
72. Z. Zhang, X. Pang, D. Xuewu, Z. Ji and Y. Jiang, Role of peroxidase in anthocyanin degradation in litchi fruit pericarp, *Food Chem.*, 2005, **90**, 47–52.
73. K. P. Ingle, A. G. Deshmukh, A. D. Padole, S. Mahendra, S. M. Dudhare, P. M. Moharil and C. V. Khelurkar, Phytochemicals: Extraction methods, identification and detection of bioactive compounds from plant extracts, *J. Pharmacogn. Phytochem.*, 2017, **6**(1), 32–36.
74. G. Micalizzi, F. Alibrando, F. Vento, E. Trovato, M. Zoccali, P. Guarnaccia, P. Dugo and L. Mondello, *Molecules*, 2021, **26**, 1588.
75. P. Tonthubthimthong, S. Chuaprasert, P. Douglas and W. Luewisutthichat, *J. Food Eng.*, 2011, **47**(4), 289–293.
76. I. E. Popova, C. Hall and A. Kubátová, *J. Chromatogr. A*, 2008, **1216**(2), 217–229.
77. W. Kemp, Energy and electromagnetic spectrum, in *Organic Spectroscopy*, Macmillan Press, London, 3rd edn, 1991.
78. T. L. Eberhardt, X. Li, T. F. Shupe and C. Y. Hse, *Wood Fiber Sci.*, 2007, **39**(2), 319–324.
79. P. Masango, Towards Understanding Steam Distillation of Essential Oils by Differential Quantification of Principal Components Using Capillary Gas Chromatography, Ph.D. Thesis, University of Surrey, Guildford, UK, 2001.
80. R. Tisserand and R. Young, Essential oil profiles, in *Essential Oil Safety A Guide for Health Care Professionals*, Elsevier, London, UK, 2nd edn, 2014, pp. 187–482.
81. T. K. T. Do, F. Hadji-Minaglou, S. Antoniotti and X. Fernandez, *Trends Anal. Chem.*, 2015, **66**, 146–157.
82. T. Mitsui, *New Cosmetic Science*, Elsevier, Amsterdam, The Netherlands, 1997.
83. A. Mosandl, *Food Rev. Int.*, 1995, **11**, 597–664.
84. J. B. Sharmeen, F. M. Mahomoodally, G. Zengin and F. Maggi, *Molecules*, 2021, **26**, 666.
85. L. Salgueiro, A. P. Martins and H. Correia, *Flavour Fragrance J.*, 2010, **25**, 253–271.
86. W. A. König and D. H. Hochmuth, *J. Chromatogr. Sci.*, 2004, **42**, 423–439.
87. A. Mosandl, *J. Chromatogr. Sci.*, 2004, **42**, 440–449.
88. F. Pellati, G. Orlandini, K. A. van Leeuwen, G. Anesin, D. Bertelli, M. Paolini, S. Benvenuti and F. Camin, *Rapid Commun. Mass Spectrom.*, 2013, **27**, 591–602.
89. V. Pieri, S. Sturm, C. Seger, C. Franz and H. Stuppner, *Metabolomics*, 2012, **8**, 335–346.
90. R. Marsili, *Sensory-Directed Flavor Analysis*, CRC Press, Boca Raton, FL, USA, 2007.

91. J. Reichling, U. Landvatter, W. Heike, K. H. Kostka and U. F. Schaefer, *J. Pharm. Biopharm.*, 2006, **64**, 222–228.
92. I. T. Carvalho, B. N. Estevinho and L. Santos, *Int. J. Cosmet. Sci.*, 2016, **38**, 109–119.
93. W. Dhifi, S. Bellili, S. Jazi, N. Bahloul and W. Mnif, *Medicine*, 2016, **3**, 1–16.
94. D. Rahmi, R. Yunilawati, B. NugrohoJati, I. Setiawati, A. Riyanto, I. Batubara and R. Indri Astuti, *Cosmetics*, 2021, **8**, 94.
95. A. M. Elgamal, R. F. Ahmed, A. M. Abd-ElGawad, A. E.-N. G. El Gendy, A. I. Elshamy and M. I. Nassar, *Plants*, 2021, **10**, 667.
96. A. Lohani, A. K. Mishra and A. Verma, *J. Cosmet. Dermatol.*, 2018, 1–8.
97. K. A. Dobos, *Colors in Cosmetics—The Importance of Dispersion*, Society of Cosmetic Chemist, Cincinnati, Ohio, 2019, https://www.scconline.org/the-importance-of-dispersion, accessed 2 June 2021.
98. B. Maktabi, M. W. Liberatore and G. Baki, *Int. J. Cosmet. Sci.*, 2021, **43**, 484–493.
99. E. Uronnachi, C. Atuegwu, C. Umeyor, C. Nwakile, J. Obasi, C. Ikeotuonye and A. Attama, *Sci. Afr.*, 2022, **16**, e01223.

CHAPTER 4

# *Toxic and Allergic Responses Caused by Secondary Metabolites Used in Cosmetic Formulations*

SWATI GAJBHIYE[a] AND KAUSHIK PAL[b*]
(0000-0003-0916-9526)

[a]Rashtrasant Tukadoji Maharaj Nagpur University, Department of Cosmetic Technology L.A.D College, Nagpur, India; [b]University Centre for Research and Development (UCRD), Department of Physics, Chandigarh University, Mohali, Gharuan, Punjab 140413, India
*E-mail: kaushikphysics@gmail.com, gajbhiyelswati@gmail.com

## 4.1 Introduction

Plants are known to synthesize numerous organic compounds, which are classified as primary and secondary metabolites, and the boundaries between these two metabolites are somewhat blurred. Primary metabolites in plants are known to play an essential role that is associated with the most important processes of the plant, including photosynthesis, respiration, growth, and development. In contrast, secondary metabolites are other phytochemicals that are accumulated in a high concentration[1] and thus we can say that the plant's secondary metabolites are distinguished from primary metabolites such as amino acids, carbohydrates, fats, *etc.*

Bioprospecting of Natural Sources for Cosmeceuticals
Edited by Deepika Kathuria, Ajay Sharma, Meenakshi Verma and Gulzar Ahmad Nayik

Published by the Royal Society of Chemistry, www.rsc.org

Many scientists also believe that secondary metabolites do not show any crucial role for living cells in normal growth, development, and reproduction, but they act as a defense mechanism to protect the plant. In contrast, primary metabolites have biological purposes across all species. Secondary metabolites are created from primary metabolites by a modified synthetic pathway.[2] Therefore, plants produce various organic compounds that do not have direct involvement in growth and development, and such substances are referred to as secondary metabolites.[3] These secondary compounds are natural products that are synthesized by plants, fungi, bacteria, algae, and animals along with marine organisms such as sponges, tunicates, coral, and snails.[4] Various medicinal and aromatic plants (MAPs) are used as traditional medicines according to the World Health Organization (WHO). These medicinal and aromatic plants synthesize various secondary metabolites such as terpenes, phenols, alkaloids, flavonoids, volatile oils, *etc*. These secondary metabolites exhibit a broad range of healthcare properties.[5] Today, plant secondary metabolites are an important source of lead molecules in drug discovery and development because of their growing diversity and also because of their high specificity in biological activities, acting as additives or in synergistic ways. Structurally different plant-derived products can have high efficacy and show a broad spectrum of activity in combination with one another.[6]

Secondary metabolites are natural products synthesized by secondary metabolism in plants that have various biological activities and they have been used in traditional medicine for centuries. In recent surveys, 25% of these secondary metabolites are used in the pharmaceutical industry. They are also used in cosmetics and nutraceuticals, and also by the chemical industry.[3] Due to the complexity of their chemical structure and also because of their biosynthetic pathways, natural products based on secondary metabolites have been perceived as biologically insignificant and received less attention from biochemists. The emerging pharmaceutical organic chemistry field has investigated the chemical properties of secondary metabolites since the 1850s with an interest in discovering novel phytochemicals. This promoted the great utility of secondary metabolites as oils, waxes, flavoring agents, perfumes, dyes, polymers, drugs, *etc*.[7] Along with the plant secondary metabolites, certain secondary metabolites are also extracted from various sources, such as animals and fungi, including subclasses that are based on the number of carbons and isoprene units, and these are also used in cosmetic industries.[8] Thus, all of the secondary metabolites are used as a primary source for drug discovery because of their beneficial biological activity, but the toxicity of some of them limits their use. Secondary metabolites are considered waste or detoxification products that are biosynthesized as a defense mechanism for survival and thus their high toxicity is an inherent characteristic for this purpose. Therefore, the toxic effects of plants, animals, and microorganisms producing a secondary metabolite that is used in cosmetic products need to be studied for their safe use in cosmetic products.[9]

Along with the plant secondary metabolites, microbial secondary metabolites are also used in cosmetics, which are intermediate products or the product of the fermentation process during metabolism. These are produced during the stationary phase and have an important role in the competition, antagonism, and self defense mechanisms. These metabolites also find wide applications in pharmaceuticals, food, cosmetics, *etc.*[10] Secondary metabolites are formed by an enzymatic pathway that occurs during free or complexed protein synthesis or by large multifunctional polypeptides that carry out a multitude of enzymatic steps such as polyketide synthases and peptide synthetases. In the secondary pathway, the important biosynthetic pathways that are involved are the formation of aromatic compounds, oligosaccharides, peptides, polyketides, isoprenes, and β-lactam rings. Knowledge of the pathways varies from the amino acid sequences of enzymes to the nucleotide sequences of genes along with those enzymatic steps that are still unknown.[11] Certain pigments are also produced as secondary metabolites from bacteria and fungi. These bacterial and fungal pigments have several biological properties, such as antioxidant, and antimicrobial, apart from only giving color. These pigments also find their use in cosmetics, for example carotenoid that is produced from the bacteria *Pedobacter*, which is known to have antioxidant properties. These carotenoids have recently been used for nutraceutical, cosmetic and pharmaceutical purposes. Another example is the production of melanin from a wide variety of microorganisms, which is an indolic polymer that has been used in cosmetics as it can resist ultra violet (UV) light by absorbing a broad range of the electromagnetic spectrum and is thus used extensively in cosmetic photoprotective creams.[12] Secondary metabolites extracted from cyanobacteria are used in cosmetics in the form of mycosporines, such as amino acids, polysaccharides, pigments, scytonemin, carotenoids, and antioxidants. These are used in skin care products for anti-wrinkle products, moisturizers, sunscreens, *etc.*[13] Cyanobacteria and algae are most widely used to produce secondary metabolites that are used in cosmetics. They share a biosynthesis pathway. Some are considered toxins, but these compounds are not considered toxic for vertebrates, and they are not studied since the toxins may be allelopathic compounds. Further research is needed to clarify the allelopathic effect, whether it is safe for dermal use and most importantly what kind of secondary metabolites are produced by the particular bacteria. Various allelopathic compounds are not often considered because their effect is target sensitive and their interaction is affected in various ways. The presence of allelopathic compounds in secondary metabolites given out by bacteria needs to be studied in detail for their safe use in cosmetic products.[14] The secondary metabolites that are produced by microbial species and through their biosynthetic pathways have led to the discovery of novel molecules or enzymes that find numerous applications in cosmetics,[15] but due to these new and novel discoveries, their adverse effects and the presence of various secondary metabolites, along with allelopathic compounds, need to be taken into consideration before they can be used safely in cosmetic preparations.

## 4.2 Toxic and Allergenic Plant Secondary Metabolites Used in Cosmetics

The following are a few examples of plant secondary metabolites according to their classification, which are toxic and allergenic in nature but are still used in cosmetic products. The following chapter discuss the toxic and allergic reactions occurring from the use of secondary metabolites in cosmetic products.

### 4.2.1 Terpene Secondary Metabolites

Terpenes are the major class of secondary metabolite product synthesized from isoprene and 2-methyl-1,3-butadiene, which is polymerized. They are classified as monoterpenes, sesquiterpenes, diterpenes, triterpenes, tetraterpenes, and polyterpenes.[16] As terpene and isoprenoid compounds are the major class of secondary metabolites, these biomolecules are said to be high-value bioproducts. In plants, the cytosolic mevalonic acid and plastid methylerythritol phosphate pathways lead to the synthesis of terpene and isoprenoid precursors. Terpenoid diversity can be attributed to the differential enzyme and secondary modifications that are acquired by terpene synthases. These secondary metabolites, like terpenes, have various applications in pharmaceuticals, cosmetics, fragrances, *etc.*[17] Most of the essential oil components represent plant secondary metabolites that are used in the pharmaceutical and cosmetic industries. These can easily penetrate the skin up to a measurable concentration in the blood. Several toxicity studies have also stated that monoterpenes and sesquiterpenes exhibit liver toxicity due to the reactive metabolite formation with an increased number of reactive species and impaired antioxidant defense. Thus, there is a probability that other terpenes also affect humans leading to liver toxicity and other related toxicities. In particular, terpenes, which are present in essential oils, can be a risk to the functioning of the liver as these are used in cosmetic preparations.[18] The following are a few examples of terpenes that are used in cosmetics, which produce toxic and allergic reactions in the human body.

#### *4.2.1.1 Limonene and Linalool*

Limonene and linalool (as shown in Figure. 4.1) are not allergenic in nature, but as terpenes they can oxidize on air exposure forming allergenic products. Oxidized linalool and its hydroperoxide fraction are found to be contact allergens. Autoxidation of fragrance terpenes, which are used in commercial formulations,[19] leads to fragrance allergies. Limonene and linalool are fragrance terpenes that have low sensitizing potential, but when exposed to air, oxidation occurs which leads to the production of different oxidation products that have a high prevalence of contact allergies towards hydroperoxides of limonene and linalool. A study also found that the patients who

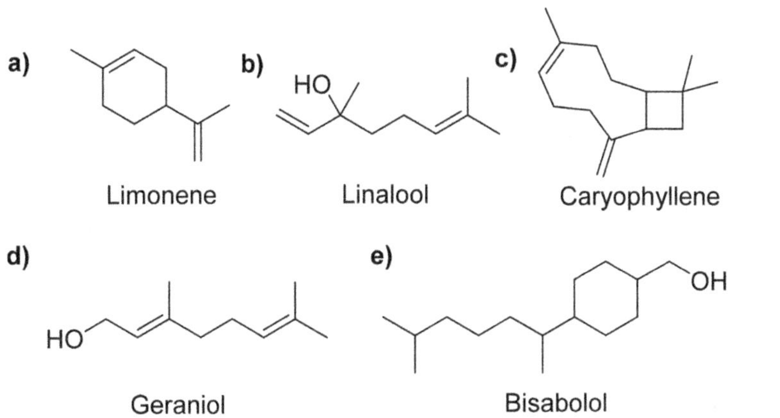


**Figure 4.1** Chemical structures of (a) limonene, (b) linalool, (c) caryophyllene, (d) geraniol and (e) bisabolol.

had allergic reactions are mostly females with a less frequent history of atopic dermatitis. The study also revealed that 71% of patients responded positively towards a rinse-off product like soap and shampoo and 42.9% responded positively towards leave-on products like cosmetics and cream, including cleaning agents, detergents, deodorants, and perfumes containing these secondary metabolites.[20] This has also been confirmed by a study done on 281 patients showing contact allergies. In dermatitis patients, 5.2% showed an allergic reaction to oxidized *R*-limonene and 6.9% to oxidized linalool. For this study, oxidized *R*-limonene 3% and oxidized linalool 6% were used in petrolatum containing limonene hydroperoxide (0.33%) and linalool hydroperoxide (1%). In total, 25% of patients showed a reaction to both the compounds, and 29% showed a reaction towards oxidized *R*-limonene and 46% towards oxidized linalool, so it was concluded that one of the mixtures leads to the specificity of the reaction.[21] Non-oxidized terpenes are mostly non-irritating but linalool and *R*-limonene are found to be irritating, especially *R*-limonene, when compared to pure compounds. Thus, autoxidation of the fragrance increases irritation.[22] In another study that investigated contact allergies of oxidized limonene and linalool, hydroperoxides of limonene and linalool were found to be an irritant.[23] Limonene and linalool behave as prehaptens as they are slowly autoxidized when exposed to air. This oxidization forms a cocktail of oxidized molecules like epoxides, diols, and allylic hydroperoxides. These allylic hydroperoxides, when tested with a local lymph node assay, were found to be a strong sensitizer as allylic hydroperoxides react with skin proteins that are recognized by the immune system. Thus, it was hypothesized that terpenes and hydroperoxides induce oxidative stress, which leads to the activation of DCs (dendritic cells). Therefore, the GSH/GSSG (reduced glutathione/oxidized glutathione) ratio, and Nrf2 (nuclear factor erythroid 2-related factor 2) and MAPK (Mitogen-activated protein kinase) pathways, were studied in the THP-1 cell line (a human monocytic leukemia cell line), which is a surrogate of DCs and validated with the h-CLAT (human Cell Line Activation Test) in an *in vitro* study. The result indicated that with both limonene and linalool oxidized cell surface thiols, the oxidative stress induced by linalool was stronger with a decrease in the GSH/GSSG ratio and strong ROS (reactive oxygen species) production. This also induces strong Nrf2 accumulation and correlates with NQO1 and HO-1 gene expression, which are two Nrf2 target genes. This is similar to the cellular defense against oxidation or electrophilic stress, which involves activation of the Nrf2 signaling pathway. Interestingly, in this study, only the linalool-induced Nrf2 pathway activation showed a high response, meaning the terpenes act differently for Nrf2 activation. The overall study underlines that autooxidation, which forms allylic hydroperoxide, does not show equal reactivity towards both, but showed strong activation towards linalool with respect to oxidative stress and Nrf2 activation.[24]

#### 4.2.1.2 Caryophyllene

Along with limonene and linalool, caryophyllene (see Figure 4.1) is also known to causes allergic contact dermatitis. β-Caryophyllene, which is a sesquiterpene used in fragrance cosmetics, was investigated for autooxidation and β-caryophyllene showed immediate oxidization upon air exposure, with caryophyllene oxide being the major oxidation product, while hydroperoxides of β-caryophyllene were not detected in the mixture in a local lymph node assay. Allergen activity was found to involve autooxidation, but to a lesser extent in comparison to *R*-limonene and linalool.[25]

#### 4.2.1.3 Geraniol

Geraniol (see Figure. 4.1) is a natural monoterpene, as per its classification, that is known to cause skin sensitivity, as geraniol reacts with oxygen and this has been studied with density functional theory (DFT) (B3LYP). Oxidation is initiated by abstracting hydrogen, which forms an allylic radical when combined with an oxygen molecule giving an intermediate peroxyl radical.[26] Geraniol is known as an important fragrance terpene and is considered an allergenic compound. Unlike other studies related to limonene and linalool, geraniol also autoxidizes upon air exposure, and thus it was investigated using a local lymph node assay. This study showed that the allylic alcohol geraniol has a different oxidation pattern from linalool and limonene, which forms hydroperoxides upon oxidation as the primary oxidation products. Geraniol follows two paths upon autoxidation; from geraniol, first hydrogen peroxide is formed along with the aldehyde geranial, and then neral forms, which is the hydroxy hydroperoxide. The formation of a hydroperoxide is a major contributor to allergenic activity along with the aldehydes geranial and neral.[27] Geraniol gets auto-oxidized and forms allergenic oxidation products causing contact sensitization. Sensitization can be caused by low molecular weight components that penetrate the skin and bind to skin proteins forming reactive species, either by autoxidation upon air exposure or by metabolic activation in the skin. Studies also show that geraniol is metabolically activated to geranial. This was further investigated with cytochrome P450 (CYP)-mediated activation of geraniol. Geranial was the major metabolite formed and was identified by a lymph node assay. Whereas geranial, neral, and 6,7-epoxygeraniol were found to be moderate sensitizers, 6,7-epooxygeranial is a strong sensitizer. In the isoenzyme study, CYP1A1 and CYP3A5 were found to be responsible for the metabolic activation of geraniol in the skin. Thus, it was concluded that geraniol is activated by both autoxidation and metabolism, and the allergens geranial and neral play a major role in sensitization to geraniol.[28] Along with this, another study has also found that dioxolan hydroperoxide and 6,7-epoxygeranial are also responsible for skin sensitization upon air-exposure of geranial. This study showed that the dioxolan derivative is formed in an acid-catalyzed closed shell reaction with 6,7-epoxygeranial and geranial, which is responsible for skin sensitization.[29]

#### *4.2.1.4 Tea Tree Oil*

Tea tree oil is used in many pharmaceutical and cosmetic preparations, but the secondary metabolites that are present in it can cause an allergic effect on the skin. The reaction of tea tree oil in cosmetics depends upon the product's use. Studies have indicated that dermatitis is less severe due to the low concentration of allergens in cosmetic products, but the side effects cannot be neglected. For example, periorbital/eyelid dermatitis can be caused by the use of soap, cream, and shampoo, dermatitis around the beard area can be due to the use of shaving oil, and dermatitis on the face and hands can be due to face cream and hand lotion. It has also been stated that stomatitis has resulted due to a contact allergy to toothpaste containing tea tree oil. Approximately three quarters of cases are known to cause allergic dermatitis due to the use of undiluted oil or oil of high concentration. The reacting sensitizers that are present in tea tree oil are reported to be ascaridole, terpinolene, α-terpinene, 1,2,4-trihydroxymenthane, and α-phellandrene. Allergenic compounds include myrcene, aromadendrene, D-carvone, L-carvone, terpinen-4-ol, viridiflorene and to a less extent sabinene, 1,8-cineole, and *p*-cymene. The components in tea tree oil change in the presence of atmospheric oxygen and also due to light, humidity, and a higher temperature. Upon oxidation, the levels of α-terpinene, γ-terpinene, and terpinolene decrease, and that of *p*-cymene increases 10 fold, and thus it can be said that *p*-cymene is a major degradation product. Oxidation of oil also forms peroxides, endoperoxides, and epoxides like ascaridole. Thus, with all the benefits of tea tree oil, the formation of metabolites can result in toxic topical preparations.[30] As tea tree oil is mostly used in cosmetic and skincare products because of its antioxidant properties, the main component that is responsible for the anti-oxidant properties is α-terpinene, which is a monoterpene also present in different essential oils used as fragrance compounds. This α-terpinene is similar to limonene, which is known to autoxidize when exposed to air and form allergenic compounds. Several studies have also investigated the auto-oxidation of α-terpinene at room temperature using a murine local lymph node assay. This showed that α-terpinene rapidly degrades and forms major oxidation products such as allylic epoxides and *p*-cymene along with hydrogen peroxide, and thus the α-terpinene sensitization potency is increased upon air exposure. In the fraction of allylic epoxides, only α- and β-unsaturated aldehydes showed strong sensitization in this assay. Overall, even if α-terpinene is known for its antioxidation properties, it gets easily autoxidized forming allergenic compounds and thus its use should be limited for topical application.[31]

#### *4.2.1.5 Bisabolol*

α-Bisabolol (as shown in Figure 4.1) is a monocyclic sesquiterpene alcohol that can be found in the essential oil of chamomile. This is used in cosmetics, such as perfumes, soaps, detergents, and personal hygiene products.[32] This essential oil is used in cosmetics because of its anti-inflammatory properties.[33] It exhibits an anti-inflammatory effect by reducing levels of TNF-α (tumor necrosis factor α), IL-1β (interleukin-1β), IL-6 (interleukin 6), iNOS (inducible nitric oxide synthase), and COX-2 (cyclooxygenase-2). Additionally, it suppresses the activation of ERK1/2 (extracellular signal-regulated protein kinase), JNK (c-Jun N-terminal kinases), NF-κB (nuclear factor kappa B), and p38 (mitogen-activated protein kinases).[34] α-Bisabolol is also known for its toxic effect. In a toxicity study done on the human epithelial cell line using an MTT (3-(4,5-dimethylthiazolyl-2)-2,5-diphenyltetrazolium bromide) assay, it was found to be a toxic compound along with nerolidol.[35] Thus, the use of such essential oils containing bisabolol should be taken into consideration for cosmetic safety.

Overall, the secondary metabolites that are present in essential oils, especially in fragrance components, can be toxic and cause allergic responses because many fragrance components act as haptens, which are low molecular weight chemicals and are immunogenic when attached to skin carrier proteins and can react directly with skin proteins. Fragrance components that act as prehaptens or prohaptens form allergens that are activated in the skin by abiotic or biotic activation and result in skin sensitization. Prehaptens are non-sensitizing and transferred into a hapten outside the skin by air oxidation without an enzymatic system. In contrast, prohaptens are also non-sensitizing and transferred into haptens in the skin but by enzymatic catalysis. Thus, it is not possible to know which particular allergen acts as a prehapten or prohapten, or both, because air oxidation and bioactivation can give the same product, but the mechanism is different. For example, geraniol and α-terpinene can act as both a prehapten and prohapten. Some substances can be moderate sensitizers but can be activated by autooxidation, such as geranial, which acts as a hapten with moderate sensitization but can be a more potent sensitizer upon air oxidation as a prehapten. Thus, we can state that components that can easily autoxidize on exposure to air can form potent sensitizers and can be a source of contact allergies from fragrance products used in cosmetics.[36]

### 4.2.2 Phenolic Secondary Metabolites

Next in the classification of secondary metabolites is the phenolic compounds that are used in cosmetics and many cause toxicity and allergic reactions when used in cosmetic preparations. Phenolic compounds are said to be widely distributed secondary metabolites present in plants. The terms phenol and polyphenol are defined as aromatic rings containing one phenol or polyphenol, but in terms of plant phenolics, this definition is not satisfactory. A plant polyphenol refers to a water-soluble phenolic compound possessing properties such as protein and alkaloid precipitation, and it is referred to as a plant secondary natural metabolite. Phenolics are uncommon in bacteria, fungi, and algae.[37] Phenolic compounds are considered an important group in plant secondary metabolites because of their morphological development, reproduction, and physiological processes. Phytochemicals are usually synthesized due to the pentose phosphate, shikimate, and phenylpropanoid pathways, and are mostly known for a wide variety of biological properties because of their molecular structure.[38] Thus, various phenolic compounds are used in the cosmetic industry as they are said to possess various properties, such as anti-aging, sun screening, depigmenting, exfoliating, anti-polluting, skin conditioning, *etc.* However, some contradictory effects obtained from non-enzymatic and enzymatic polyphenols should be taken into consideration due to their toxic effect.[39] Phenolics can act as redox-sensitive signaling cascades, which can inhibit DNA damage and thus apart from having beneficial effects they can act as mutagens. These harmful effects are due to the pro-oxidant rather than the antioxidant action. Therefore, it is important to evaluate their toxic effects before their use in cosmetics.[40]

Phenolic compounds are a diverse group of chemical compounds which consist of heterogeneous structures and they are classified in many different ways. The classification of the phenolic compounds that are used in cosmetics is presented in Scheme 4.1.

Unlike tannins, phenolic compounds are also said to be widely distributed secondary metabolites in plants. Phenolic compounds are mainly distributed in plant tissues and the concentration may vary from 0.5 to 5 g per 100 g dry weight of plant tissue. Mostly the phenolic compounds are considered a by-product of the metabolism of the aromatic amino acid phenylalanine and a large concentration of phenolic compounds in plants is due to the presence of cinnamic acid derivatives. Cinnamic acids play a key role in the formation of complex phenolic compounds.[41] The following are examples of phenolic compounds that result in toxic and allergic reactions, but are still used in cosmetics products.

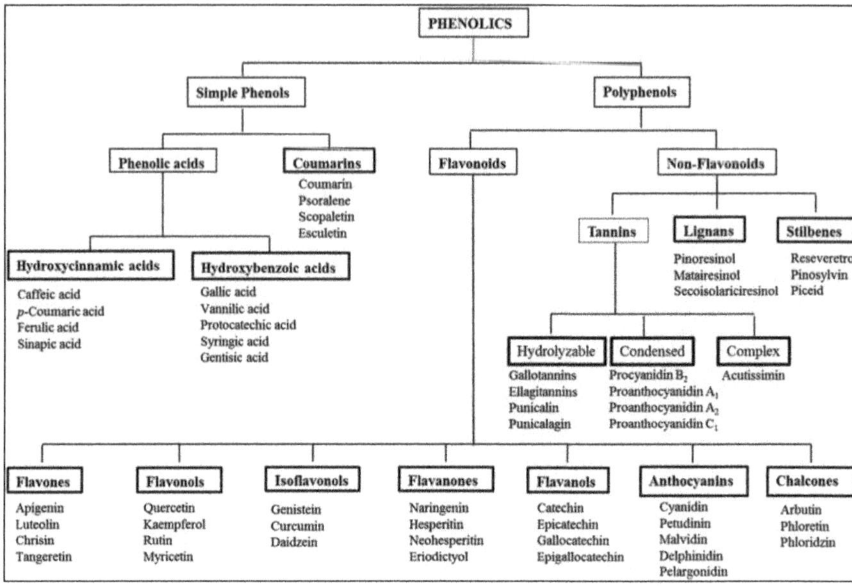


**Scheme 4.1** Classification of phenolic compounds.

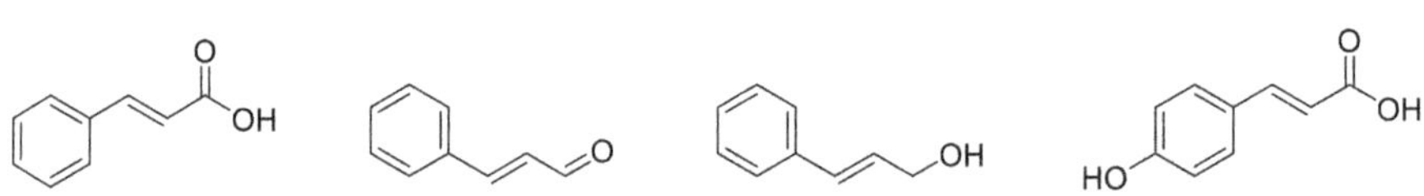


cinnamic acid    cinnamaldehyde    cinnamyl alcohol    p-coumaric acid

**Figure 4.2** Structure of cinnamic acid and its derivatives.

### *4.2.2.1 Cinnamic Acid*

Cinnamic acid is the first molecule in the phenylpropanoid pathway and is mostly a source of hydrocinnamic acid derivatives like coumaric, caffeic, ferulic, and sinapic acid, which are present in plants in either the ester or glycosidic forms.[42] The cinnamic acid (3-phenylprop-2-enoic acid and 3-phenylacrylic acid) and its derivatives that are used in cosmetics include cinnamic acid, cinnamaldehyde, cinnamyl alcohol, 4-hydroxycinnamic acid (*p*-coumaric acid), 3,4-dihydroxy cinnamic acid (caffeic acid), and 4-hydroxy-3-methoxy cinnamic acid (ferulic acid) (see Figure 4.2). These are used in cosmetics as perfuming and masking agents, UV filters, UV-absorbing agents, antioxidants, skin conditioning agents, hair conditioning agents, *etc*. These cinnamic acid derivatives are also widely used in skin-lightening and anti-aging preparations. Cinnamic acid derivatives are known to possess some allergic and photoallergic potential and are thus restricted to being used in concentrations not exceeding certain fixed values in cosmetic products. Cinnamic acid derivatives used in cosmetic as a UV protectant have been restricted for use in higher concentrations because of their adverse effects, such as photo contact allergies, phototoxic contact dermatitis, estrogenic modulation, and generation of reactive oxygen species.[43]

### *4.2.2.2 Ferulic Acid*

Ferulic acid is commonly found in various fruits and vegetables and is a phenolic acid compound with antioxidant properties. These antioxidant properties depend on a number of hydroxyl and methoxy groups attached to the phenyl ring. Ferulic acid is readily absorbed into the body and stays for a longer duration in the blood than any other phenolic acid. It has low toxicity but possesses many physiological functions, such as antimicrobial, anti-inflammatory, *etc*. In cosmetics, it is used in skin care formulations as a sunscreen and brightening component. It has a protective role in the skin structure and is thus mostly used in anti-aging cosmetics and also in an anti-blemish cosmetic formulation as it can inhibit melanogenesis enzymes.[44] Ferulic acid is generally considered safe but people with sensitive skin or with an inflammatory condition such as acne or eczema should do a patch test before using cosmetics that contain ferulic acid. People allergic to food containing ferulic acid should also do a patch test as they might develop skin reactions with such compounds, which could be in the form of an itchy rash.[45]

### *4.2.2.3 Coumarin*

Coumarin and its derivatives are natural compounds that have cosmetic use; these are used not only as a fragrance, but also as a UV absorber and antioxidant, and thus their photosensitizing properties should be evaluated for safe administration in cosmetic preparations.[46] Furanocumarins, which are the result of prenylation in the 6- or 8-position of 7-hydroxy-coumarin, are known to be phototoxic in nature when exposed to the sun, and such compounds can result in acute dermatitis along with blisters and vesicles followed by hyperpigmentation.[47] Coumarin is a naturally occurring constituent in many plants and essential oils, such as sweet clover, lavender, sweet woodruff, *etc.* and its toxicity profile depends on its purity.[48] Coumarins are known as phenolic secondary metabolites consisting of naturally occurring benzopyrone, which is found in many microorganisms, plants, and animals. Coumarin (2*H*-1-benzopyran-2-one) is the most widely used in the fragrance cosmetic industry at a concentration of 0.5–6.4%. An earlier study has shown the toxic effect of coumarin intake in animals and still, it is used in cosmetic skin care preparations such as soaps, detergents, lotions, *etc*. Thus, cosmetics containing coumarin cause skin allergies in individuals with sensitive skin and can possibly have skin irritation/toxic effects. The absorption of coumarin and its derivatives, specifically osthole, daphnoretin, coumarin, byakangelicin,

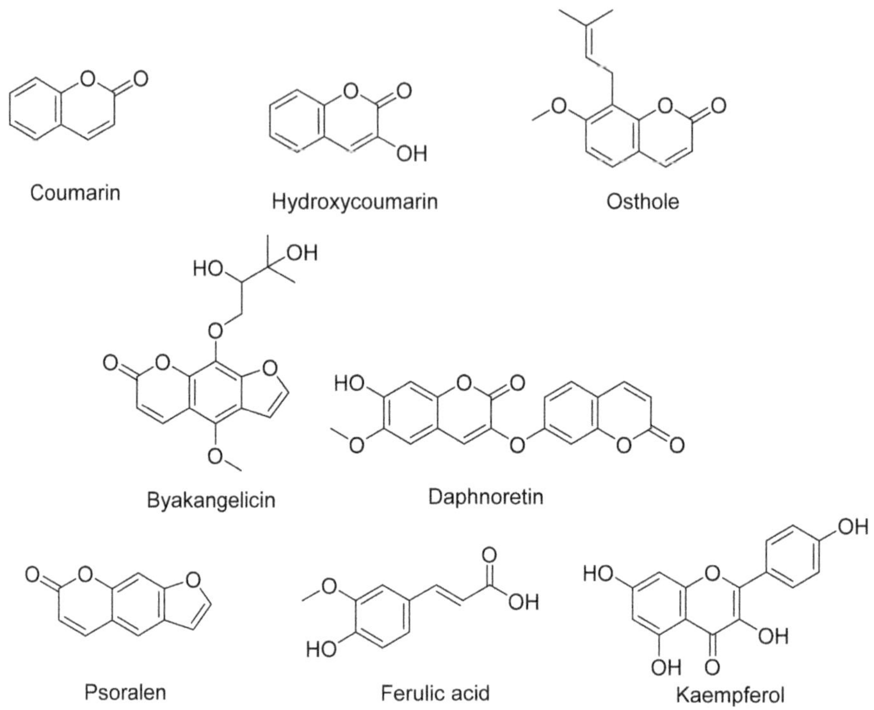


**Figure 4.3** Examples of coumarin-based chemical constituents in cosmetics.

and 7-hydroxycoumarin, *via* the topical route of application, was examined for dermal irritation, skin penetration, and toxicity (as shown in Figure 4.3).

The experimental results showed that chemical structure affects delivery and skin irritation because coumarins have an aromatic ring which joins to condense the lactone ring, which makes it soluble in ethanol. Similarly, osthole comprises a coumarin structure alongside an alkyl chain and ether group. Daphnoretin is known for its coumarin analog and lastly, byakangelicin is similar to the osthole compound with more hydroxyl and ether groups. Thus, it has increased aqueous ethanol solubility in comparison to osthole. In contrast, the hydroxyl groups in 7-hydroxycoumarin do not increase solubility in comparison to coumarin. Overall dermal absorption is shown by osthole, coumarin, and 7-hydrocoumarin and the highest accumulation was observed by osthole followed by coumarin. Due to the high molecular weight, daphnoretin has limited skin permeability along with byakangelicin. After being absorbed by the skin this coumarin gets stored in large amounts in the skin inducing cytotoxicity. Osthole demonstrated considerable inhibition, primarily affecting the viability of skin cells, followed by daphnoretin and then coumarin. These effects were observed in keratinocytes and fibroblast cells. In contrast, byakangelicin and 7-hydroxycoumarin did not show any cytotoxicity towards keratinocytes and fibroblasts. Further in the study, a dermal proteomics study was done with the expression of protein biomarkers to evaluate skin sensitization of topical application of coumarins, with proteins such as 14-3-3, COX-2, heat shock proteins, epidermal growth factor receptors, and cytokeratins, and it was found that they are involved in skin sensitization. The results showed many major influences on cell viability as it was assumed that coumarin might disrupt the skin first and then it can hyperproliferate; therefore, further study is needed to understand the mechanism. In all the other studies, coumarin showed weak sensitization and so the presence of coumarin and its derivatives in cosmetic products can account for irritation and the use of coumarin and its derivatives should be reduced to ensure the safe administration of cosmetic products.[49]

#### 4.2.2.4 Psoralen

Psoralens are secondary metabolites that are naturally found in many plants and food. The psoralens bergapten (5-methoxypsoralen, 5-MOP) and xanthotoxin (8-methoxypsoralen) are found to be toxic in nature.[50] Bergapten is a psoralen present in many citrus oils, which are used in various skin care products (sunscreen), and increase skin cancer risk.[51] In an acute derma toxicity study in animals, it showed signs of skin irritation with edema. In a human study, various citrus-derived peel oils (undiluted) were tested and they tended to induce photosensitization due to the presence of furocoumarin. Thus, to reduce the side effects, it should be used in diluted form and should be used in low concentrations in cosmetic preparations to avoid skin sensitization. A cosmetic ingredient review has given the concentration limit of these oils that can be used safely to facilitate the safety assessment of these ingredients.[52] The safe level of bergapten is 0.0015% and thus 100 g of oil should contain 0.0015 g of bergapten (International Fragrance Association). The maximum safe strength of this essential oil is around 0.5 g in 100 g of carrier and not more than that as it is carcinogenic in nature. Care should be taken that other oils used in the composition do not contain any photosensitizing components.[53] It has also been reported that bergamot oil containing bergapten is photocarcinogenic. Studies show that application of bergapten to hairless mouse skin results in tumours in the presence of UV light even at the lowest concentration (5, 15, and 50 ppm). Therefore, the use of bergamot oil in cosmetics should be limited or restricted to 15 ppm of bergapten, equivalent to a 0.4% dilution of commercial bergamot oil.[54]

#### 4.2.2.5 Kaempferol

Kaempferol (3,5,7-trihydroxy-2-(4-hydroxyphenyl)-4*H*-1-benzopyran-4-one, also known as kaempferol-3, kaempferide or kaempferol flavonol) is a type of flavonoid with a molecular weight of 286.23 and has hydrophobic properties due to its diphenyl propane structure. It is a natural flavonol that can be isolated from tea, various vegetables and fruits, and medicinal plants.[55] Kaempferol is used in skin care products as it has anti-collagenase and anti-tyrosinase activities. It can also work as an anti-aging agent as it has also shown anti-wrinkle activity against UV-induced photoaging and is thus used in an anti-aging topical formulations.[56] A study was done on the deposition of flavonoids within the skin and flux across the skin to predict the flavonoid's penetration and reach of the systemic circulation. It was also found that of flavonoids in mouse skin and pig skin, kaempferol showed the greatest deposition.[57] Therefore, kaempferol is known for its toxic effect and has been found to be genotoxic. Although kaempferol is antioxidant in nature, it sometimes acts as a pro-oxidant, which is said to be key to its genotoxic effect. When the phenoxy radical resulting from the interaction with flavonoids gives up a hydrogen atom, it can function as an antioxidant when it reacts with another radical. However, if it encounters oxygen species, it takes on a pro-oxidant role. In this pro-oxidant state, it facilitates the reduction of iron and copper ions, leading to lipid peroxidation and the generation of hydroxyl radicals. Various studies have also shown that changes in antioxidant and pro-oxidant activities by different enzymes are mediated to give a kaempferol pro-oxidant effect. The mutagenic effect of kaempferol is because of the conversion of kaempferol into genotoxic quercetin by CYP1A1 enzyme action. It has also been known for its carcinogenic effect as well. It can also reduce cellular uptake and interfere with bioavailability.[58]

### 4.2.3 Alkaloid Secondary Metabolites

Last in the classification of plant secondary metabolites that are used in cosmetics is alkaloids, which are plant-based secondary metabolites used in creams, lotions hair masks, *etc.* Higher doses of these are known to exhibit toxic properties.[59] Alkaloids are organic compounds containing nitrogen atoms in their structure, which causes alkalinity of these compounds.[60] These are divided into groups like true alkaloids, polyamine alkaloids, protoalkaloids, pseudoalkaloids, peptides, and cyclopeptide alkaloids.[61] Many of the alkaloids are considered toxic in nature because of the presence of various chemical components such as caffeine, pyrrolizidine, *etc.* The following are examples of a few toxic alkaloids that are used in cosmetics.

#### *4.2.3.1 Caffeine*

Caffeine is a secondary metabolite that is biosynthesized by the plant genus *Coffea*. This alkaloid belongs to the family methylxanthine.[62] The alkaloid caffeine present in *Coffea* plant extracts is largely used in cosmetics because of its anti-fat activity and its ability to penetrate the skin barrier.[63] This alkaloid caffeine has antioxidant properties mostly used in various creams and lotions as it is believed to lower the photo-aging process in the skin by UV radiation absorption, which can prevent skin cancer. It has also been used in anti-cellulite products as it is said to be effective in preventing excessive accumulation of fat in cells. It is also believed to improve the microcirculation of blood in the skin and stimulate hair growth by inhibiting 5-α-reductase

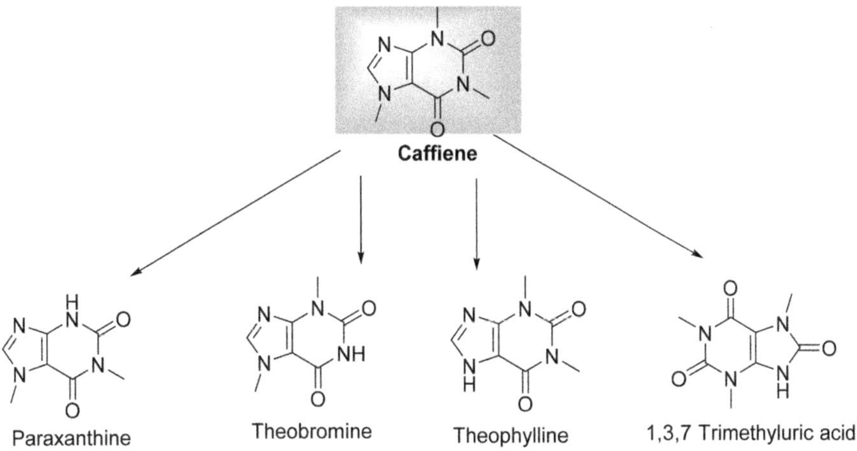


**Figure 4.4** Metabolic conversion of caffeine.

activity.[64] Caffeine in low doses is considered to be completely metabolized. The four primary metabolic conversions are *N*3-demethylation to paraxanthine, *N*1-demethylation to theobromine, *N*7-demethylation to theophylline and C-8 hydroxylation to form 1,3,7-trimethyluric acid (see Figure 4.4).[65]

Caffeine metabolism tends to be affected by many factors such as age, hormones, liver disease, smoking, *etc*. The CYP1A2 isoform of cytochrome P450 tends to metabolize it. The variability of the pharmacokinetics in different individuals can be seen in the polymorphism of this isoform. Various research has also stated that several loci can be located that are involved in caffeine consumption, which has consequences for sleep, anxiety, and neurodegenerative disorders. Caffeine is used for the acute treatment of migraines, but it could potentially be the underlying reason for migraines and can cause headaches. Thus, caffeine use must be limited, especially for those suffering from underlying diseases.[66] Caffeine is completely absorbed by the stomach with peak plasma concentration occurring in 20–40 minutes. Thus, a toxic level can be reached quickly. As the liver metabolizes caffeine by *N*-demethylation, acetylation, and oxidation, substances such as alcohol that are metabolized in the same manner can prolong the half-life of caffeine by 72%.[67] Caffeine is usually referred to as lipophilic, but is a more amphiphilic molecule. Because of the lipophilic moieties, it can partition into the lipid bilayer and can diffuse across the cell; thus, caffeine is also known to accumulate in tissues. In general, concentrations above 15 mg $L^{-1}$ can be considered toxic with mild psychological side effects such as irritability and nervousness and thus the use of caffeine should be taken into account by those with pre-existing medical conditions. Caffeine is considered as a less complex mechanism in the case of normal use but in the case of a lethal dose, it can be a more complex molecule interacting with various molecular targets that can lead to side effects such as antagonism of GABA(A) (gamma-aminobutyric acid), which can result in seizures. Caffeine toxicity is related to the concentration of caffeine at high doses, which leads to higher peak plasma concentration, and thus the use of caffeine-containing cosmetics or any other dermal application should be kept limited especially for people with pre-existing medical conditions.[68]

### 4.2.3.2 Capsaicin

Another alkaloid that is used in cosmetics is capsaicin (as shown in Figure. 4.5), a major component in chili peppers, and its secondary metabolites are capsaicinoids, which are irritating to body tissues. Capsaicin is used in cosmetics as it prevents the photo-aging of skin and has a strong potential as a cosmetic ingredient in anti-photoaging applications.[69] Capsicum-derived ingredients are used in skin conditioning agents and also in fragrance components in cosmetics. These ingredients are used in cosmetics at as high as 5%, which include capsaicin. Studies have also shown that capsicum fruit extract induces cytotoxicity in the human buccal mucosa fibroblast cell line. Capsaicin is not recognized as a safe drug administration for fever. *In vitro* percutaneous absorption of capsaicin is shown in humans. Skin permeation enhancement of anti-inflammatory agents in the presence of capsaicin has also been demonstrated. Capsaicin contains a vanillyl moiety that produces its sensory effects on sensory neurons and thus capsaicin is a known activator of vanilloid receptor 1. It has been shown that dermal application inhibited sensitization in a mouse study. The skin irritant effects of capsaicin appear to be the same as vanilloid receptors. This mechanism of action means that it is considered a skin irritant. Thus, there should be limitations over the use of capsaicin, which can reduce skin irritants. As capsaicin enhances the penetration of anti-inflammatory agents through the skin, the use of this ingredient in cosmetics should be limited.[70] Skin reactions are known to be induced by capsaicin. This depends on the time of action and concentration. Capsaicin activation of TRPV1 (transient receptor potential cation channel subfamily V member 1) on nociceptive nerve endings results in nociceptive signal transmission to the central nervous system, which leads to a burning sensation. Further activation releases neuropeptides such as SP (substance P) and CGRP (calcitonin gene-related peptide) leading to neurogenic inflammation. For dermal toxicity, a lethal dose level is recommended of not more than 512 mg $kg^{-1}$ in mice, and it is highly irritant in humans leading to a burning sensation. It has been reported that long-term exposure leads to

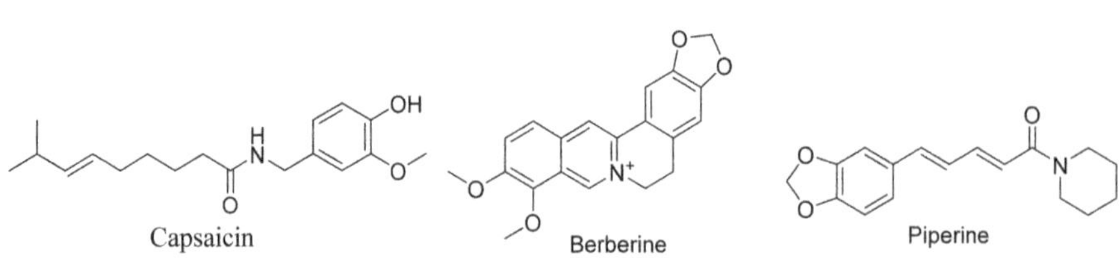


**Figure 4.5** Chemical structures of alkaloids predominantly used in cosmetics.

blistering and rashes, which is proof of dermal toxicity in humans. If topically applied it results in burning pain and it also increases the temperature and blood flow. Investigation of dermal micro-vessels also resulted in a skin neurogenic vasodilator reaction by capsaicin.[71]

#### *4.2.3.3 Berberine*

Berberine (see Figure. 4.5), which is a key component in *Coptis japonica*, is used in many skin conditions, like skin sagging, swelling, cellulite, *etc*. Thus, *Coptis japonica* root extract has wide applications in cosmetics. It shows potential inhibition of triglyceride accumulation and shows down-regulation of mRNA expression in various differentiation factors, which are derived from subcutaneous preadipocytes.[72] Berberine is known to suppress the interleukin release in eosinophil cultures, which decreases tyrosinase expression.[73] It is mostly used in cosmetics because of its antioxidant and anti-wrinkle properties. It reduces ROS production and promotes hyaluronic acid production, and is also known to decrease the protein and mRNA expression of MMP-2 and MMP-9, which are related to wrinkle formation by increasing the expression of TIMP-1 and TIMP-2 (tissue inhibitors of metalloproteinases) along with proinflammatory cytokines, which inhibiting TNF-$\alpha$ in a concentration-dependent manner.[74] Berberine has various pharmacological effects, including antioxidant, which is why it is used in dermatological products. It can be applied directly to the skin to treat burns but it has some toxicity issues and thus it has short-term use and is not suitable for children and during pregnancy and breastfeeding.[73] Berberine is also known as a natural antioxidant and was investigated for its hepatoprotective effect. It was found to induce liver damage, decreasing the PQ-induced cell death and diminishing PQ-induced liver function enzymes, formation of ROS, and the release of lactate dehydrogenase. It was also found to inhibit cellular glutathione depletion in PQ-treated hepatocytes.[75] The oral administration of powdered root *Berberis vulgaris* resulted in diarrhea in 40% of rats affecting the gastrointestinal tract. A study done on cats also resulted in vomiting, as well as death. Oral administration of berberine can cause hemorrhagic inflammatory problems in the small and large intestines. It has also been found to be poisonous in dogs, resulting in nausea, diarrhea, muscular tremors, and sometimes paralysis. In a sub-acute toxicity study, it was found to induce chronic arthritis, liver and kidney enlargement, and decreased bilirubin protein binding. Some studies have reported an immunotoxic effect. It was also reported to decrease leukocytes, neutrophils, lymphocytes, *etc*., decreasing B and T cells and influencing the proliferation of lymphocytes, suppressing both cellular and humoral immune function. Even in sub-chronic toxicity, it has been shown to be damaging to the lungs and liver through the high level of alanine aminotransferase and aspartate aminotransferase.[76] It is known for its antioxidant effect, but it can also lead to the production of ROS in certain conditions; thus, its use in cosmetics should be limited to avoid its toxic effect as it can affect the immune cells present in the human body.

### 4.2.3.4 Piperine

Piperine (see Figure 4.5) is another alkaloid that belongs to the largest class of plant secondary metabolites, which is termed alkaloids. It is a natural pungent secondary metabolite alkaloid extracted from the plant genus *Piper*. It is obtained from the oleoresin of *P. nigrum* or *P. longum*. Piperine is a natural spice with antioxidant, and anti-inflammatory properties and it is used mostly for treating and preventing the aging process and related conditions. It is used in skin aging and alopecia as a cosmetic ingredient.[77] Iit is estimated that over 700 members of this class have been researched to date. These alkaloids possess a piperidine nucleus, which is is known to cause toxicity. Piperine can be hydrolyzed by alkali into piperidine.[78] Cosmoperine is isolated from piperine and used in cosmetics as a natural bio-enhancer improving the active compound's permeability through the skin.[79] Tetrahydropiperine (THP) has been recently developed and introduced into the cosmetic market, it is used in topical bioavailability enhancers and is a derivative of the pungent alkaloid piperine, which is extracted from the black pepper plant. Along with its beneficial effect, the use of THP in the range 0.01–0.1% is considered effective but topical irritation also occurs.[80] As this is linked with topical irritation its use in cosmetics should be restricted.

### 4.2.3.5 Pyrrolizidine Alkaloids

Pyrrolizidine alkaloids are known to be esters of necic acid and 1-hydroxymethylpyrrolizidines, *i.e.*, necine acid, and these originate from L-arginine or L-ornithine. They have two conjoined 5-membered rings with a bridged nitrogen atom and are considered to be hepatotoxic. The hepatotoxicity of the alkaloids is determined by the C-1 and C-2 double bond positions.[81] Pyrrolizidine found in plant extracts such as *Senecio vulgaris*, *Symphytum officinalis*, and *Tussilago farfara* produces pyrrolizidine alkaloids, which are used in cosmetics. These alkaloids are also known to be mutagenic and carcinogenic.[82] Along with this, they are known to cause abdominal pain, nausea, and vomiting and can also damage the pulmonary region as well.[83] The absorption of pyrrolizidine alkaloids through the skin was found to be low (up to 5%) when compared to oral intake. Topical application showed low absorption and 0.1–0.4% was also recovered in urine as well. Cosmetic exposure to these alkaloids can have a risk of cancer when present in shaving products that have 5% comfrey extract in them. Leave-on face, body lotion, and body wash containing *Symphytum officinale* at 4% in the product showed the worst effects through skin penetration. Thus, cosmetics containing comfrey extracts are prohibited. Along with pyrrolizidine, alkaloids showed a low amount of systemic absorption with *Symphytum officinale* topical application, but its potential for health effects should be considered and pyrrolizidine-free extracts should be used in cosmetic preparations.[84] Retrorsine is the major toxic alkaloid found in *Senecio vulgaris*, and jacobine found in *Symphytum officinale* is known to be the most toxic. Pyrrolizidines are toxic towards the liver, where they cause veno-occlusive disease, which results in lesions. The formation of pyrrole-type metabolites is related to a toxic mechanism of action. These pyrrolic dehydro alkaloids are formed *via* dehydrogenation by hepatic monooxygenases. These metabolites are reactive alkylating agents that react with cellular nucleophiles, causing DNA alkylation, DNA cross-linking, and liver cell necrosis.[85] As pyrrolizidine alkaloids are considered hepatotoxic, mutagenic and carcinogenic for humans, their use for dermal administration should be below 0.01 ppm as they can sometime show 100% skin penetration if the skin barrier is not intact.[86] It should be noted that only 1,2-dehydropyrrolizidine alkaloids in the pyrrolizidine alkaloid structures are considered to be toxic; without the presence of this structure they are considered non-toxic. Thus, toxicity requires metabolic activation of reactive compounds.[87] Therefore for the safe use of extracts containing pyrrolizidine alkaloids in cosmetics, one should check for the presence of 1,2-dehydropyrrolizidine alkaloids in the extract and then they can be incorporated in the cosmetic formulation from a safety point of view.

## 4.3 Secondary Metabolites Derived from Microorganisms and Animals That Cause Toxic and Allergic Responses

Microorganisms produce secondary metabolites when one or more nutrients in the culture medium are depleted. This has an important ecological function because it provides diverse survival functions in nature, but is not essential to the growth of microorganisms. However, they are important to humans in terms of health and nutrition.[88] Secondary metabolites are a result of a later phase of fermentation called the idiophase, followed by the trophophase, which is an active growth phase. Compounds resulting in idiophase do not have any direct relationship with the growth of microorganisms. Secondary metabolites are formed during fermentation after the growth of microorganisms is completed. A single microorganism can produce different secondary metabolites. There are a few secondary metabolites that are toxic in nature and efforts are being made to reduce their production.[89] These microbial secondary metabolites, such as antibiotics, pigments, antioxidants, *etc.* are not important for the growth and development of the microorganism, but are of great importance for human and animal health.[90] Even in cosmetic industries, microorganisms producing low molecular weight chemical compounds such as organic acids, alcohols, proteins, and polysaccharides are used. These can be used as UV protectors, anti-aging agents, immunomodulators, binding agents, thickening agents, coloring agents, and stabilizing agents.[91] Microorganisms are therefore highly used in the production of metabolites because of they have a reasonable cost and are rich in fatty acids, enzymes, peptides, lipopolysaccharides, pigments, *etc.*, which are used in cosmetic applications. Along with microorganisms, compounds such as ceramides, mycoporine-like amino acids, and carotenoids are extensively used in the cosmetics industry. There has been an increasing trend in the use of biological and eco-friendly products in the cosmetics industry. Some such biological products can be responsible for adverse effects, and therefore methodical and systematic rigorous assessment is required in terms of clinical research to understand the potential of such secondary metabolites obtained from microorganisms before any validation.[92] The following are a few examples of secondary metabolites produced by microorganisms that can be toxic in normal environmental defense systems, as these secondary metabolites are common compounds produced by the microorganism that need to be considered before using them in cosmetics. Furthermore, the presence of such compounds obtained by the biotechnological process also needs to be evaluated. This is because several secondary metabolites are obtained by a single microorganism; thus, the isolation of toxic compounds if present in the extract along with the valuable secondary metabolites needs to be considered. Cyanobacteria are known to have various functional ingredients and are widely used in cosmetics as they produce pigments and have bioactive potential. They have great potential for health applications, as they produce compounds such as polyphenols, peptides, polysaccharides, and pigments. Pigments obtained from cyanobacteria, such as carotenoids and phycobiliproteins, are used as a cosmetic enhancers for moisturizers and as stabilizing agents in natural cosmetics. Carotenoids are also used as sunscreen, anti-aging, and antioxidant agents. Phycobiliproteins are used as a natural dye. *Arthrospira platensis* (*Spirulina*) is also used in the cosmetics industry. Extracts obtained from cyanobacteria species are known to have great potential as antioxidant and anti-inflammatory agents.[93] Recently, the use of microalgae and cyanobacteria in cosmetic applications has increased in products containing pigments, proteins, essential fatty acids, polysaccharides, vitamins, and minerals. Cyanobacteria, known as blue–green algae (*Arthrospira*, *Spirulina*), have been used since historical times. The microalgae *Haematococcus pluvialis* is used in the production of astaxanthin, and is known to have excellent antioxidant properties. Species such as *Chlorella*, *Nannochloropsis*, *Porphyridium*, *Nostoc*, and *Dunaliellaare* are used in anti-aging products, and as protection from UV damage and oxidative stress, but there should be more research conducted to evaluate them before they can be stated as safe natural ingredients as most of them are labeled as a natural cosmetic product.[94] In general, cyanobacteria are used as a source of several secondary products but cyanotoxins are also produced as a result of a chemical defense mechanism as a secondary product and this role of producing toxic secondary metabolites should be considered in the biotechnological procedure of obtaining secondary metabolites from bacteria along with their mode of action. As cyanobacteria are known to produce a wide range of toxic secondary compounds, which are chemically classified as cyclic peptides, alkaloids, and lipopolysaccharides, and can result in hepatotoxins, neurotoxins, cytotoxin, dermatotoxins, and irritant toxins, these should be evaluated in the bacterial extract that is obtained from the biotechnological process. For example, Aplysia toxins produced by *Lyngbia*, and lipopolysaccharides produced by *Anabaena*, *Anacystis*, *Microsystis*, *Oscillatoria*, *Spirulina*, and almost all cyanobacteria are known to be dermatoxic. Therefore, such bacteria when used to extract valuable secondary metabolites for cosmetic use should be evaluated for the presence of toxic secondary metabolites along with the mechanism of action.[95] Most of the algae metabolites are obtained by the bioreactor which is used for algae cultivation, but it increases the cost and therefore some use natural habitats to harvest algae. Culturing algae in natural habitats such as tanks or pounds means that they can be easily contaminated. Open cultivation can interfere with biotic and abiotic environments and results in low-quality algae production. Therefore, the quality of the raw material should be evaluated before use, especially for the side effect of being phototoxic. Photoallergens include chlorophyll, which can be classified into four types, chlorophyll a in cyanobacteria and algae, chlorophyll b in green algae, chlorophyll c in brown algae and dinoflagellates, and chlorophyll d in red algae. Most of the time consumers are not aware of natural cosmetic products that may develop side effects due to the complex mixture of natural and chemical compounds and therefore a safety evaluation of the cosmetic should be performed, especially if such raw materials are used in a cosmetic formulation. Cyanotoxins are metabolites that are reported to have toxic effects on immune and brain cells and therefore clinical studies should be determined for the safety and efficacy of such compounds on humans.[96] Along with these, there are some natural dyes that are used in cosmetics and are derived from insects, such as cochineal beetles and lac scale insects, and animals such as mollusks and shellfish. Being a natural dye, it has some toxicity issues, like for example carminic acid.[97]

### 4.3.1 Carminic Acid

Color pigments that are obtained from the cochineal extracted from the dried bodies of the female insect *Dactylopius coccus* are used in colored cosmetics that contain carminic acid (see Figure 4.6) as the coloring principle is known to be toxic in nature. The study showed that people working in extracting carmine from insects as a cosmetic coloring agent suffered from asthma and the bronchial provocation test was found to be positive. It also

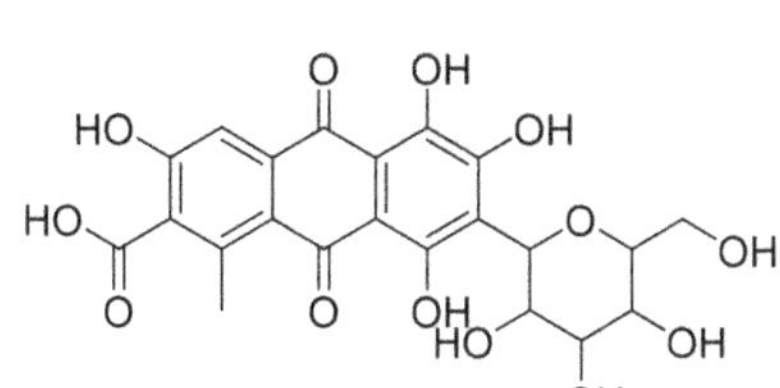


**Figure 4.6** The structure of carminic acid.

has dermal toxicity issues; when carmine-containing lip salve was used by women, they developed soreness of the lips with cracking and bleeding. The same was also the case when an oil-soluble carmine preparation was used. Even patients allergic to carmine in food and drink have reported reactions to cosmetics containing carmine in them.[98] Their use in cosmetics is limited with today's trend towards green cosmetics, but their presence in colored cosmetics is unavoidable.

### 4.3.2 Bee Products

Bee products are extensively used in cosmetics, especially honey, which is regenerative and has antimicrobial properties because of its high osmolarity due to the presence of hydrogen peroxide and lysozyme. Manuka honey is a potent antiseptic agent due to the presence of methylglyoxal. Propolis is also very commonly used in cosmetics; it is a bee product that is rich in phenolic compounds which possess antimicrobial, antioxidant, UV protectant, and regenerative properties. Royal jelly is known to have jelleine peptides that are used in collagen production, as it contains 10-hydroxy-*trans*-2-decenoic acid. Another bee product is bee pollen, which is rich in flavonoids, a phenolic compound, along with hydroxy acids, vitamins, *etc.* and is widely used in cosmetics as an emulsifier. Last but not least is bee venom, a bee product that is used in cosmetics because of its natural toxin content, which is rich in peptides. Sometimes when these bee products are used in cosmetics, they can cause some allergic reactions. A honey allergy is rare, but frequent honey allergens can cause a hypersensitive reaction, *i.e.*, bee pollen and bee proteins in honey cause allergies. Propolis contains more than 26 allergenic substances and allergic reactions are frequently caused by the esters of caffeic acid and cinnamic acid derived from poplar buds, which are phenolic secondary metabolic compounds. These cause contact allergic reactions in hypersensitive people because of the presence of these esters. Potent allergen compounds include 3-methyl-2-butyl-caffeate (54.2%), 3-methyl-3-butyl-caffeate (28.3%), 2-methyl-2-butylcaffeate (4.3%), caffeic acid (1.3%), benzyl caffeate (1.0%), caffeic acid phenethyl ester (CAPE, 7.9%) and benzyl salicylate. Although propolis allergies are rare, the allergic response can be more frequent after topical application than oral administration. Royal jelly in ointment form can cause skin rashes and eczema *via* topical application. Moreover, the allergic reaction and irritation caused buy bee venom is due to the presence of phospholipase A2, melittin, and hyaluronidase. It can be toxic in large amounts but topical application is safe, as has been confirmed by a previous study.[99] Another study was done on propolis, which is commonly used in cosmetics such as lip balms, lotions, ointments, shampoos, conditioners, *etc.* and it was found to be allergenic; 1.2% to 6.6% of patients patch-tested for dermatitis were sensitive to propolis and thus dermatologists need to consider patch tests before using cosmetic products that contain propolis.[100]

**Table 4.1** Toxic materials used in fragrance products.

| S. no. | Examples (fragrance material) | Allergenic potential |
|---|---|---|
| 1 | D-Limonene (monoterpene) | III |
| 2 | Linalool (monoterpene) | III |
| 3 | Geraniol (monoterpene) | II |
| 4 | Citronellol (monoterpenoid-modified terpenes) | II |
| 5 | Citral (aliphatic aldehyde of the terpene) | II |
| 6 | Farnesol (acyclic sesquiterpenes) | I |
| 7 | Coumarin (phenolic compound – benzopyrone derivatives) | II |
| 8 | Eugenol (phenolic compound) | II |
| 9 | Benzyl alcohol (phenolic compound) | III |
| 10 | Benzyl benzoate (phenolic compound) | III |
| 11 | Benzyl salicylate (phenolic compound) | III |
| 12 | Cinnamyl alcohol (phenolic compound) | II |
| 13 | Isoeugenol (phenolic compound – phenylpropene) | I |
| 14 | Cinnamal (phenolic compound) | I |

### 4.3.3 Toxicological Profile

The following are a few examples of secondary metabolites that are present in plant extracts that are used as a fragrance material in personal care products that are known to have a toxicological profile and some are also known as sensitizers for allergic contact dermatitis. The examples are listed in Table 4.1 with their allergenic potential: [Group I: important allergens, Group II: clearly allergenic, but less important in terms of sensitization frequency, Group III: (extremely) rare sensitizers or even non-sensitizers].[101]

The examples listed in Table 4.2 are prohibited substances used in cosmetic products due to the presence of toxic secondary metabolites that are present in plant and animal extracts.[102]

## 4.4 Conclusions and Future Perspectives

The possibility of secondary metabolite toxicity and allergic reactions is an important issue because consumers use cosmetic products containing several types of secondary metabolites and their derivatives, so they are regularly exposed and their toxic components can lead to synergistic or additive effects on the human body. No doubt secondary metabolites have a remarkable place in cosmetic preparation, but knowledge about their various derivatives which have inherent toxicity should be evaluated for the safe use of secondary metabolites in cosmetic preparations, along with their safe concentration level. Many natural substances have been classified as hazardous substances that are used in personal care products, but when used as an essential oil or as a plant extract containing those toxic chemical compounds, they should be classified first before being used in cosmetics. In the future, the cosmetic

**Table 4.2** Prohibited substances used in cosmetic products.

| S. no. | Examples | Description |
|---|---|---|
| 1 | *Atropa belladonna* | It's a poisonous plant. Its roots, leaves and fruits contain toxic alkaloids such as atropine, hyocyamine and scopolamine. It is used as a facial cosmetic.[103] |
| 2 | *Cantharides, Cantharis vesicatoria* (Spanish fly-beetle) | It is used in hair care products because of the hair growth-stimulating composition found in *Cantharis* extracts.[104] It is known to have simple monoterpenes, and is derived from degradation of sesquiterpene farnesol.[105] |
| 3 | *Claviceps purpurea* | It contains toxic ergot alkaloids. Mostly used in colored cosmetics.[106] |
| 4 | Hemlock *(Conium maculatum)* | Known as a poisonous plant because it contains toxic secondary metabolite piperidine alkaloids.[107] |
| 5 | *Prunus laurocerasus* L. (cherry laurel water) | It is mostly use as an antioxidant.[108] It is considered toxic because of the presence of cyanogenic glycoside secondary metabolites.[109] |
| 6 | *Veratrum* spp. (false hellebore) | Its oil from the beans was used in facial cosmetics.[110] It is prohibited because it contains toxic steroidal alkaloids.[111] |
| 7 | *Pyrethrum album* L. | It is mostly used as an insecticide for scabies, head lice, and crab lice in shampoo preparations. It causes eye irritation.[112] |
| 8 | *Chenopodium ambrosioides* L. | Leaf extract is used in skin disorders (skin itching, eczema, and pimples) and is also used for making soaps.[113] It is known to have chronic toxicity.[114] |
| 9 | *Evernia prunastri* (oakmoss extract) | This extract is mostly used in fragrances and is known to cause contact allergy.[115] |
| 10 | *Ficus carica Linn* | Leaf extract contains polyphenolic compounds and is known to be toxic only at higher doses.[116] |

ingredients that are derived from plant sources should be reviewed by an expert panel to determine their safe use in cosmetics, which could raise concerns about toxicity and allergens. This is because the final cosmetic formulation can contain multiple botanicals with the presence of similar toxic constituents and can reach a toxic level that can be hazardous to consumers. Therefore, the industry must follow good manufacturing practices to limit the toxic effect in the final cosmetic formulation. This is essential for the fast growing economy of the cosmetics industry, which requires safe and efficient raw ingredients that are natural and do not cause toxic or allergic reactions.

## References

1. R. Croteau, T. M. Kutchan and N. G. Lewis, Natural products (secondary metabolites), *Biochem. Mol. Biol. Plants*, 2000, **24**, 1250–1319.

2. J. N. Kabera, E. Semana, A. R. Mussa and X. He, Plant secondary metabolites: biosynthesis, classification, function and pharmacological properties, *J. Pharm. Pharmacol.*, 2014, **2**(7), 377–392.
3. R. Tiwari and C. S. Rana, Plant secondary metabolites: a review, *Int. J. Eng. Res. Gen. Sci.*, 2015, **3**(5), 661–670.
4. P. Vaishnav and A. L. Demain, Unexpected applications of secondary metabolites, *Biotechnol. Adv.*, 2011, **29**(2), 223–229.
5. N. Kaur and T. Ahmed, Bioactive secondary metabolites of medicinal and aromatic plants and their disease-fighting properties, in *Medicinal and Aromatic Plants*, 2021, pp. 113–142.
6. S. R. Leicach and H. D. Chludil, Plant secondary metabolites: Structure–activity relationships in human health prevention and treatment of common diseases, *Stud. Nat. Prod. Chem.*, 2014, **42**, 267–304.
7. S. N. Jimenez-Garcia, M. A. Vazquez-Cruz, R. G. Guevara-Gonzalez, I. Torres-Pacheco, A. Cruz-Hernandez and A. A. Feregrino-Perez, Current approaches for enhanced expression of secondary metabolites as bioactive compounds in plants for agronomic and human health purposes-A review, *Pol. J. Food Nutr. Sci.*, 2013, **63**(2), 67–78.
8. H. H. Agus, Terpene toxicity and oxidative stress, in *Toxicology*, Academic Press, 1 January 2021, pp. 33–42.
9. A. Madariaga-Mazón, R. B. Hernández-Alvarado, K. O. Noriega-Colima, A. Osnaya-Hernández and K. Martinez-Mayorga, Toxicity of secondary metabolites, *Phys. Sci. Rev.*, 2019, **4**(12), 20180116.
10. V. Kumar, V. Ahluwalia, S. Saran, J. Kumar, A. K. Patel and R. R. Singhania, Recent developments on solid-state fermentation for production of microbial secondary metabolites: Challenges and solutions, *Bioresour. Technol.*, 2021, **323**, 124566.
11. B. Ruiz, A. Chávez, A. Forero, Y. García-Huante, A. Romero, M. Sánchez, D. Rocha, B. Sánchez, R. Rodríguez-Sanoja, S. Sánchez and E. Langley, Production of microbial secondary metabolites: regulation by the carbon source, *Crit. Rev. Microbiol.*, 2010, **36**(2), 146–167.
12. M. P. N. Rao, M. Xiao and W. J. Li, Fungal and bacterial pigments: secondary metabolites with wide applications, *Front. Microbiol.*, 2017, **8**, 1113.
13. B. Nowruzi, G. Sarvari and S. Blanco, The cosmetic application of cyanobacterial secondary metabolites, *Algal Res.*, 2020, **49**, 101959.
14. J. P. Leflaive and L. O. Ten-Hage, Algal and cyanobacterial secondary metabolites in freshwaters: a comparison of allelopathic compounds and toxins, *Freshwater Biol.*, 2007, **52**(2), 199–214.
15. J. Zhang, R. Zeng and A. Labes, Marine microbial-derived molecules and their potential medical and cosmetic applications, *Front. Microbiol.*, 2021, **12**, 706152.
16. P. Noriega, Terpenes in Essential Oils: Bioactivity and Applications, in *Terpenes and Terpenoids—Recent Advances*, 10 October 2020.
17. S. D. Tetali, Terpenes and isoprenoids: a wealth of compounds for global use, *Planta*, 2019, **249**(1), 1–8.

18. T. Zárybnický, I. Boušová, M. Ambrož and L. Skálová, Hepatotoxicity of monoterpenes and sesquiterpenes, *Arch. Toxicol.*, 2018, **92**(1), 1–3.
19. M. Matura, M. Sköld, A. Börje, K. E. Andersen, M. Bruze, P. Frosch, A. Goossens, J. D. Johansen, C. Svedman, I. R. White and A. T. Karlberg, Selected oxidized fragrance terpenes are common contact allergens, *Contact Dermatitis*, 2005, **52**(6), 320–328.
20. D. Dittmar and M. L. Schuttelaar, Contact sensitization to hydroperoxides of limonene and linalool: Results of consecutive patch testing and clinical relevance, *Contact Dermatitis*, 2019, **80**(2), 101–109.
21. J. B. Christensson, A. T. Karlberg, K. E. Andersen, M. Bruze, J. D. Johansen, B. Garcia-Bravo, A. G. Arnau, C. L. Goh, R. Nixon and I. R. White, Oxidized limonene and oxidized linalool–concomitant contact allergy to common fragrance terpenes, *Contact Dermatitis*, 2016, **74**(5), 273–280.
22. J. B. Christensson, P. Forsström, A. M. Wennberg, A. T. Karlberg and M. Matura, Air oxidation increases skin irritation from fragrance terpenes, *Contact Dermatitis*, 2009, **60**(1), 32–40.
23. H. Audrain, C. Kenward, C. R. Lovell, C. Green, A. D. Ormerod, J. Sansom, M. M. Chowdhury, S. M. Cooper, G. A. Johnston, M. Wilkinson and C. King, Allergy to oxidized limonene and linalool is frequent in the UK, *Br. J. Dermatol.*, 2014, **171**(2), 292–297.
24. C. Raffalli, E. Clouet, S. Kuresepi, M. H. Damiens, J. P. Lepoittevin, M. Pallardy, P. J. Ferret, E. Giménez-Arnau and S. Kerdine-Römer, Editor's Highlight: Fragrance Allergens Linalool and Limonene Allylic Hydroperoxides in Skin Allergy: Mechanisms of Action Focusing on Transcription Factor Nrf2, *Toxicol. Sci.*, 2018, **161**(1), 139–148.
25. M. Sköld, A. T. Karlberg, M. Matura and A. Börje, The fragrance chemical β-caryophyllene—air oxidation and skin sensitization, *Food Chem. Toxicol.*, 2006, **44**(4), 538–545.
26. C. Bäcktorp, L. Hagvall, A. Börje, A. T. Karlberg, P. O. Norrby and G. Nyman, Mechanism of air oxidation of the fragrance terpene geraniol, *J. Chem. Theory Comput.*, 2008, **4**(1), 101–106.
27. L. Hagvall, C. Bäcktorp, S. Svensson, G. Nyman, A. Börje and A. T. Karlberg, Fragrance compound geraniol forms contact allergens on air exposure. Identification and quantification of oxidation products and effect on skin sensitization, *Chem. Res. Toxicol.*, 2007, **20**(5), 807–814.
28. L. Hagvall, J. M. Baron, A. Börje, L. Weidolf, H. Merk and A. T. Karlberg, Cytochrome P450-mediated activation of the fragrance compound geraniol forms potent contact allergens, *Toxicol. Appl. Pharmacol.*, 2008, **233**(2), 308–313.
29. L. Hagvall, C. Bäcktorp, P. O. Norrby, A. T. Karlberg and A. Börje, Experimental and theoretical investigations of the autoxidation of geranial: a dioxolane hydroperoxide identified as a skin sensitizer, *Chem. Res. Toxicol.*, 2011, **24**(9), 1507–1515.
30. A. C. de Groot and E. Schmidt, Tea tree oil: contact allergy and chemical composition, *Contact Dermatitis*, 2016, **75**(3), 129–143.

31. J. Rudbäck, M. A. Bergström, A. Börje, U. Nilsson and A. T. Karlberg, α-Terpinene, an antioxidant in tea tree oil, autoxidizes rapidly to skin allergens on air exposure, *Chem. Res. Toxicol.*, 2012, **25**(3), 713–721.
32. M. R. Youssefi, A. Nikpay, N. Hassanpour, A. Mirzapour, P. S. Tabari, R. Pavela, F. Maggi and R. Petrelli, In vitro scolicidal activity of the sesquiterpenes isofuranodiene, α-bisabolol and farnesol on Echinococcus granulosus protoscoleces, *Molecules*, 2020, **25**(16), 3593.
33. Z. Naz, M. S. Faisal, A. Muheem, S. Singh and J. F. Ahmad, Determination of (-) α-BISABOLOL in Matricaria chamomile oil and in nanoformulation by HPTLC mhod: it's application in ex vivo studies, *Indo Am. J. Pharm. Res.*, 2015, **5**(6), 2190–2197.
34. E. Ramazani, M. Akaberi, S. A. Emami and Z. Tayarani-Najaran, Pharmacological and biological effects of alpha-bisabolol: An updated review of the molecular mechanisms, *Life Sci.*, 2022, 120728.
35. G. P. Kamatou and A. M. Viljoen, A review of the application and pharmacological properties of α-bisabolol and α-bisabolol-rich oils, *J. Am. Oil Chem. Soc.*, 2010, **87**(1), 1–7.
36. A. T. Karlberg, A. Börje, J. Duus Johansen, C. Lidén, S. Rastogi, D. Roberts, W. Uter and I. R. White, Activation of non-sensitizing or low-sensitizing fragrance substances into potent sensitizers–prehaptens and prohaptens, *Contact Dermatitis*, 2013, **69**(6), 323–334.
37. V. Lattanzio, P. A. Kroon, S. Quideau and D. Treutter, Plant phenolics—secondary metabolites with diverse functions, *Recent Adv. Polyphenol Res.*, 2009, **1**, 1–35.
38. M. Działo, J. Mierziak, U. Korzun, M. Preisner, J. Szopa and A. Kulma, The potential of plant phenolics in prevention and therapy of skin disorders, *Int. J. Mol. Sci.*, 2016, **17**(2), 160.
39. M. L. Soto, E. Falqué and H. Domínguez, Relevance of natural phenolics from grape and derivative products in the formulation of cosmetics, *Cosmetics*, 2015, **2**(3), 259–276.
40. A. Svobodová, J. Psotová and D. Walterová, Natural phenolics in the prevention of UV-induced skin damage. A review, *Biomed. Pap. Med. Fac. Univ. Palacky Olomouc Czech Repub.*, 2003, **147**(2), 137–145.
41. B. G. Swanson, Tannins and polyphenols, in *Encyclopaedia of Food Sciences and Nutrition*, ed. B. Caballero, L. C. Trugo, and P. M. Finglas, 2nd edn, 2003, pp. 5729–5733.
42. Y. Desjardins, Physiological and ecological functions and biosynthesis of health-promoting compounds in fruit and vegetables, in *Improving the Health-Promoting Properties of Fruit and Vegetable Products*, ed. B. F. A. Tomás, and M. I. Gil, Elsevier, USA, 1 January 2008, vol. 23, pp. 201–247.
43. A. Gunia-Krzyżak, K. Słoczyńska, J. Popiół, P. Koczurkiewicz, H. Marona and E. Pękala, Cinnamic acid derivatives in cosmetics: current use and future prospects, *Int. J. Cosmet. Sci.*, 2018, **40**(4), 356–366.
44. K. Zduńska, A. Dana, A. Kolodziejczak and H. Rotsztejn, Antioxidant properties of ferulic acid and its possible application, *Skin Pharmacol. Physiol.*, 2018, **31**(6), 332–336.

45. https://www.medicalnewstoday.com/articles/ferulic-acid.
46. K. Kasperkicwicz, A. Erkiert-Polguj and E. Budzisz, Sunscreening and photosensitizing properties of coumarins and their derivatives, *Lett. Drug Des. Discovery*, 2016, **13**(5), 465–474.
47. A. Sallegger, Bio-guided isolation of phytotoxic coumarin derivatives of Thamnosma montana Torr. & Frém., a Rutaceae native to the southwestern United States/vorgelegt von Aaron Sallegger, PhD thesis, University of Graz, 2009.
48. F. Floc'h, F. Mauger, J. R. Desmurs, A. Gard, F. Bagneris and B. Carlton, Coumarin in plants and fruits: implications in perfumery, *Perfum. Flavor.*, 2002, **27**(2), 32–36.
49. T. L. Pan, P. W. Wang, I. A. Aljuffali, Y. L. Leu, Y. Y. Hung and J. Y. Fang, Coumarin derivatives, but not coumarin itself, cause skin irritation via topical delivery, *Toxicol. Lett.*, 2014, **226**(2), 173–181.
50. M. M. Diawara and P. J. Kulkosky, Reproductive toxicity of the psoralens, *Pediatr. Pathol. Mol. Med.*, 2003, **22**(3), 247–258.
51. J. Bruneton, Principles of herbal pharmacology, in *Principles and Practice of Phytotherapy: Modern Herbal Medicine*, ed. K. Bone, and S. Mills, Churchill Livingstone, Elsevier, 2012, pp. 45–82.
52. C. L. Burnett, M. M. Fiume, W. F. Bergfeld, D. V. Belsito, R. A. Hill, C. D. Klaassen, D. C. Liebler, J. G. Marks Jr, R. C. Shank, T. J. Slaga and P. W. Snyder, Safety assessment of citrus-derived peel oils as used in cosmetics, *Int. J. Toxicol.*, 2019, **38**(2_suppl), 33S–59S.
53. S. Clarke, Families of compounds that occur in essential oils, *Essential Chemistry for Aromatherapy*, 2008, vol. 3, pp. 41–77.
54. R. Tisser and R. Young, *The Skin, Essential Oil Safety*, Churchill Livingstone, 2nd edn, 2014, pp. 69–98.
55. J. I. Ren, Y. Lu, Y. Qian, B. Chen, T. A. Wu and G. Ji, Recent progress regarding kaempferol for the treatment of various diseases, *Exp. Ther. Med.*, 2019, **18**(4), 2759–2776.
56. E. S. Mostafa, A. Maher, D. A. Mostafa, S. S. Gad, M. A. Nawwar and N. Swilam, A Unique Acylated Flavonol Glycoside from Prunus persica (L.) var. Florida Prince: A New Solid Lipid Nanoparticle Cosmeceutical Formulation for Skincare, *Antioxidants*, 2021, **10**(3), 436.
57. S. Y. Chuang, Y. K. Lin, C. F. Lin, P. W. Wang, E. L. Chen and J. Y. Fang, Elucidating the skin delivery of aglycone and glycoside flavonoids: How the structures affect cutaneous absorption, *Nutrients*, 2017, **9**(12), 1304.
58. W. Alam, H. Khan, M. A. Shah, O. Cauli and L. Saso, Kaempferol as a dietary anti-inflammatory agent: Current therapeutic standing, *Molecules*, 2020, **25**(18), 4073.
59. A. Stępniowska, P. Cieplińska, W. Fac and J. Górska, Selected Alkaloids Used in the Cosmetics Industry, *J. Cosmet. Sci.*, 2021, **72**(2), 229–245.
60. J. Kurek, in *Alkaloids: Their importance in Nature and Human life*, BoD–Books on Demand, 13 November 2019.
61. I. Chiocchio, M. Mandrone, P. Tomasi, L. Marincich and F. Poli, Plant secondary metabolites: An opportunity for circular economy, *Molecules*, 2021, **26**(2), 495.

62. R. Pech-Kú, J. A. Muñoz-Sánchez, M. Monforte-González, F. Vázquez-Flota, B. A. Rodas-Junco and S. T. Hernández-Sotomayor, Caffeine extraction, enzymatic activity and gene expression of caffeine synthase from plant cell suspensions, *J. Visualized Exp.*, 2018, (140), e58166.
63. A. Barbulova, F. Apone and G. Colucci, Plant cell cultures as source of cosmetic active ingredients, *Cosmetics*, 2014, **1**(2), 94–104.
64. M. Gajewska, A. Paini, J. S. Benito, J. Burton, A. Worth, C. Urani, H. Briesen and K. W. Schramm, In vitro-to-in vivo correlation of the skin penetration, liver clearance and hepatotoxicity of caffeine, *Food Chem. Toxicol.*, 2015, **75**, 39–49.
65. J. G. Bessems, A. Paini, M. Gajewska and A. Worth, The margin of internal exposure (MOIE) concept for dermal risk assessment based on oral toxicity data–a case study with caffeine, *Toxicology*, 2017, **392**, 119–129.
66. I. W. Altabakhi, J. Anderson and P. M. Zito, Acetaminophen/Aspirin/Caffeine, in *StatPearls*, StatPearls Publishing, Treasure Island (FL), 2022.
67. A. Murray and J. Traylor, Caffeine toxicity, in *StatPearls*, StatPearls Publishing, Treasure Island (FL), 2022.
68. C. Willson, The clinical toxicology of caffeine: A review and case study, *Toxicol. Rep.*, 2018, **5**, 1140–1152.
69. Q. Wu, P. Bai, H. Guo, M. S. Guo, Y. Xia, X. Gao, X. Wang, J. Wu, T. T. Dong and K. W. Tsim, Capsaicin, a Phytochemical from Chili Pepper, Alleviates the Ultraviolet Irradiation-Induced Decline of Collagen in Dermal Fibroblast via Blocking the Generation of Reactive Oxygen Species, *Front. Pharmacol*, 2022, **13**, 872912.
70. W. Johnson Jr, Final report on the safety assessment of capsicum annuum extract, capsicum annuum fruit extract, capsicum annuum resin, capsicum annuum fruit powder, capsicum frutescens fruit, capsicum frutescens fruit extract, capsicum frutescens resin, and capsaicin, *Int. J. Toxicol.*, 2007, **26**(Suppl 1), 3–106.
71. C. Căruntu, C. Negrei, M. A. Ghiță, A. Căruntu, A. I. Bădărău, I. Buraga, D. Boda, A. Albu and D. Brănişteanu, Capsaicin, a hot topic in skin pharmacology and physiology, *Inflammation*, 2015, **8**(40), 21.
72. K. Yashiki, A. Kiso, Y. Y. Zhou, D. Iwasaki, T. Kambara and K. Mizutani, Abstracts: The effects of Coptis japonica root extract and its key component, berberine, on human subcutaneous adipocytes, *Int. J. Cosmet. Sci.*, 2010, **32**(5), 392.
73. M. Dulić, P. Ciganović, L. Vujić and M. Zovko Končić, Antidiabetic and cosmeceutical potential of common barbery (Berberis vulgaris L.) root bark extracts obtained by optimization of 'green'ultrasound-assisted extraction, *Molecules*, 2019, **24**(19), 3613.
74. Y. A. Jang and J. T. Lee, Anti-wrinkle effect of berberine by inhibition of MMP-2 and MMP-9 activity in fibroblasts, *J. Appl. Biol. Chem.*, 2018, **61**(1), 9–15.
75. A. Eftekhari, A. Hasanzadeh, R. Khalilov, H. Hosainzadegan, E. Ahmadian and M. A. Eghbal, Hepatoprotective role of berberine against

paraquat-induced liver toxicity in rat, *Environ. Sci. Pollut. Res.*, 2020, **27**(5), 4969–4975.
76. N. Singh and B. Sharma, Toxicological effects of berberine and sanguinarine, *Front. Mol. Biosci.*, 2018, **5**, 21.
77. B. Chopra, A. K. Dhingra, R. P. Kapoor and D. N. Prasad, Piperine and its various physicochemical and biological aspects: A review, *Open Chem. J.*, 2016, **3**(1), 75–96.
78. A. Thawabteh, S. Juma, M. Bader, D. Karaman, L. Scrano, S. A. Bufo and R. Karaman, The biological activity of natural alkaloids against herbivores, cancerous cells and pathogens, *Toxins*, 2019, **11**(11), 656.
79. A. K. Srivastava and V. K. Singh, Biological action of Piper nigrum-the king of spices, *Eur. J. Biol. Res.*, 2017, **7**(3), 223–233.
80. V. Badmaev and M. Majeed, Skin and its role in providing nourishment to the body-Pungent principles of black pepper, alkaloid piperine, and tetrahydropiperine (THP) as Enhancers of Nutrient and Drug Bioavailability, *Cosmet. Sci. Technol.*, 2004, 1–9.
81. S. Badal and R. Delgoda, *Pharmacognosy*, Elsevier, 2017.
82. T. D. Bucheli, B. W. Strobel and H. C. Hansen, Personal care products are only one of many exposure routes of natural toxic substances to humans and the environment, *Cosmetics*, 2018, **5**(1), 10.
83. J. K. Aronson, in *Meyler's Side Effects of Drugs: the International Encyclopedia of Adverse Drug Reactions and Interactions*, Elsevier, 13 October 2006.
84. https://www.mattilsynet.no/kosmetikk/stoffer_i_kosmetikk/risk_profile_symphytum_officinale.9962/binary/Risk%20Profile%20Symphytum%20officinale.
85. M. J. Martena, *Safety of Herbal Preparations on the Dutch Market*, Wageningen University and Research, 2010.
86. C. Allgaier and S. Franz, Risk assessment on the use of herbal medicinal products containing pyrrolizidine alkaloids, *Regul. Toxicol. Pharmacol.*, 2015, **73**(2), 494–500.
87. M. Habs, K. Binder, S. Krauss, K. Müller, B. Ernst, L. Valentini and M. Koller, A balanced risk–benefit analysis to determine human risks associated with pyrrolizidine alkaloids (PA)—The case of tea and herbal infusions, *Nutrients*, 2017, **9**(7), 717.
88. *Food Biosynthesis*, ed. A. M. Grumezescu, and A. M. Holban, Academic Press, 19 June 2017, p. 25.
89. P. S. Nigam, A. Singh and D. Lyngby, Production of secondary metabolites–Fungi, *FEBS Lett.*, 2014, 570.
90. B. P. Singh, M. E. Rateb, S. Rodriguez-Couto, M. D. Polizeli and W. J. Li, Microbial secondary metabolites: recent developments and technological challenges, *Front. Microbiol.*, 2019, **10**, 914.
91. H. A. El-Enshasy, M. A. Hamid, R. A. Malek, E. Nagib and M. R. Sarmidi, Microbial metabolites in the cosmetics industry, in *The Handbook of Microbial Bioresources*, 2016, pp. 388–405.

92. P. L. Gupta, M. Rajput, T. Oza, U. Trivedi and G. Sanghvi, Eminence of microbial products in cosmetic industry, *Nat. Prod. Bioprospect.*, 2019, **9**(4), 267–278.
93. F. Pagels, C. Almeida, V. Vasconcelos and A. C. Guedes, Cosmetic Potential of Pigments Extracts from the Marine Cyanobacterium Cyanobium sp, *Mar. Drugs*, 2022, **20**(8), 481.
94. M. L. Mourelle, C. P. Gómez and J. L. Legido, The potential use of marine microalgae and cyanobacteria in cosmetics and thalassotherapy, *Cosmetics*, 2017, **4**(4), 46.
95. R. P. Rastogi, D. Madamwar and A. Incharoensakdi, Bloom dynamics of cyanobacteria and their toxins: environmental health impacts and mitigation strategies, *Front. Microbiol.*, 2015, **6**, 1254.
96. K. Thiyagarasaiyar, B. H. Goh, Y. J. Jeon and Y. Y. Yow, Algae metabolites in cosmeceutical: An overview of current applications and challenges, *Mar. Drugs*, 2020, **18**(6), 323.
97. S. S. Affat, Classifications, advantages, disadvantages, toxicity effects of natural and synthetic dyes: A review, *Univ. Thi-Qar J. Sci.*, 2021, **8**(1), 130–135.
98. J. B. Greig, *WHO Food Additives Series 46: Cochineal Extract, Carmine, and Carminic Acid*, Food Standards Agency, 2012.
99. A. Kurek-Górecka, M. Górecki, A. Rzepecka-Stojko, R. Balwierz and J. Stojko, Bee products in dermatology and skin care, *Molecules*, 2020, **25**(3), 556.
100. S. E. Walgrave, E. M. Warshaw and L. A. Glesne, Allergic contact dermatitis from propolis, *Dermatitis*, 2005, **16**(4), 209–215.
101. U. Klaschka, Natural personal care products—analysis of ingredient lists and legal situation, *Environ. Sci. Eur.*, 2016, **28**(1), 1–4.
102. https://ec.europa.eu/growth/toolsdatabases/cosing/pdf/COSING_Annex%20II_v2.pdf.
103. M. A. Berdai, S. Labib, K. Chetouani and M. Harandou, Atropa Belladonna intoxication: a case report, *Pan Afr. Med. J.*, 2012, **11**(1), 72.
104. S. C. Chi and S. Y. Kim, inventors, Kwang Dong Pharm Co Ltd, assignee, Composition for Improving Hair Growth Comprising an Non-Polar Solvent-Soluble Extract of Cantharis as an Active Ingredient, *US pat.*, US11/814437, 28 August 2008.
105. M. Gronquist and F. C. Schroeder, *Insect Natural Products*, 2010, pp. 67–108.
106. L. Neubauer, J. Dopstadt, H. U. Humpf and P. Tudzynski, Identification and characterization of the ergochrome gene cluster in the plant pathogenic fungus Claviceps purpurea, *Fungal Biol. Biotechnol.*, 2016, **3**(1), 1–4.
107. A. E. Al-Snafi, Pharmacology and toxicology of Conium maculatum-A review, *Pharm. Chem. J.*, 2016, **3**(2), 136–142.
108. I. T. Karabegović, S. S. Stojičević, D. T. Veličković, Z. B. Todorović, N. Č Nikolić and M. L. Lazić, The effect of different extraction techniques on the composition and antioxidant activity of cherry laurel

(Prunus laurocerasus) leaf and fruit extracts, *Ind. Crops Prod.*, 2014, **54**, 142–148.

109. I. Demirbolat and M. Kartal, Prulaurasin content of leaves, kernels and pulps of Prunus lauracerasus L.(Cherry Laurel) during ripening, *J. Res. Pharm.*, 2019, **23**(1), 69–75.
110. V. Borgia, The prehistory of poison arrows, in *Toxicology in Antiquity*, Academic Press, January 2019, vol. 1, pp. 1–10.
111. K. D. Welch, B. T. Green, D. R. Gardner, C. A. Stonecipher and D. Cook, Toxic plants, in *Reproductive and Developmental Toxicology*, Academic Press, January 2022, vol. 1, pp. 933–953.
112. F. T. Fraunfelder, Herbal medicine and dietary supplement induced ocular side effects, *Clin. Ocul. Toxicol.*, 2008, 307–313.
113. F. M. Kasali, J. Tusiimire, J. N. Kadima and A. G. Agaba, Ethnomedical uses, chemical constituents, and evidence-based pharmacological properties of Chenopodium ambrosioides L.: extensive overview, *Future J. Pharm. Sci.*, 2021, **7**(1), 1–36.
114. V. Kuete, Physical, hematological, and histopathological signs of toxicity induced by African medicinal plants, in *InToxicological Survey of African Medicinal Plants*, Elsevier, 1 January 2014, pp. 635–657.
115. J. B. Christensson, L. Hagvall and A. T. Karlberg, Fragrance allergens, overview with a focus on recent developments and understanding of abiotic and biotic activation, *Cosmetics*, 2016, **3**(2), 19.
116. Z. Li, Y. Yang, M. Liu, C. Zhang, J. Shao, X. Hou, J. Tian and Q. Cui, A comprehensive review on phytochemistry, bioactivities, toxicity studies, and clinical studies on Ficus carica Linn. leaves, *Biomed. Pharmacother.*, 2021, **137**, 111393.

CHAPTER 5

# The Wonders of Plant Secondary Metabolites as Cosmeceuticals

P. SHAH[a], Y. SONI[b], M. RAVAL[a] AND N. LUNAGARIYA[b*]

[a]Department of Pharmacognosy and Phytochemistry, Ramanbhai Patel College of Pharmacy, Charotar University of Science and Technology (CHARUSAT), CHARUSAT campus, Changa 388421, India; [b]Department of Pharmaceutical Chemistry and Analysis, Ramanbhai Patel College of Pharmacy, Charotar University of Science and Technology (CHARUSAT), CHARUSAT campus, Changa 388421, India
*E-mail: nitinlunagariya.ph@charusat.ac.in, nitin21286@gmail.com

## 5.1 Introduction

The term cosmeceutical is a blend of two words, *i.e.* "cosmetics" and "pharmaceutical". The word cosmeceutical was first introduced by Raymound Reed, a cosmetic chemist. However, the term came into vogue after Dr Albert Kligman described cosmeceuticals as cosmetics with therapeutic importance. Still, the term is not officially recognized in the USA, EU, or India.[1,2]

The cosmeceutical industry has been constantly blooming in the last decade. In 2021, the global market size of cosmeceuticals was valued at 45.07 billion USD and it has been estimated to reach 95.09 billion USD by 2030.[3] Due to the increase in awareness about the harmful effects of synthetic chemicals used in cosmetics, consumers are turning to natural ingredients, which are safer than synthetic chemicals.[4] Nature is full of biodiversity and chemo diversity. The constantly growing cosmeceutical

Bioprospecting of Natural Sources for Cosmeceuticals
Edited by Deepika Kathuria, Ajay Sharma, Meenakshi Verma and Gulzar Ahmad Nayik

Published by the Royal Society of Chemistry, www.rsc.org

industry is in need of different natural products with cosmeceutical potential and is one of the frontiers in bioprospecting. Plants contain various secondary metabolites, which have different pharmacological actions, responsible for their cosmeceutical potential.[4] This chapter attempts to summarize some of the wonders of plant secondary metabolites used in the cosmeceutical industry.

## 5.2 Cultural History of Natural Product-based Cosmeceuticals

Natural products for cosmetic purposes have been used since ancient times.[5] There are historical written and pictorial records along with much physical evidence obtained during the excavation of historical sites, which has revealed the use of different natural products not only for beauty enhancement, but also for the treatment of various skin and hair-related problems.[5,6] In ancient times, kohl was used in eye makeup, for protecting the eyes from hot and dry winds. Eye-painting was used not only to obtain attractive eyes, but also for the treatment of some visible or non-visible eye-related pathological conditions.[6] Henna has been used since ancient times for coloring hair, skin and nails, and making tattoos on the body. Henna is extracted from the leaves of *Lawsonia inermis or L. alba* and it forms a non-permanent color. The Egyptians used scented oils for softening and protecting the skin against the hot and dry weather. The Egyptians also used a mixture containing freshly squeezed olive oil to prevent wrinkles. There are reports from Biblical times about the usage of different oils, such as olive oil, almond oil and sesame oil, as emollients and for protection against the sun's rays. The Greeks liked to have saffron-yellow hair, which can be found from their statues. In ancient Rome, women used walnut extract to darken their hair.[7] In India, women are still using sandalwood and turmeric for skin care and henna for hair-color. Coconut oil, olive oil, shea butter, jojoba oil and various essential oils are still being used for different cosmeceutical purposes.[5]

## 5.3 Major Secondary Metabolites with Cosmetic Applications

Plants contain chemically and biologically diverse classes of phyto-constituents, such as alkaloids, flavonoids, terpenoids, glycosides, tannins, *etc.*[4,8] Compounds that have an effect on the skin, hair, nails and oral health can be used for cosmeceutical purposes. Due to the increase in consumer preference towards natural products over synthetic chemicals, a large number of secondary metabolites have been extensively studied for their cosmeceutical use.[4] Here, we have tried to cover various wonders of secondary metabolites (see Figure 5.1 and 5.2) from plants, which have been proven for cosmeceutical purposes.

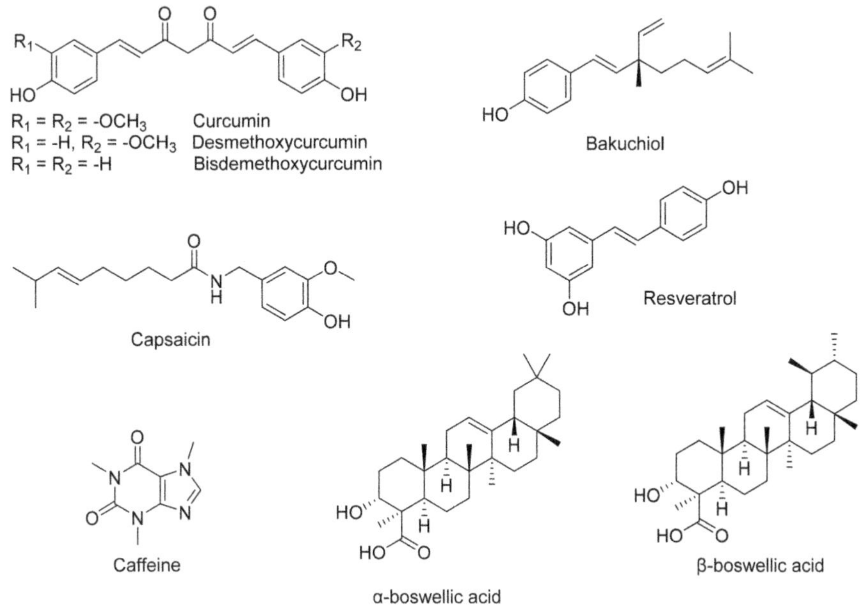


**Figure 5.1** Structures of some of the secondary metabolites present in cosmeceuticals.

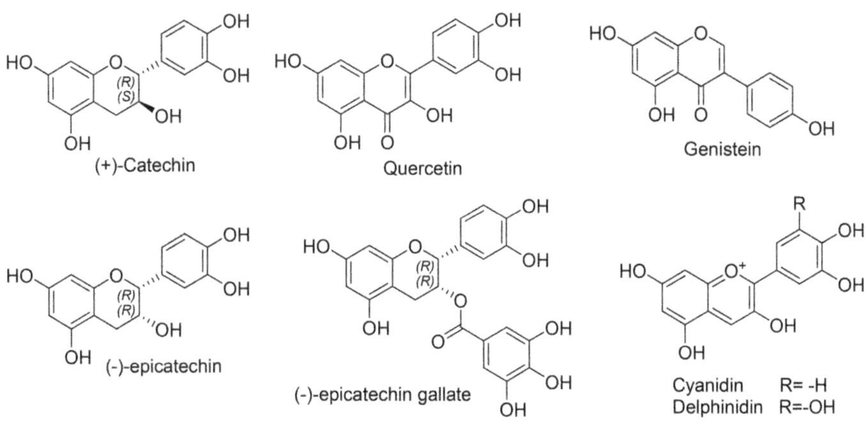


**Figure 5.2** Structures of some of the flavonoids present in cosmeceuticals.

## 5.3.1 Curcuminoids

Curcuminoids are naturally occurring phenylpropanoids, present in the rhizomes of *Curcuma longa* Linn. They are coloring pigments giving the yellow color to turmeric. Curcumin is the active ingredient from *C. longa* rhizomes, first isolated and named by Vogel and Pelletier. Chemically, curcumin is diferuloylmethane, or (1*E*,6*E*)-1,7-bis(4-hydroxy-3-methoxyphenyl)hepta-1,6-diene-3,5-dione.[9]

For the extraction of curcumin from plants of the *Curcuma* genus, there are different conventional (such as Soxhlet extraction, maceration, and solvent extraction) and modern (such as super-critical fluid extraction, microwave-assisted extraction, ultrasound-assisted extraction, and enzyme-assisted extraction) extraction methods.[10] There is a report of subcritical solvent extraction of curcuminoids from *C. longa*, with a yield of curcuminoids of 13.58% at 135 °C in 5 min with a mixture of water and ethanol (50 : 50, v/v) as a solvent, in comparison to 13.96% yield with 100% ethanol under atmospheric conditions (60 °C/120 min). This study showed that sub-critical solvent extraction is a much faster and more efficient extraction technique with the potential to be utilized commercially.[11]

Curcuminoids have been well studied for different biological activities, such as anti-oxidant, anti-inflammatory, anti-microbial, antimutagenic, anti-diabetic, hepatoprotective, immunomodulatory and expectorant. They have also been reported to be effective for use in the treatment of Alzheimer's disease and cancer.[12] Curcumin has been reported to inhibit various inflammatory mediators, such as phospholipase, lipooxygenase, cyclooxygenase-2 (COX-2), leukotrienes, thromboxane, prostaglandins, nitric-oxide, collagenase, elastase, hyaluronidase, monocyte chemoattractant protein-1 (MCP-1), interferon-inducible protein, tumor necrosis factor, and interleukin-12 (IL-12).[13] Significant anti-oxidant and anti-inflammatory activities make curcuminoids more attractive to researchers for use in cosmetic formulations. However, their poor aqueous solubility and instability (*e.g.*, photo-degradation and chemical instability) are the main hurdles for their direct application to the skin. To overcome these issues, different drug delivery systems have been developed with improved bioavailability as well as stability. In 2007, a group of researchers from Thailand reported the study of a facial cream containing curcuminoid-loaded solid lipid nanoparticles (SLNs), which showed effective anti-ageing activity with acceptable safety. It improved skin wrinkles, hydration, melanin content and skin elasticity without any sign of skin irritation.[14] In 2020, Espinosa-Olivares *et al.* developed nano-structured lipid carriers (NLCs) with a prolonged release and enhanced penetration of curcuminoids into the skin.[15] Curcuminoid-layered double hydroxide nanohybrids were developed as a novel slow release biomaterial for application as next-generation anti-microbials.[16] The skin-permeation of curcuminoids can be improved with niosome formulation for effective trans-dermal delivery.[17] A group of researchers from Thailand developed a hydrogel patch of curcuminoids for cosmeceutical purposes, using chitosan from different sources.[18] Curcumin elastic vesicles were reported for anti-photoageing activity in mouse models.[19] Curcuminoids, being an effective protein kinase C modulator, might be of therapeutic value in the treatment of scleroderma. Curcumin can be developed as an effective treatment of psoriasis by reducing keratinocyte proliferation through decreasing the expression of pro-inflammatory cytokines such as IL-6 and IL-8. The significantly higher levels of phosphorylase kinase in psoriatic epidermis were reduced to normal skin levels by curcumin, it being a selective inhibitor of phosphorylase kinase. Curcumin, being a photosensitizing compound, can be utilized for phototherapy of psoriasis. Curcumin also inhibited a number of tumors in 7,12-dimethylbenz[*a*]anthracene (DMBA)-induced and tetradecanoylphorbol-13-acetate (TPA)-induced models of carcinogenesis. Topical application of curcumin inhibits epidermal hyperplasia. In India, turmeric has been used for wound healing since ancient times. Reports suggest that curcumin speeds up the process of reepithelization of the epidermis, increases the migration of myofibroblasts, fibroblasts and macrophages in the wound, and leads to extensive neovascularization and greater collagen deposition. The enhanced wound-healing activity of turmeric is due to increased expression of transforming growth factor-β1. Studies have also shown the positive effect of curcumin in the treatment of wounds whose healing is impaired by diabetes as well as radiation. Tumor angiogenesis is hindered by curcumin; thus, it inhibits carcinogenesis in the skin.[20,21] Some of the cosmeceutical products containing turmeric are listed in Table 5.1.

The major drawback of curcuminoids is yellow staining on the skin after application. To overcome this staining problem, the cosmeceuticals market is now moving towards the use of tetrahydrocurcumin, a hydrogenated form of curcumin, which is off-white in color. This solves the problem without compromising the efficacy.[22]

### 5.3.2 Caffeine

Chemically, caffeine is a xanthine class of alkaloid. It is a central nervous stimulant in coffee, tea and some soft drinks available on the market. It is a 1,3,5-trimethyl xanthine, extracted from the mature ripe seeds of *Coffea arabica* (coffee) or leaves of *Camellia sinesis* (tea).[23] For isolation of caffeine, tea leaves are extracted with hot boiling water along with alkali salt, which forms alkaline salts of tannins, which will remain soluble in water, whereas caffeine can be separated by extraction with dichloromethane. On evaporation of dichloromethane, caffeine can be separated.[24] Caffeine can also be obtained during the production of decaffeinated teas. Valorization of spent coffee grounds and coffee wastewater can also be used for environmentally friendly extraction of caffeine.[25–27] Caffeine can also be obtained from the coffee silver skin, which detaches during the coffee roasting process.[28]

As caffeine is extensively used in day-to-day life through various beverages and dietary intake, many research groups have investigated its effect on the cardiovascular, nervous and metabolic systems of the body.[29] It has been studied for its potential as an antioxidant and anticarcinogenic. On the oral administration of black and green tea, caffeine was found to inhibit the development of cancer. When caffeine was orally administered, it showed inhibition of UVB-induced carcinogenesis, through *in vivo* upregulation of tumor suppressor genes. Later on, topical application of caffeine was investigated, which revealed that caffeine showed inhibition of carcinogenesis and decreased

**Table 5.1** A list of some of the cosmeceutical products containing turmeric available on the market for various uses.

| S. no. | Product | Manufacturer (country) | Site of application | Label claims |
|---|---|---|---|---|
| 1 | SNEC-G Oral Gel (Nanoemulsifying curcumin gel) | SNEC-30 (India) | Inside the oral cavity | It can be used to treat mouth ulcers, gingivitis and periodontitis, and prevent the foul odor of the mouth. It also helps in decreasing the inflammation and burning sensation in the mouth. It is an instant oral pain relief gel |
| 2 | Vicco turmeric WSO cream | Vicco laboratories (India) | On the skin | It acts on the skin as an antiseptic, exfoliator, and protective layer against pollution damage and ultraviolet (UV) rays. It also adds glow to the skin |
| 3 | Turmeric cream | Ziva Ayurveda (India) | On the skin | It works on acne, pimples and zits, and promotes even skin tone. It also helps in minute wound healing on the skin |
| 4 | Turmeric lip balm | Ilaa botanicals (India) | On the lips | It moisturizes and heals chapped lips |
| 5 | Vicco turmeric face wash | Vicco laboratories (India) | On the face | It fights pimples and skin infections. It clears the pores, removes blackheads and accelerates skin cell renewal. It removes tan and lightens dark circles. Its antifungal properties help in curing minor wounds, cuts and small skin injuries as well |
| 6 | WOW skin science turmeric clay face mask | Kapco International Limited (India) | On the face | It repairs the skin, minimizes blemishes, and fades acne scars. It also protects the skin from acne flare-ups and helps in the skin-renewal process |

tumors through increasing apoptosis in sunburnt cells of hairless SKH-1, UVB-1-treated mice. Caffeine also shows a photo-protective effect. Therefore, topical applications containing caffeine may inhibit sunlight-induced skin cancer in humans.[30] Caffeine has also shown inhibition of phosphodiesterase and agonistic activity towards ryanodine as well as adenosine receptors. Due to its various actions at the molecular level, it is used for the treatment of various dermatologic disorders.[29,31] It stimulates hair growth through the inhibition of 5α-reductase.[31] In various clinical trials, it was found that caffeine promoted hair growth in patients diagnosed with androgenic alopecia (AGA).[29] Topical sun-protective cosmeceuticals containing 0.1% caffeine can act as hair follicle photoprotectants against UV radiation-induced damage.[32] Its anti-oxidant and anti-inflammatory activity might be beneficial for the reduction of facial redness, suggesting its use in the treatment of rosacea.[29]

Caffeine was reported as a strong inhibitor of collagenase and elastase, whereas a weak inhibitor of tyrosinase. These enzymes are responsible for UV-induced skin ageing. Therefore, caffeine showed anti-ageing activity on the skin through inhibition of these enzymes, which is also supported with *in silico* data.[33] Higher binding affinity of caffeine with keratin justifies scientifically the presence of caffeine in hair growth stimulating and skin brightening cosmeceutical products.[34] It acts as an anti-cellulite by preventing excessive fat accumulation in the skin through the stimulation of lipolysis, *via* inhibition of phosphodiesterase activity. It also promotes lymphatic drainage, which helps in its anti-cellulite action. It increases the blood circulation around the eyes, which reduces swelling around puffy eyes.[31] Being hydrophilic in nature, the penetration of caffeine into the skin is very difficult. Therefore, it requires several modifications in its formulation to ease its penetration into the skin.[35] For targeted drug delivery of caffeine, various lipid and polymer-based nano-carriers (such as SLNs, NLCs, and chitosan), liposomes and niosomes have been developed and studied to improve bioavailability and efficacy.[36–38] Caffeine-containing micro-emulsion sticks (a semisolid dosage form) were developed, which showed higher penetration into the skin.[39] A bioadhesive emulgel formulation was made containing 1% caffeine, which showed good skin adhesion properties as well as better delivery of caffeine to the skin, and is a promising cosmeceutical formulation.[40] Due to its beneficial effect, a lot of cosmeceutical products available on the market contain caffeine (see Table 5.2).

### 5.3.3 Capsaicin

Capsaicin, (*E*)-*N*-[(4-hydroxy-3-methoxyphenyl)methyl]-8-methylnon-6-enamide, is the active ingredient of fruits of plants of the *Capsicum* genus, such as *Capsicum annuum 'Cayenne'*, *C. annuum 'Jalapneo'*, and *C. chili.* Capsaicin was first purified and named in 1876 by Thresh. In 1919, Nelson and Dawson resolved its molecular structure.[41] Capsaicin acts as an analgesic when used in the right dose and the right frequency.[42] Being a lipophilic substance, it cannot be solubilized in water. There are many extraction methods using

**Table 5.2** A list of some of the cosmeceutical products containing caffeine available on the market.

| S. no. | Product | Manufacturer (country) | Site of application | Label claims |
|---|---|---|---|---|
| 1 | Coffee hydrogel under eye patches for dark circles with caffeine & hyaluronic acid | mCaffeine (India) | Under the eyes | Caffeine rich extract reduces puffiness under the eye, leaving it refreshed and awake |
| 2 | Coffee under eye cream | mCaffeine (India) | Under the eyes | Caffeine rich extract relieves dark circles and tones the skin |
| 3 | Coffee face wash | mCaffeine (India) | On the face | It reduces redness, inflammation and puffiness |
| 4 | Coffee lip scrub | mCaffeine (India) | Over the lips | Caffeine rich coffee exfoliates away the dead skin, and removes chapped and flaky skin |
| 5 | Coffee body polishing oil | mCaffeine (India) | All over the body | Caffeine rich coffee reduces cellulite, and polishes and tones the skin |
| 6 | Coffee lotion sunscreen SPF 50 PA++ | mCaffeine (India) | Face and other areas of the body | It protects the skin from UV damage, fights free radicals and prevents premature ageing |
| 7 | The ordinary caffeine solution 5% + EGCG | The Ordinary (Canada) | Under the eyes | Caffeine combats the appearance of pigmentation and puffiness in the eye area and reduces the appearance of cellulite |
| 8 | REVOX B77 just caffeine 5% eye contour serum | Revox B77 (Europe) | Around the eyes | It reduces the appearance of puffy eyes and dark circles |
| 9 | Mario Badescu caffeine eye cream | Maria Badescu (USA) | Around the eyes | It hydrates, softens and smoothens while de-puffing to promote more radiant-looking skin under the eyes |
| 10 | Serum shot 5% caffeine | Eveline Cosmetics (Poland) | Under the eyes | It reduces dark circles and puffiness under the eyes and adds radiance and nourishes the skin |
| 11 | Revolution skincare revitalizing caffeine eye cream | Revolution Skincare (London) | Under the eyes | It brings down the puffiness and dark circles under the eyes |
| 12 | Revolution skincare caffeine 5% + hyaluronic acid under eye serum | Revolution Skincare (London) | Under the eyes | It reduces the appearance of dark circles and puffiness |
| 13 | SR cosmetics caffeine series AC fine cream | SR Cosmetics (Israel) | On skin | Skin rejuvenation |

| | | | | |
|---|---|---|---|---|
| 14 | WOW SKIN SCIENCE caffeine face serum | WOW SKIN SCIENCE (India) | On the face | The serum is infused with pure caffeine which is rich in antioxidants and has skin calming properties that reduce puffiness and boost lymphatic circulation. It helps to tone and firm the skin. Coffee extract is rich in antioxidants and helps to boost collagen levels in the skin, which reduces fine lines and wrinkles visible on the skin |
| 15 | Reduxcell body slimming caffeine gel | ReduxCell (Paris) | On belly, thighs, arms | Anti-cellulite firming cream |

different solvents that have been reported in the literature. Amongst them, the pressurized liquid extraction technique using methanol as a solvent was found to be the most effective for extracting maximum capsaicin from *C. annuum* samples.[43] Capsaicin has been reported for a variety of biological activities, such as antioxidant, anticarcinogenic and anti-inflammatory, but its use is limited due to its pungency and irritation on long-term usage.[44] Capsaicin is also reported to promote weight loss, and thus is beneficial in the treatment of obesity.[42] Topical application of creams, bandages and patches containing capsaicin has been used to alleviate the pain from a wide range of chronic conditions. On initial application to the skin, capsaicin binds to transient receptor potential vanilloid 1 (TRPV1), which results in an increase in intracellular calcium, triggering the release of neuropeptides such as substance P and calcium gene related peptide (CGRP), which causes enhanced sensitivity to noxious stimuli, and results in pain, inflammation and heat-sensation. On the repetitive use of capsaicin, it results in desensitization due to a decrease in the levels of substance P, and it acts as an analgesic.[45,46] Capsaicin, being a TRPV1 agonist, can adjust the translation of hypoxia-inducible factor-1α (HIF-1α), and consequently induces normal differentiation by obstructing hyperproliferation of psoriatic epidermis. On the topical application of capsaicin, there is inhibition of cutaneous vasodilation and depletion of substance P, which makes it effective in the treatment of psoriasis. A study was reported on the development of SLNs and NLCs of capsaicin, which resulted in higher skin permeation and greater skin retention with no skin irritation.[47] A carbopol-based hydrogel formulation containing capsaicin was reported with good anti-microbial and anti-fungal activities.[48] Cosmeceutical products containing capsaicin for pain-relief are listed in Table 5.3.

### 5.3.4 Boswellic Acids

Chemically, boswellic acids (BAs) are pentacyclic triterpenes, major constituents of the resin derived from the genus *Boswellia*. The oleo-gum resin obtained from *Boswellia* species is known as frankincense or olibanum. Frankincense was used for various religious rituals amongst various communities. In ancient Egypt, it was also used in the process of mummification.[49,50] For commercial purposes, BAs are obtained from *Boswellia serrata*. Scientists have developed various procedures for the extraction of BAs using various techniques such as solvent extraction, steam distillation, hydrodistillation, microwave-assisted extraction, ultrasonic extraction, and supercritical fluid extraction.[51] In Ayurveda, *B. serrata* resin was reported for the treatment of mouth sores, diarrhea, dysentery, fever, vaginal discharge, hemorrhoids, hair loss, syphilis, and skin and blood related disorders.[52] In traditional Chinese medicine, *B. serrata* resin is used in the treatment of rheumatic disorders, menstrual disorders and bruises.[50] *B. serrata* has been extensively studied for its anti-inflammatory and antiarthritic activities.[53] BAs have been reported for their anti-inflammatory and anti-cancer potential. BAs have also been

**Table 5.3** A list of some of the cosmeceutical products containing capsaicin available for pain-relief.

| S. no. | Product | Manufacturer (country) | Site of application | Label claims |
|---|---|---|---|---|
| 1 | Hansaplast lion heat plaster (capsicum plaster) | Essity Operations Goa Limited (India) | On the calf, neck, back | It provides relief from neck, back and muscular pain. It induces warmth around the affected area for effective and faster pain relief |
| 2 | Medigrip capsicum plaster | Precision coatings Pvt. Ltd (India) | On undamaged skin | It is used as a counter-irritant to produce heat on undamaged skin. It is used in the treatment of backaches, swelling of nerves, stiffness of muscles and swollen joints |
| 3 | Mendwell capsicum plaster (pain relief patch) | Prominence healthcare Pvt. Ltd. (India) | On undamaged skin | It is used as a counter-irritant and gives relief from pain by producing heat in the treatment of backaches, swelling of nerves, stiffness of muscles, swollen joints, and boils |

reported for their therapeutic potential against neurodegenerative disorders, such as Alzheimer's disease.[54]

Frankincense essential oil was reported for anti-inflammatory and antibacterial activities, which can be utilized in the treatment of acne on the skin. It improves the skin, reduces wrinkles and elevates the process of generation of new cells, making it a promising anti-ageing agent.[53] It has also been evaluated for its inhibitory action on elastase and collagenase through *in vitro* and *in silico* studies, which again makes it a promising candidate for anti-ageing cosmeceuticals.[55] A group of scientists developed enriched Boswellia extract (Olibose®) and evaluated its anti-inflammatory activity in human dermal fibroblast cells. Olibose® was found to inhibit the levels of pro-inflammatory markers TNF-α and nitric oxide (NO), providing scientific evidence for its anti-inflammatory potential and use for the treatment of various dermatological disorders.[56] BAs have also been proved to be an effective leukotriene inhibitor, when applied topically, which is beneficial for rheumatic patients.[57] A double-blind clinical study revealed that Bosexil®, a phytosome formulation containing BAs, improves scales, itching and erythema, without any case of worsening. This clinical study showed that this topical formulation may be used for the treatment of skin-related disorders such as psoriasis and erythematous eczema. Due to its anti-inflammatory properties, it could also be used for the treatment of other inflammatory skin disorders.[58] A single blind evaluation showed the protective effects of an emulsion containing Bosexil® (BAs phytosome) against UV damage on the skin of healthy volunteers.[59] A sport cream containing *B. serrata* extract as one of the ingredients showed improvement in neck/shoulder pain.[60] A topical proniosomal gel containing BAs was developed, which showed anti-inflammatory activity with enhanced bioavailability.[61] Cosmeceutical products containing boswellic acids on the market for different purposes are listed in Table 5.4.

### 5.3.5 Resveratrol

Chemically, resveratrol is 3,4′,5-trihydroxy-*trans*-stilbene, which is a naturally occurring polyphenol of the stilbene class of compounds. The biologically active *trans*-isomer is converted to its *cis*-form on exposure to light.[62] In 1939, it was first isolated from the roots of white hellebore (*Veratrum grandiflorum*) by Takaoka and thereafter it was considered as a phytoalexin, produced by plants in response to environmental stress or pathogenic attack.[63–66] It is widely distributed in nature, found present in red grapes, various berries (raspberry, strawberry, blueberry, and cranberry), peanuts and Japanese knotweed root (*Polygonum cuspidatum* var. *Japonicus*). However, natural sources are not sufficient to meet the industrial need for resveratrol due to its enormous biological activities, as well as its application in the pharmaceutical and cosmeceutical fields. Moreover, multi-step isolation and purification processes make it difficult, time-consuming and impracticable to extract resveratrol from plants. It can also be synthesized chemically; however, it is accompanied by the formation of many unwanted side products, which makes the purification complicated. In addition to this, it also affects environment sustainability due to use of harmful chemicals and solvents on a large scale. Therefore, to fulfil the industrial need, it is generally obtained by biotechnological synthesis from various microbial hosts, such as *Escherichia coli*, *Corynebacterium glutamicum*, *Saccharomyces cerevisiae*, and *Pichia pastoris*.[63,66]

In 1992, Siemann and Creasy reported the presence of resveratrol in conspicuous concentrations in red wine, which explained the French paradox, *i.e.* the low incidence of morbidity and mortality related to coronary heart disease (CHD) might be due to regular consumption of red wine.[67] Thereafter, Jang *et al.* reported resveratrol as an effective chemo-preventive agent against cancer in 1997.[68] Since then, resveratrol has been reported to have various biological activities such as anti-cancer, anti-oxidant, anti-inflammatory, and protection against cardiovascular diseases, neuronal diseases and metabolic disorders.[64,65,69]

Resveratrol is one of the potential skin care agents with scientific proof. It has been extensively studied for the treatment of various skin disorders, such as skin cancer, photo-ageing, allergies, dermatitis, melanogenesis and microbial infection. Topical administration is a preferable method to treat skin-related diseases. Because of its anti-oxidant and anti-inflammatory activity, it is a key ingredient of many of the cosmeceutical products with anti-ageing and skin-brightening effects.[70] Being a phytoestrogen, resveratrol activates estrogen receptors and helps to slow down the ageing process of the skin, by increasing the number of fibroblasts and glycosaminoglycans,

**Table 5.4** A list of some of the cosmeceutical products containing boswellic acids available on the market.

| S. no. | Product | Manufacturer (country) | Site of application | Label claims |
|---|---|---|---|---|
| 1 | Psorolin B Ointment | Dr JRK's Research and Pharmaceuticals Pvt Ltd (India) | Over the affected area | It contains boswellic acid which has steroid like strong anti-inflammatory activity and reduces scaling, itching and redness |
| 2 | Verdura Boswell Cream | Verdura Care (India) | Over the affected area | It reduces itching, redness and scaling |
| 3 | Boswellia Cream | Gapdang Herb (Republic of Korea) | On the skin, face, neck, joints | Skin care moisturizer. It helps in soothing dry and itchy skin. It also helps in easing rheumatic pain |
| 4 | Boswellia Cream | Sol The Lab (Republic of Korea) | On the skin, face, neck, joints | Skin care moisturizer. It helps in soothing dry and itchy skin. It also helps in easing rheumatic pain |
| 5 | Arnica Boswella Cream | WiseWays Herbals (USA) | On the skin | It gives relief from stiff and sore muscles |
| 6 | V-Cense™ Revitalising Night Cream | REN Clean Skincare (USA) | On the skin | It smooths and hydrates the skin overnight. It reduces the appearance of fine lines and wrinkles, and prevents premature ageing |
| 7 | Natural Pain Relieving Cream | The Feel Good Lab (USA) | On the muscles and joints | For arthritis, back pain, sore muscles and joints |
| 8 | Boswella cream | Life Extension (USA) | On the joints | It may provide localized relief from fluid imbalance |

and thus the amount of collagen and hyaluronic acid. Thus, it can be incorporated in cosmetic formulations such as anti-ageing creams.[71] Resveratrol is reported to inhibit human tyrosinase enzyme and melanogenesis at the cellular level, responsible for its skin whitening effect.[72] There are several *in vitro* and *in vivo* studies reported claiming that the topical application of resveratrol showed strong protection against UVB radiation, resulting in anti-proliferative and chemopreventive effects against skin carcinogenesis.[71] Resveratrol was also found to be protective against UVA radiation-induced damage on retinal pigment epithelial (RPE) cells through various mechanisms, such as suppressing intracellular hydrogen peroxide ($H_2O_2$) production, lowering mitogen-activated protein kinase (MAPK) activation, and a reduction in COX-2 expression, which might be responsible for the increase in RPE cell viability even after exposure to UVA radiation. Resveratrol can be taken further as a chemopreventive agent against age-related macular degeneration (AMD).[73] Topical application of resveratrol promoted hair growth through stimulation of the transition of hair follicles from the telogen to anagen phase and delayed catagen phase progression. It can be developed as a potential molecule for the prevention and treatment of hair loss.[74]

Despite its usefulness in topical formulations, its topical application is accompanied by various challenges because of its poor solubility and low photo-stability. Resveratrol tends to undergo photo-degradation and oxidative discoloration in solutions, limiting its use in cosmetics.[75] Different photo-stabilizers can be added to topical formulations to decrease the photodegradation of resveratrol.[70] Different formulation techniques have been employed to deal with the problems related to its solubility and stability.[76] In 2020, Sharkawy *et al.* developed a Pickering emulsion of resveratrol using chitosan/gum Arabic nanoparticles, which showed higher levels of resveratrol retention in cutaneous layers of the skin, with increased photostability against UV radiation.[77] A cream containing resveratrol was found to be effective in increasing skin hydration.[78] A company named Medskin Solutions Dr Suwelack AG developed an emulsion containing 2% bioavailable, solubilized *trans*-resveratrol, which showed a strong skin rejuvenating effect *in vivo* through increasing skin elasticity and skin firmness.[79] Dendrimer nanostructures have been developed entrapping resveratrol with enhanced solubility and stability in aqueous solution, to combat the pharmacokinetic and formulation-related challenges for resveratrol.[80] Various cosmeceutical products containing resveratrol available on the market are listed in Table 5.5.

### 5.3.6 Bakuchiol

Bakuchiol is in the meroterpenoid class of naturally occurring molecules, and was first isolated from the seeds of *Psoralea corylifolia* by a group of scientists from India.[81] It is also reported to be present in several other plants, however the seeds of *P. corylifolia*, containing about 6.24% bakuchiol, are used for its large-scale isolation. Isolation of bakuchiol can be done by extraction with a suitable solvent under cold or hot conditions, followed by column

**Table 5.5** A list of some of the cosmeceutical products containing resveratrol available on the market.

| S. no. | Product | Manufacturer (country) | Site of application | Label claims |
|---|---|---|---|---|
| 1 | Timeless Antioxidant Serum | Raw beauty Natural body care products (India) | On the face, neck and decolletage area | It reduces signs of ageing, boosts glow, lightens pigmentation and strengthens skin |
| 2 | SkinCeuticals Prevent Resveratrol BE | SkinCeuticals (USA) | On the skin of the face, neck and decolletage area | It improves nighttime repair and helps promote natural skin repair. It helps to correct the damage caused by daily external aggressions, while restoring the natural luminosity to the skin, and reduces visible signs of ageing – wrinkles, lines and blemishes |
| 3 | Caudalie Resveratrol-Lift Firming Cashmere Cream | Caudalie (France) | On the face and neck | It corrects wrinkles, and regenerates, firms and nourishes the skin |
| 4 | Dr Althea Resveratrol 345NA_Intensive Repair Cream | Kolmar Korea Co. Ltd (South Korea) | On the face and neck | It provides young and firm looking skin with its highly effective anti-ageing and anti-inflammatory function |
| 5 | Rejusure Resveratrol & Ferulic Acid Facial Oil | Rejusure Skin Sciences (India) | On the face | It can be used for reducing hyperpigmentation, dark spots, freckles, and eye dark circles |

chromatography.[82] A patented technique was reported for the isolation of bakuchiol from the seeds of *P. corylifolia*, after alkaline treatment, which removes unwanted furanocoumarins by chemical modification, with subsequent extraction of bakuchiol using solvent extraction.[83] In comparison with conventional methods of extraction, ultrasonic extraction and supercritical fluid extraction have been reported with a higher yield of bakuchiol.[82] *In vitro* grown *P. drupacea* was reported to contain up to 11% bakuchiol, and thus it can be a potential source to meet industrial demand.[84]

*P. corylifolia* seeds were used in Ayurveda for the treatment of various skin-related disorders such as leukoderma, leprosy, psoriasis, eczema, ringworm, and scabies.[85] Bakuchiol is reported to have a broad spectrum of biological activities, such as anti-oxidant, anti-inflammatory, anticancer, anti-bacterial, anti-microbial, hepatoprotective, cardioprotective, hypolipidemic and hypoglycemic potential.[82] Despite structural dissimilarity to retinol, it is a functional mimic of retinol. It acts through a similar gene expression pattern to retinol, but not completely identical.[86] Thus, it acts as a naturally occurring plant-based alternative for retinol, with less undesired effects.[82,86]

Bakuchiol was reported to exhibit inhibitory effects on melanin production in B16 mouse melanoma cells, which might be useful for a skin-whitening effect.[87] Its potent inhibitory action on anti-tyrosinase enzyme also supports its depigmenting action.[88] It improves skin photo-ageing and suppresses the proliferation of skin cancer cells, and can be incorporated in sun-protective products.[89,90] A randomized, double-blind clinical study of a facial cream containing 0.5% bakuchiol reported improvement of facial wrinkles and pigmentation, responsible for its effects against photo-ageing of the skin.[91] An open-label, single center clinical study of a facial serum containing 1.5% bakuchiol showed anti-ageing activity with increased skin-firmness and radiance.[92] Another clinical study was reported on the anti-ageing effect of a bakuchiol-containing emulsion for sensitive skin.[93] Given its usefulness, the cosmetics industry has been inspired to incorporate it as natural plant-based retinol in a number of products for anti-ageing effects.[82] This compound is also reported to have an anti-acne effect by down-regulating the synthesis of 5α-reductase. Acne skin is prone to *Staphylococcus* and *Candida* infection. Therefore, antibacterial and antimicrobial treatment is also required along with anti-acne treatment. Bakuchiol showed promising inhibitory activity against proliferation of *Propionibacterium acnes* as well as antimicrobial activity against *Staphylococcus* and *Candida* micro-organisms.[94] It was also reported to have anti-fungal activity against *C. albicans* and *non-albicans Candida* (NAC), except *C. glabrata*, suggesting its use in anti-fungal formulations.[95] Some of the cosmeceutical products containing bakuchiol are listed in Table 5.6.

### 5.3.7 Other Natural Products as Wonders in Cosmeceuticals

Polyphenolics are present in the outer skin of various tropical fruits, such as pomegranate, pineapple, apple, grapes and passion fruit, and protect them against UV radiation through their anti-oxidant effect.[96] Various

**Table 5.6** A list of some of the cosmeceutical products containing bakuchiol available on the market.

| S. no. | Product | Manufacturer (country) | Site of application | Label claims |
|---|---|---|---|---|
| 1 | Bakuchiol Slip | 82°E (India) | On the face and neck | It reduces hyper-pigmentation and simultaneously soothes the skin |
| 2 | Bakuchiol Plant Retinol Creme | Lotus Organics[+] (India) | On the face | Anti-ageing face cream, restores the firmness and elasticity of the skin and provides intense hydration and nourishment |
| 3 | Bakuchiol Plant Retinol Miracle Face Oil | Lotus Organics[+] (India) | On the face | It protects the skin from free radical damage, restores firmness of the skin and gives an anti-ageing effect to the skin |
| 4 | Bakuchiol Plant Retinol Night Recovery Creme | Lotus Organics[+] (India) | On the face | Anti-ageing night cream |
| 5 | Bakuchiol Plant Retinol Super Serum | Lotus Organics[+] (India) | On the face | It reduces wrinkles and illuminates the skin |
| 6 | 0.5% Retinol Anti-Ageing Night Cream | Plum Goodness (India) | On the face | It prevents fine lines and wrinkles, and soothes the skin |
| 7 | Bio-retinol cream | Aminu (India) | On the face | It provides firmness and smoothness to the skin and prevents skin-ageing |
| 8 | Daily Dew Phyto-Retinol & Peptide Eye Cream | TrueKind (India) | Around the eyes | It moisturizes and plumps the skin |
| 9 | Bakuchiol Natural Anti-ageing elixir | St Botanica (India) | On the face | Rejuvenating face oil with anti-ageing benefits that help firm, tighten and tone the skin |
| 10 | Revolution 1% Bakuchiol Serum | Revolution Skincare London (UK) | On the face | It helps reduce the appearance of lines and encourages an even skin tone |
| 11 | Real Bakuchiol Firming Serum | NEOGEN DERMALOGY (Republic of Korea) | On the face | It helps create firm, smooth and radiant skin |
| 12 | DermaCult Bakuchiol Serum | O3+ (India) | On the face | Anti-ageing |

polyphenolic-rich plant extracts are used in a range of cosmeceutical products mainly due to their anti-oxidant effect.[96] Flavonoids are a major class of compounds under the polyphenolics category of secondary metabolites. They are further classified as flavanols (*e.g.* catechins), flavonols (*e.g.* quercetin), isoflavones (*e.g.* genistein), flavanones (*e.g.* silymarin), and anthocyanidins (*e.g.* cyanidin and delphinidin), based on their chemical structures (see Figure 5.2). Due to their phenolic structures, they interfere at different phases of the oxidation process in living cells. Therefore, their structural differences affect their anti-oxidant activity. In cosmeceutical products, polyphenolic enriched extracts are used to utilize the synergistic effect of all polyphenolics present in the plant extract.[97] Different conventional and modern extraction techniques have been reported in the literature, using different solvents. Due to their abundant presence in the peels of tropical fruits, wastage and by-products from the food industry (*e.g.* apple pomace, grape skins and seeds, and pineapple peel) can be used as an interesting and low-value source of polyphenolics.[96] They are reported to have a large number of biological activities, such as anti-oxidant, anti-inflammatory, anti-microbial, anti-viral, analgesic, anti-pyretic, cardioprotective and hepatoprotective activities.[96] Due to their prevalent antioxidant activity, they are scientifically proven to exhibit anti-ageing, anti-inflammatory and skin-protective effects.[98] As they also exhibit anti-microbial activity, their inclusion in the cosmeceutical product minimizes microbial growth in the product, and thus increases the shelf-life of the product.[99] Green tea extract is rich in polyphenolics and extensively used in cosmeceuticals, due to its anti-oxidant, anti-inflammatory, photoprotective and anti-hyaluronidase properties.[100]

In the market, a number of cosmetic products are present containing essential oils as one of the ingredients for multiple purposes, such as to provide fragrance, as an active ingredient, as a preservative, *etc*. Essential oils can be extracted from different parts of the plant using different extraction techniques.[101] They exhibit various biological activities, such as anti-oxidant, anti-inflammatory, and antimicrobial. From the dermatological and trichological perspectives, various essential oils are incorporated in a vast range of cosmeceutical products due to their variety of actions, such as anti-ageing, skin-whitening, anti-acne and protection against damage by UV rays. Due to their antimicrobial properties, they can also be used as a natural preservative in cosmeceutical products, which makes it possible to avoid the use of harmful chemical preservatives.[101] Tea tree oil, rosemary oil, eucalyptus oil, lavender oil, chamomile oil, bergamot oil, peppermint oil and geranium oil are amongst several examples.[102,103] However, they do not always only provide beneficial effects. Topical application of essential oils containing a higher concentration of allergenic constituents has been reported to cause various allergic reactions.[101] Apart from essential oils, active constituents isolated from them are also used in different cosmetics for different biological activities, such as menthol, eugenol, anethole, carvone, geraniol, citral, linalool, and farnesol.[102]

Since antiquity, *Aloe vera* has been used for different medicinal purposes. Historical evidence states that Cleopatra and Nefertiti, Egyptian queens, used *A. vera* in regular practice to maintain and enhance beauty. In the ancient time of the Greek king Alexander the Great, it was used in the treatment of the wounds of his soldiers in the war.[104] *A. vera* gel is one of the most commonly used ingredients in skin care formulations across the globe. Various scientific reports of *A. vera* are available for different biological activities, including those with cosmeceutical potential such as, anti-oxidant, anti-inflammatory, wound-healing, antiseptic, and anti-ageing effects.[104,105]

## 5.4 Conclusion and Future Prospects

The market for cosmeceuticals is increasing in leaps and bounds. Due to the availability of the internet, all information is just one touch away. Nowadays, consumers are more aware about the use of cosmetic products with therapeutic benefits. Since antiquity, natural products have been used for the treatment of various dermatological conditions. Cosmetics containing chemicals are prone to severe side effects, and consumers are turning towards the use of natural cosmeceuticals. As there are no stringent regulatory guidelines for this class of products, more and more start-up companies are launching cosmeceutical products containing natural ingredients to attract consumers with different label claims. Instead of going to dermatologists, consumers also prefer to use these cosmetic products with therapeutic benefits to avoid the payment of doctor's fees. Everybody wants to look good in the current time of social media. Skin care products are leading the cosmetic market. Manufacturers are attracting consumers by including natural ingredients in their products, along with attractive packaging containers with mind-influencing advertisements of their usage benefits. As they are available primarily as non-prescription products, their market is completely dependent on publicity. Detailed scientific studies have to be carried out to test the claims regarding the therapeutic effects of traditionally used natural products. Though these products might have been clinically tested, clinical studies are too confined with very little data. Clinical studies should be done in large populations to check for efficacy and toxicity. Some consumers might be allergic to some of the natural ingredients or their degradation products. Therefore, standardization and stability data should also be checked for herbal cosmeceuticals. Regulatory authorities should make more stringent guidelines for cosmeceutical products looking to their therapeutic efficacy. Initially, this will slow down the growth of the cosmeceutical market, but later on more reliable cosmeceuticals with standardized natural actives will be available for consumers.

To meet the demand of the cosmeceutical industry, large scale cultivation and harvesting of natural products is required. As a green alternative to this, valorification of by-products of the food industry and agro-waste can be done to extract bioactive constituents of cosmeceutical importance. Plant cell culture techniques can also be utilized for the production of cosmeceutically important secondary metabolites at an industrial scale. This technology is not yet explored much for the cosmetic industry. Formulation scientists are also working on the development of effective formulations to make cosmeceutical active ingredients more bioavailable, giving rise to the field of nano-cosmeceuticals. Adulteration by inferior plant material or species is a big concern. In essence, standardization of the formulation and clinical effectiveness must be very well validated in order to have safe and effective cosmeceuticals.

## Acknowledgements

The authors would like to thank Charotar University of Science and Technology (CHARUSAT) and Ramanbhai Patel College of Pharmacy (RPCP), Gujarat for their support.

## References

1. K. Sri Manjari, D. Chakraborty, A. Kumar and S. Singh, in *Bioprospecting of Plant Biodiversity for Industrial Molecules*, ed. S. K. Upadhyay, and S. P. Singh, John Wiley & Sons Ltd, USA, 2021, p. 189.
2. A. Pandey, G. K. Jatana and S. Sonthalia, Cosmeceuticals, *StatPearls*, StatPearls Publishing, USA, 2023, https://www.ncbi.nlm.nih.gov/books/NBK544223/, accessed 07 August 2023.
3. Global Cosmeceutical Market Size & Share Report, 2019–2025, https://www.grandviewresearch.com/industry-analysis/cosmeceutical-market, accessed 4 November 2022.
4. N. Amberg and C. Fogarassy, *Resources*, 2019, **8**, 137.
5. N. R. Desam and A. J. Al-Rajab, in *Bioactive Natural Products for Pharmaceutical Applications*, ed. D. Pal, and A. K. Nayak, Springer, Switzerland, 2021, p. 643.
6. J. Murube, *Ocul. Surf.*, 2013, **11**, 2.
7. L. C. Parish and J. T. Crissey, *Clin. Dermatol.*, 1988, **6**, 1.
8. R. N. Raji, A. Vysakh, D. Suma, M. K. Preetha and M. S. Latha, *Phytochemicals from Medicinal Plants: Scope, Applications, and Potential Health Claims*, ed. H. A. Rasul Suleria, M. R. Goyal, and M. Sadiq Butt, Apple Academic Press, USA, 2019, vol. 3.
9. S. C. Gupta, S. Patchva, W. Koh and B. B. Aggarwal, *Clin. Exp. Pharmacol. Physiol.*, 2012, **39**, 283.
10. T. Jiang, R. Ghosh and C. Charcosset, *Trends Food Sci. Technol.*, 2021, **112**, 419.
11. H. L. Kwon and M. S. Chung, *Food Chem.*, 2015, **185**, 58.
12. V. Krup, L. H. Prakash and A. Harini, *J. Homeopathy Ayurvedic Med.*, 2013, **2**, 133.
13. N. Chainani-Wu, *J. Altern. Complement. Med.*, 2003, **9**, 161.
14. P. Plianbangchang, W. Tungpradit and W. Tiyaboonchai, *Naresuan Univ.: J. Sci. Technol. (NUJST)*, 2013, **15**, 73.

15. M. A. Espinosa-Olivares, N. L. Delgado-Buenrostro, Y. I. Chirino, M. A. Trejo-Márquez, S. Pascual-Bustamante and A. Ganem-Rondero, *Eur. J. Pharm. Sci.*, 2020, **155**, 105533.
16. C. Madhusha, K. Rajapaksha, I. Munaweera, M. de Silva, C. Perera, G. Wijesinghe, M. Weerasekera, D. Attygalle, C. Sandaruwan and N. Kottegoda, *ACS Omega*, 2021, **6**, 9600.
17. N. Rungphanichkul, U. Nimmannit, W. Muangsiri and P. Rojsitthisak, *Pharmazie*, 2011, **66**, 570.
18. P. Boriwanwattanarak, K. Ingkaninan, N. Khorana and J. Viyoch, *Int. J. Cosmet. Sci.*, 2008, **30**, 205.
19. R. Agrawal and I. P. Kaur, *Rejuvenation Res.*, 2010, **13**, 397.
20. R. L. Thangapazham, A. Sharma and R. K. Maheshwari, in *The Molecular Targets and Therapeutic Uses of Curcumin in Health and Disease*, ed. B. B. Aggarwal, Y. J. Surh, and S. Shishodia, Springer, USA, 2007, p. 343.
21. L. Vollono, M. Falconi, R. Gaziano, F. Iacovelli, E. Dika, C. Terracciano, L. Bianchi and E. Campione, *Nutrients*, 2019, **11**, 2169.
22. D. Nagore and P. Pawar, presented in part at 14th World Conference on Cosmetic Dermatology & Skin Allergies, Athens, Greece, November 2018.
23. A. M. Hui, J. R. Jagdeo, N. Brody, and R. Rupani, in *Cosmeceuticals and Active Cosmetics*, ed. R. K. Sivamani, J. R. Jagdeo, P. Elsner, and H. I. Maibach, CRC Press, USA, 3rd edn, 2015, p. 19.
24. T. Onami and H. Kanazawa, *J. Chem. Educ.*, 1996, **73**, 556.
25. A. M. Ferreira, H. M. D. Gomes, J. A. P. Coutinho and M. G. Freire, *Sustainability*, 2021, **13**, 7509.
26. G.-C. Mitraka, K. N. Kontogiannopoulos, M. Batsioula, G. F. Banias and A. N. Assimopoulou, *Sustainability*, 2021, **13**, 8818.
27. L. S. Torres-Valenzuela, A. Ballesteros-Gomez, J. Serna, A. Arango and S. Rubio, *Environ. Sci.: Water Res. Technol.*, 2020, **6**, 757.
28. S. H. Xuan, K. S. Lee, H. J. Jeong, Y. M. Park, J. H. Ha and S. N. Park, *Biomater. Res.*, 2019, **23**, 2.
29. M. J. Visconti, W. Haidari and S. R. Feldman, *J. Dermatol. Dermatol. Surg.*, 2020, **24**, 18.
30. A. F. Stallings and M. P. Lupo, *J. Clin. Aesthet. Dermatol.*, 2009, **2**, 36.
31. A. Herman and A. P. Herman, *Skin Pharmacol. Physiol.*, 2013, **26**, 8.
32. J. Gherardini, J. Wegner, J. Chéret, S. Ghatak, J. Lehmann, M. Alam, F. Jimenez, W. Funk, M. Böhm, N. V. Botchkareva, C. Ward, R. Paus and M. Bertolini, *Int. J. Cosmet. Sci.*, 2019, **41**, 164.
33. K. E. Lee, S. Bharadwaj, U. Yadava and S. G. Kang, *J. Enzyme Inhib. Med. Chem.*, 2019, **34**, 927.
34. D. Gajalakshmi and E. Kavitha, *Mater. Today: Proc.*, 2022, **69**, 1408.
35. L. Luo and M. E. Lane, *Int. J. Pharm.*, 2015, **490**, 155.
36. S. A. Abosabaa, M. G. Arafa and A. N. ElMeshad, *J. Drug Delivery Sci. Technol.*, 2020, **60**, 102084.
37. Y. Rahimpour and H. Hamishehkar, *Expert Opin. Drug Delivery*, 2012, **9**, 443.

38. P. Ganesan and D. K. Choi, *Int. J. Nanomed.*, 2016, **11**, 1987.
39. R. Iswandana, W. Chalista and E. S. Sitepu, *J. Appl. Pharm. Sci.*, 2018, **8**, 43.
40. Y. de Lafuente, A. Ochoa-Andrade, M. E. Parente, M. C. Palena and A. F. Jimenez-Kairuz, *Int. J. Cosmet. Sci.*, 2020, **42**, 548.
41. E. K. Nelson and L. E. Dawson, *J. Am. Chem. Soc.*, 1923, **45**, 2179.
42. V. Fattori, M. S. N. Hohmann, A. C. Rossaneis, F. A. Pinho-Ribeiro and W. A. Verri, *Molecules*, 2016, **21**, 844.
43. M. Lu, C. T. Ho and Q. Huang, *J. Food Drug Anal.*, 2017, **25**, 27.
44. M. De Lourdes Reyes-Escogido, E. G. Gonzalez-Mondragon and E. Vazquez-Tzompantzi, *Molecules*, 2011, **16**, 1253.
45. A. T. E. Jolayemi, in *Aromatic and Medicinal Plants - Back to Nature*, ed. H. A. El-Shemy, IntechOpen, UK, 2017, p. 239.
46. R. Arora, N. S. Gill, G. Chauhan and A. C. Rana, *Int. J. Pharm. Sci. Drug Res.*, 2011, **3**, 280.
47. U. Agrawal, M. Gupta and S. P. Vyas, *Artif. Cells, Nanomed., Biotechnol.*, 2015, **43**, 33.
48. E. Goci, E. Haloci, A. Di Stefano, A. Chiavaroli, P. Angelini, A. Miha, I. Cacciatore and L. Marinelli, *Biomolecules*, 2021, **11**, 432.
49. F. Iram, S. A. Khan and A. Husain, *Asian Pac. J. Trop. Biomed.*, 2017, **7**, 513.
50. E. Basch, H. Boon, T. Davies-Heerema, I. Foppo, S. Hashmi, J. Hasskarl, D. Sollars and C. Ulbricht, *J. Herb. Pharmacother.*, 2004, **4**, 63.
51. S. Mishra, R. S. Bishnoi, R. Maurya and D. Jain, *Asian J. Pharm. Clin. Res.*, 2020, **13**, 33.
52. H. Hussain, A. Al-Harrasi, R. Csuk, U. Shamraiz, I. R. Green, I. Ahmed, I. A. Khan and Z. Ali, *Expert Opin. Ther. Pat.*, 2017, **27**, 81.
53. B. G. Alraddadi and H.-J. Shin, *Cosmetics*, 2022, **9**, 119.
54. H. Haghaei, S. Soltani, S. Arefhosseini, M.-R. Rashidi and S. Karima, *Pharm. Sci.*, 2021, **27**, 14.
55. S. Hourfane, H. Mechqoq, F. Errajouani, J. M. Rocha and N. El Aouad, *Cosmetics*, 2022, **9**, 131.
56. A. Sudhakar, J. Shantakumar and M. V. Joseph, *Am. J. Phytomed. Clin. Ther.*, 2021, **9**, 6.
57. S. Singh, A. Khajuria, S. C. Taneja, R. K. Johri, J. Singh and G. N. Qazi, *Phytomedicine*, 2008, **15**, 400.
58. S. Togni, G. Maramaldi, F. Di Pierro and M. Biondi, *Clin., Cosmet. Invest. Dermatol.*, 2014, **7**, 321.
59. L. Rigano, G. Maramaldi, A. Bonfigli, S. Togni, M. Meneghin and S. Brancato, *Esper. Dermatol.*, 2015, **17**, 139.
60. S. Hu, G. Belcaro, M. R. Cesarone, B. Feragalli, R. Cotellese, M. Dugall, C. Scipione, V. Scipione, C. Maione, G. Maramaldi, S. Togni and A. Riva, *Minerva Med.*, 2021, **112**, 255.
61. M. Mehta, H. Dureja and M. Garg, *Drug Delivery*, 2016, **23**, 3072.
62. J. Fidalgo, A. N. Barros and A. Casas, in *Resveratrol - Recent Advances, Application, and Therapeutic Potential*, ed. A. Imran, IntechOpen, UK, 2022, DOI: 10.5772/intechopen.107860.

63. S. B. Thapa, R. P. Pandey, Y. I. Park and J. K. Sohng, *Molecules*, 2019, **24**, 2571.
64. J. M. Pezzuto, T. P. Kondratyuk and T. Ogas, *Expert Opin. Ther. Pat.*, 2013, **23**, 1529.
65. C. Li, X. Xu, Z. Tao, C. Sun and Y. Pan, *Expert Opin. Ther. Pat.*, 2016, **26**, 1189.
66. A. Ratz-Łyko and J. Arct, *J. Cosmet. Laser Ther.*, 2019, **21**, 84.
67. E. H. Siemann and L. L. Creasy, *Am. J. Enol. Vitic.*, 1992, **43**, 49.
68. M. Jang, L. Cai, G. O. Udeani, K. V. Slowing, C. F. Thomas, C. W. Beecher, H. H. Fong, N. R. Farnsworth, A. D. Kinghorn, R. G. Mehta, R. C. Moon and J. M. Pezzuto, *Science*, 1997, **275**, 218.
69. S. Intagliata, M. N. Modica, L. M. Santagati and L. Montenegro, *Antioxidants*, 2019, **8**, 244.
70. M. H. Lin, C. F. Hung, H. C. Sung, S. C. Yang, H. P. Yu and J. Y. Fang, *J. Food Drug Anal.*, 2021, **29**, 15.
71. M. A. Nunes, F. Rodrigues and M. B. P. P. Oliveira, in *Handbook of Grape Processing By-Products: Sustainable Solutions*, ed. C. M. Galanakis, Academic Press, USA, 2017, p. 267.
72. J. Park and Y. C. Boo, *Evid. Based Complement. Alternat. Med.*, 2013, 645257.
73. C. M. Chan, C. H. Huang, H. J. Li, C. Y. Hsiao, C. C. Su, P. L. Lee and C. F. Hung, *Int. J. Mol. Sci.*, 2015, **16**, 5789.
74. Y. Zhang, C. Ni, Y. Huang, Y. Tang, K. Yang, X. Shi, Y. Zhang, Z. Li, J. Wang, Y. Zhu, H. Li, Y. Ma, J. Lin, J. Wang, Q. Liu and W. Wu, *Clin., Cosmet. Invest. Dermatol.*, 2021, **14**, 1805.
75. F. Silva, A. Figueiras, E. Gallardo, C. Nerin and F. C. Domingues, *Food Chem.*, 2014, **145**, 115.
76. B. Tian and J. Liu, *J. Sci. Food Agric.*, 2020, **100**, 1392.
77. A. Sharkawy, F. M. Casimiro, M. F. Barreiro and A. E. Rodrigues, *Int. J. Biol. Macromol.*, 2020, **147**, 150.
78. J. Igielska-Kalwat, M. Firlej, A. Lewandowska and B. Biedziak, *Acta Biochim. Pol.*, 2019, **66**, 371.
79. A. S. Brinke, C. Janssens-Böcker and M. Kerscher, *J. Cosmet., Dermatol. Sci. Appl.*, 2021, **11**, 155.
80. A. S. Chauhan, *Ann. N. Y. Acad. Sci.*, 2015, **1348**, 134.
81. G. Mehta, U. R. Nayak and S. Dev, *Tetrahedron*, 1973, **29**, 1119.
82. T. P. A. Krishna, B. Edachery and S. Athalathil, *RSC Adv.*, 2022, **12**, 8815.
83. Q. Jia and M. F. Hong, *US pat.*, US2006/0251749A1, 2006.
84. K. Lystvan, V. Belokurova, Y. Sheludko, J. L. Ingham, V. Prykhodko, O. Kishchenko, E. Paton and M. Kuchuk, *Plant Cell, Tissue Organ Cult.*, 2010, **101**, 99.
85. F. Alam, G. N. Khan and M. H. H. B. Asad, *Phytother. Res.*, 2018, **32**, 597.
86. R. K. Chaudhuri and K. Bojanowski, *Int. J. Cosmet. Sci.*, 2014, **36**, 221.
87. O. Ohno, T. Watabe, K. Nakamura, M. Kawagoshi, N. Uotsu, T. Chiba, M. Yamada, K. Yamaguchi, K. Yamada, K. Miyamoto and D. Uemura, *Biosci., Biotechnol., Biochem.*, 2010, **74**, 1504.

88. N. P. Ariantari and E. S. P. Ratnasantasyacitta, *J. Trop. Pharm. Chem.*, 2022, **6**, 177.
89. D. Petric, *Food Health*, 2022, **4**, 16.
90. J. E. Kim, J. H. Kim, Y. Lee, H. Yang, Y. S. Heo, A. M. Bode, K. W. Lee and Z. Dong, *Oncotarget*, 2016, 7, 14616.
91. S. Dhaliwal, I. Rybak, S. R. Ellis, M. Notay, M. Trivedi, W. Burney, A. R. Vaughn, M. Nguyen, P. Reiter, S. Bosanac, H. Yan, N. Foolad and R. K. Sivamani, *Br. J. Dermatol.*, 2019, **180**, 289.
92. D. Bacqueville, A. Maret, M. Noizet, L. Duprat, C. Coutanceau, V. Georgescu, S. Bessou-Touya and H. Duplan, *Clin., Cosmet. Invest. Dermatol.*, 2020, **13**, 359.
93. Z. D. Draelos, H. Gunt, J. Zeichner and S. Levy, *J. Drugs Dermatol.*, 2020, **19**, 1181.
94. R. K. Chaudhuri and F. Marchio, *Cosmet. Toiletries*, 2011, **126**, 502.
95. M. A. F. Nordin, F. A. Razak and W. H. Himratul-Aznita, *Evid. Based Complement. Alternat. Med.*, 2015, **2015**, 918624.
96. M. K. Khan, L. Paniwnyk and S. Hassan, in *Plant Based "Green Chemistry 2.0"*, ed. Y. Li, and F. Chemat, Springer, Singapore, 2019, p. 197.
97. O. V. Zillich, U. Schweiggert-Weisz, P. Eisner and M. Kerscher, *Int. J. Cosmet. Sci.*, 2015, **37**, 455.
98. D. J. de Lima Cherubim, C. V. B. Martins, L. O. Fariña and R. A. da Silva de Lucca, *J. Cosmet. Dermatol.*, 2020, **19**, 33.
99. M. Daglia, *Curr. Opin. Biotechnol.*, 2012, **23**, 174.
100. W. Koch, J. Zagórska, Z. Marzec and W. Kukula-Koch, *Molecules*, 2019, **24**, 4277.
101. E. Guzmán and A. Lucia, *Cosmetics*, 2021, **8**, 114.
102. A. Sarkic and I. Stappen, *Cosmetics*, 2018, **5**, 11.
103. U. S. Abelan, A. C. de Oliveira, É. S. P. Cacoci, T. E. A. Martins, V. M. Giacon, M. V. R. Velasco and C. R. R. de Castro Lima, *J. Cosmet. Dermatol.*, 2022, **21**, 1407.
104. K. Manvitha and B. Bidya, *J. Pharmacogn. Phytochem.*, 2014, **2**, 85.
105. A. Surjushe, R. Vasani and D. G. Saple, *Indian J. Dermatol.*, 2008, **53**, 163.

CHAPTER 6

# Applications of Mushrooms in Cosmeceuticals

CHASHEEN FIZZA[a], WASEEM KHALID[b*], SAFURA KAUSAR[b], ZAHRA MAQBOOL[b], SHANZA MUKHTAR[c], EMAN ALI[d], BLESSING TITYLAYO AJALA[a], MUHAMMAD ZEESHAN AHMAD[d] AND MUHAMMAD IRFAN[a]

[a]School of Food Science and Engineering, South China University of Technology, Guangzhou, China; [b]Department of Food Science, Government College University, Faisalabad, Pakistan; [c]Department of Nutrition and Dietetics, The University of Faisalabad, Faisalabad, Pakistan; [d]National Institute of Food Science and Technology, University of Agriculture Faisalabad, Pakistan
*E-mail: waseemkhalid@gcuf.edu.pk

## 6.1 Introduction

Mushrooms are the epigeous fruiting body of land fungi. Since fungi lack chlorophyll as well as cellulose, they live in a contrasting way compared to various non-motile living forms like plants. Mushrooms are eukaryotic fungi that digest food from the external environment and entities. Mushrooms have the ability to absorb nutrients in their cells. Hence, mushrooms may live in diverse environments: above ground, on the ground, and even on different plants, for instance dead wood. More research on mushrooms has revealed that they offer several health-related advantages in the human body.[1] Mushrooms have been used as medicines as well as food for ages because of their richness in nutrients and physiologically

Bioprospecting of Natural Sources for Cosmeceuticals
Edited by Deepika Kathuria, Ajay Sharma, Meenakshi Verma and Gulzar Ahmad Nayik

Published by the Royal Society of Chemistry, www.rsc.org

active substances with several applications, including in agriculture, food, pharmacology, cosmetics, food-related businesses, and others.[2] Mushrooms have been farmed and utilized by humans for centuries as a kind of macrofungi due to their appealing sensory qualities, plentiful nutritional contents, numerous functional activities, and controllable cultivation circumstances.[3] Mushrooms are quickly evolving into one of the most successful enterprises on the planet.[4] For many years, people have consumed and enjoyed mushrooms for their flavor, economic and ecological worth, and therapeutic capabilities.[5] Many chemicals extracted from various mushroom species have been discovered, the majority of which are lectins, fatty acids, steroids, proteins, vitamins, phenols, terpenoids, calcium, polysaccharides[1] and phosphorus. Mushrooms have been shown to possess a broad range of biological actions, exhibiting antimicrobial, anticancer, hypocholesterolemic, antihypertensive, antioxidant, antidiabetic, hepatoprotective, anti-obesity, antiallergic, antiaging, and anticoagulant properties.[2] The quantity of commercially used cosmeceutical products containing extracts of mushroom or their biologically active additives, asserting positive effects against lines, skin texture, wrinkles, aging, photoprotection, and pigmentation, shows that mushroom cosmetic industry ingredients are gradually entering the cosmeceutical market.[6] Edible mushrooms are well recognized as an underutilized resource of quality nutraceuticals and food that can enhance humans' health plus life quality. Nevertheless, inedible species or poisonous ones (such as *Ramaria formosa*, *Amanita muscaria*, *Inonotus obliquus*, *Ganoderma lucidum*, *Phellinus igniarius*, and others) provide myco-pharmaceutical (nutraceuticals or dietary supplements) development, inducing bioactive secondary metabolites with health-promoting and therapeutic properties. Nutraceuticals developed from macrofungi are collected from the mycelium or reproductive body and are supplied orally in the formulation of extract pills or capsules. Many commercially found products developed from macrofungi are already available on the market, including cosmeceuticals (topically applied lotions, creams, and ointments) and nutracosmetics (given orally). Their popularity is due to the lack of regulation and safety when compared to regular medications.[7] As the world population has grown, so has production interest as well as the subsequent use of mushrooms as an origin of food and competent industrial element. Since 1990, the world has been focused on the mushroom business, resulting in a tremendous growth in production. Because of their taste, flavor, and nutritional content, edible mushrooms have grown in popularity. Mushrooms are superior to meat and dairy products as well as other animal products, as evidenced by previous research. Several types of vitamins found in mushrooms are credited for boosting human wellbeing by lowering the risks of certain illnesses.[8] In this chapter, we will be discussing the basics of mushrooms, their categories and deep insights into some prominent features (composition, bioactive components, and types), and their industrial application, most specifically to food and cosmeceuticals.

## 6.2 Types of Mushrooms

Mushrooms are roughly classified into three types: 1: edible mushrooms, such as *Agaricus bisporus*, the button mushroom; 2: therapeutic mushrooms, such as *Ganoderma lucidum* (lingzhi); and 3: dangerous mushrooms, such as *Amanita phalloides*, also known as death cap.[9–11] All species names and authorities have been confirmed and rectified using legitimate mycological databases, and a total of 1662 taxa have been summarized. There are 1020 edible mushrooms, 692 therapeutic mushrooms, and 480 deadly mushrooms among the 1662 taxa.

### 6.2.1 Poisonous Mushrooms

Poisoning from toxic mushrooms occurs worldwide each year, and a large number of cases of poisoning have been documented. Accidental-type poisoning by harvested mushrooms may have serious consequences, and can even result in death under certain circumstances.[12] Poisoning by mushrooms is an international food safety concern and one among the leading causes of food-imposed poisoning mortality. More than 100 toxic mushrooms have been discovered in recent years, with greater than 30 exhibiting lethal toxicities. Within the huge variety of deadly mushrooms, *Inocybe* is one of the most prevalent species.[13] Poisonous mushrooms can induce six types of symptoms: cytotoxicity, neurotoxicity, myotoxicity (rhabdomyolysis), metabolism disturbance (comprising endocrine plus associated toxicities), gastrointestinal disturbance, and other undesirable effects. The majority of fatal poisonings are linked to myotoxic and cytotoxic poisonous mushrooms.[14] Pezizales (Ascomycota), Agaricales, Russulales (Basidiomycota), and Boletales are the most common orders of toxic mushrooms, with the majority of species forming a fleshy fruiting body. Lethal Agaricales species include amanitin-producing mushrooms such as *Amanita bisporigera* and *A. phalloides*, *Lepiota brunneoincarnata* and *Galerina marginata*. 90% of deadly mushroom poisoning instances are caused by *Amanita* species, which are regarded to be the most hazardous mushrooms. The death cap, *A. phalloides*, is one of the most well-known species, having killed countless individuals throughout history and been named among the world's 10 most feared mushroom species. In addition to the amanitin-producing mushrooms, recognized fatal mushrooms include *Trogia*, *Russula*, *Gyromitra*, and *Paxillus* species.[15] These species carry dangerous poisons in varied concentrations that produce a variety of symptoms and syndromes. Poisonous chemicals are found in both toxic and edible mushrooms, and the level of intoxication depends on the amount taken. When there is a dearth of awareness of mushroom poisoning, it is hard to avoid accidental mushroom consumption.[16] Accurate mushroom identification is a critical step in preventing poisoning as well as facilitating effective treatments through the early discovery of poison-related symptoms.[17] Events of deadly mushroom-poisoning have previously been recorded, but the toxin features, symptoms, and toxicity pathways have not been described. As a result, it is critical to evaluate data on mushroom poisons and toxicity. Extensive clinical and experimental-based research is necessary in order to discover potential negative impacts of therapeutic edible mushrooms and also ensure their safe ingestion.

### 6.2.2 Medicinal Mushrooms

Mushrooms have been employed as food and medicines for ages. Medically used mushrooms are considered as functional foods and are available as health supplements for use in alternative plus complementary therapies. The variety of chemicals derived from mushrooms has piqued the interest of researchers as a possible root of pioneering products, with unique modes of action or potent strategies against present life-threatening illnesses. Biologically active molecules can be found as proteins, components of the cell wall (polysaccharides, *e.g.* glucans), or organic secondary metabolites (steroids, terpenes, and phenols, among others). The action of these chemicals is greatly influenced by a variety of circumstances, including the species of mushroom, its developmental stage, and the growth conditions. Extracts and/or chemicals derived from mushrooms have displayed anti-inflammatory, anticancer, hypoglycemic, immunomodulatory, antibacterial, antiviral, antioxidant, hepatoprotective, anti-neurodegenerative, hypocholesterolemic, and antiangiogenic properties.[18] A mushroom known as 'Sanghuang,' of which 14 species have been identified all over the globe and which is mostly taken as a complement to high level processed foods, has been demonstrated to be among one of the most significant therapeutic fungus groupings. The principal therapeutic effects of Sanghuang include anticancer, antioxidant, immunomodulatory, anti-inflammatory, anti-diabetic, and hepatoprotective characteristics; these are mostly mediated by the secondary metabolites. However, in animal tests, its basidiocarp powder was shown to be non-toxic.[19] *Auricularia*, *Hericium*, *Lentinus*, *Grifola*, *Tremella*, *Pleurotus*, and *Flammulina* are edible mushrooms with potential medicinal and functional characteristics. *Trametes*, *Ganoderma*, and other species are only known for their therapeutic characteristics, because they are gritty, have a tougher texture, and have a bitter flavor. Mushrooms are high in vitamins, including thiamine (B1), riboflavin (B2), ascorbic acid (vitamin C), niacin, and biotin. Vitamins A and D are uncommon; however, when exposed to UV radiation, several species have measurable amounts of vitamin D. The crude fat content of mushrooms is made up of lipids, namely monoglycerides, diglycerides, and triglycerides, and phospholipids plus sterols.[20] The therapeutic effects of mushrooms are attributable to a variety of secondary metabolites contained inside the carpophore and mycelial structures that may influence various biological activities of humans in a non-synergistic or synergistic way.[21] Biological activities are aided by complexes of polysaccharide and protein, peptidoglycans, or heteroglycans. Owing to the capacity of binding to specific receptors on the cell wall, and also triggering particular immunological responses, they are principally accountable for immunomodulatory effects.

Medicinal mushrooms are commonly used for cancer treatment as biological response modifiers (BRMs), and are beneficial for curing cancer, minimizing treatment adverse reactions, and enhancing the patient's quality of life. Terpenes, which are composed of five-carbon isoprene atoms and are formed by the summing of the functional groups, are another significant family of chemicals with bioactivity. The immune system is influenced by increasing the gene expression that codes for immune-related proteins, but they also have anticancer, antioxidant, and anti-inflammatory activities. Terpenoid levels are high in mushrooms of the species *Ganoderma P. Karst.*[22] These mushrooms are an excellent antioxidant source, although they are not widely consumed. Several mushroom types can be called functional foods due to their greater levels of antioxidants and bio-elements. Various basidiomycete mushrooms contain physiologically active polysaccharides. The overwhelming majority of the 651 species plus 7 infraspecific taxa from the 182 genera of higher hetero- as well as homobasidiomycetes exhibit pharmacologically active types of polysaccharides in the fruit bodies, cultured mycelia, or broth.[23]

### 6.2.3 Edible Mushrooms

There are about 2000 kinds of fungus in nature, yet only 25 are acknowledged as foods and just a handful are exploited commercially. Because of their nutritional, therapeutic, and organoleptic properties, edible mushrooms are considered a delicacy. *Agaricus bisporus* is the most commonly eaten fungus on this planet, followed by the species *Lentinus edodes* and *Pleurotus*. China is the world's greatest mushroom producer, followed by America, Poland, France, Spain, and the Netherlands.[24]

Mushrooms are important as a meal not only because of their pleasant organoleptic features, but also because of their high number of elements that are required in a healthy diet. Mushrooms, in reality, include amino acids, fatty acids (among them many are unsaturated or poly-unsaturated, like linolenic acids, oleic, and linoleic), trace materials, and vitamins. From carbohydrates, the most prevalent sugar found is trehalose, the 1,1-dimer of D-glucose, which is responsible for various gastrointestinal intolerances in people who have difficulties digesting it. Polyphenols found in mushrooms have a variety of applications as medical and functional food additives.[25] Chitin, also called fungal fiber, is a homopolymer of *N*-acetylglucosamine plus other sugars found in fungal cell walls and is a more complex polysaccharide. Regardless of their culinary features, edible mushrooms gathered from contaminated soils may include harmful elements such as mercury, arsenic and other heavy metals.[26] Edible mushrooms and their byproducts are heavily involved in a variety of industries with a wide range of uses. Mushrooms are used as nutritional supplements in probiotics and also fortified in RTE and RTC meals owing to their nutritional and functional properties. The bioactive compounds extracted from them have been discovered to have potent credibilities in food items. Aside from that, mushrooms have been discovered to be useful in the manufacturing of biochars, biosorbents, carbon dots, nanoparticles, media, and skincare formulations. All of these are synthesized or produced with the help of mushrooms and have shown positive outcomes in *in vitro* trials. Nevertheless, edible coatings and skincare products derived from these edible mushrooms are currently restricted and are not commercially or industrially used yet. Furthermore, edible mushrooms and their waste have enormous economic capabilities in a variety of businesses, and this may frequently lead to the synthesis of innovative goods. Nonetheless, these edible mushrooms are underutilized sources with several potential industrial applications. As a result, there is an urgent need for prudent and competent administration in the production chain in order to investigate the viability of edible mushrooms.[27]

## 6.3 Constituents of Mushrooms

Numerous scientists from across the world have examined the molecules derived from mushrooms with bioactivities, including their preparation techniques and structural characterization, which are largely related to proteins, terpenes, unsaturated fatty acids, polysaccharides, and phenolics.[6] Dietary fiber, protease inhibitors and proteases, hydrophobins and ribosome-inactivating proteins are the most notable achievements. Furthermore, studies concentrating on the roots of bioactive nutraceuticals may lead to the production of possible functional meals with high quality as well as nutritional value.[28] The following sections will provide a detailed summary of such active substances.

### 6.3.1 Carbohydrates

The carbohydrate content of mushrooms accounts for the majority of the fruity bodies, comprising 50 to 65% of the dry weight. About 11% of the total sugars are free sugars.[29] Carbohydrates are involved primarily in the structural structure of mushrooms, with the exception of sugar-free components, and are necessary in sustaining the hyperosmotic content and providing the release of energy consistent with their fast metabolic rate.[8] Some mushroom polysaccharides, like lentinan from *Lentinus edodes*, pleuran from the *Pleurotus* species, schizophyllan from *Schizophyllum*, ganoderan from *Ganoderma lucidium*, and commune calocyban from *Calocybe indica*, have been produced and exploited as functional foods, the majority of which have been proven as an outstanding presenter of D-glucans having the common (1–3) or (1–6) linkage.[3] Mannitol is the most abundant sugar in *A. bisporus*. *A. polytricha* AAPS-1 homopolysaccharides with various conformations and couplings of D-glucopyranosyl backbones have been found. Various numbers of (1–6) attached to the -(1–3) are D-glycosidic linkages. Auricularia polysaccharide linked D-glucopyranosyl backbones have been related to a variety of chemical, physical, and biological characteristics.[30] Carbohydrates derived from mushrooms possess anti-tumor and immunomodulatory properties. When coupled with proteins, they can create polysaccharide protein-complexes and peptidoglycans. Polysaccharides are the most powerful category of bioactive chemicals discovered in mushrooms. The majority of polysaccharides extracted from fungi are water insoluble; of those that are water soluble, the majority are heteropolysaccharides with a complex structure, making it difficult to determine the primary and secondary structures.[31] In general, mushroom polysaccharides are very diverse in terms of sugar content, main chain degree of branching, molecular mass, polymer structure, conformation, and other physical features, all of which have a substantial impact on the polysaccharide's bioactivity and mode of action.[32]

### 6.3.2 Proteins

Fungal immunomodulatory proteins (FIP), lectins, ribosome inactivating proteins (RIPs), laccases, antimicrobial or antifungal proteins, and ribonucleases are among the bioactive proteins and peptides produced by mushrooms. Lectins are glycoproteins that link to the cell surfaces of carbohydrates and have the potential to agglutinate cells.[33] Proteins that inhibit ribosome activity, RIPs, are enzymes that render the ribosome inactive through removing one or more adenosine residues of rRNA. RIPs from numerous mushroom species, including *Pleurotus tuber-regium*, *Flammulina velutipes*, *Calvatia caelata*, *Lyophyllum shimeiji*, and *Hypsizigus marmoreus*, have been purified over the last decade.[34] These mushroom RIPs have a variety of bioactivities, including HIV-1 reverse transcriptase inhibition, anti-proliferative and antifungal. Laccases are multicopper oxidases, a group of enzymes involved in the pathogenesis, immunogenesis, and morphogenesis of organisms, as well as the metabolic cycle of complex chemical compounds. Laccase was extracted from the bodies of *Pleurotus eryngi*. With an $IC_{50}$ of 2.2 M, laccase inhibits the activity of HIV-1 reverse transcriptase. FIPs are beneficial in tumor immunotherapy because they inhibit tumor invasion and metastasis. Lentin (27.5 kDa) is an antifungal protein extracted from the productive bodies of the edible mushroom *L. edodes*. Lentin suppresses the growth of the mycelium in different fungal species, including *Mycosphaerella arachidicola*, *Botrytis cinerea*, and *Physalospora piricola*.[35] Bioactive peptides extracted from mushrooms possess antifungal, antibiotic and antimicrobial activities, and antihypertensive, antiviral, anticancer, ACE inhibitory and antioxidant potential.[36] Overall, their potential biotechnological/medical uses should increase the hunt for novel BPs that have yet to be found, given that fungal species constitute a rich repository of these compounds.[37]

### 6.3.3 Minerals

In comparison to vegetables, mushrooms have a respectable mineral content (6–10.5%). The principal elements of the ash are phosphorus, potassium, or calcium, magnesium, copper, zinc, and iron, depending on the mushroom species. Selenium levels in mushrooms are typically high, varying from 0.46 to 5.63 mg $kg^{-1}$ dry matter.[38] This is a trace mineral that is a cofactor in several enzyme systems and has a greater anti-oxidative influence. It is crucial to remember that the manner of growing and the compost type used might impact the mineral elements and the trace element concentrations in mushrooms.[39] Furthermore, because these compounds are typically present in the culture of substrates, a buildup of trace heavy metals, particularly hazardous elements like cadmium, lead, and mercury, can be discovered in them.[40,41] Some research suggests that the cleaning and peeling processes of mushrooms can lessen the concentration of heavy metals in the edible sections.[42]

### 6.3.4 Vitamins

Mushrooms are excellent sources of many vitamins, including vitamin B2, vitamin B3, and folate. Some also include vitamin C, vitamin B1, vitamin D, carotene (vitamin A precursor), vitamin E, and vitamin B12.[43]

### 6.3.5 Lipids

Edible mushrooms have a low total fat content ranging from 1.3 to 5.8%. Unsaturated fatty acids, particularly palmitic acid (C16:0), linoleic acid (C18:2), and oleic acid (C18:1), predominate over the saturated fatty acids in general. The lipid profile of *A. bisporus* also contains glycerides, free fatty acids, and phospholipids.[44] Mushroom lipids are extremely suited for humans as they pose little danger of plaque development in blood passages. The crude fat in medicinal and edible mushrooms contains lipids, sterol esters, monoglycerides, diglycerides, triglycerides, and other compounds.[45,46] Fatty acid compositions vary among the fungal species and are key elements of organelles accounting for 30–70% of fungi organelles. A survey of the fat content in *Pleurotus* species was conducted, which reported the content ranging between 0.9 and 7.5%. *Pleurotus eryngii* (PE) and *Pleurotus tuber-regium* (PTR0) had the highest (7.50%) and lowest (1.06%) fat levels, respectively.[47] The agro-waste employed in the growing process might be the cause of this diverse variance in fat content. *Pleurotus* species are low in lipids and high in fatty acids, namely linoleic acid and oleic acid. Previous studies have shown that *Pleurotus* species are excellent options for hypocholesterolemia and anti-inflammation in modern nutrition.[31,44] Kutaiba revealed that the *Pleurotus ostreatus* diet suppressed the low density lipoproteins as well as very low density lipoprotein formation, which dramatically lowered the total cholesterol levels in humans.[31,48]

## 6.4 Phytochemicals

Macrofungi are recognized to contain a wide range of secondary metabolites. Medicinal mushrooms include chemical components that have a distinct physiological effect on the human body. These bioactive substances consist of phenols, flavonoids, alkaloids, terpenoids, tannins, and steroids. Many compounds, such as kejoic acid, L-ergothioneine, *p*-coumaric acid, ellagic acid, betulin, caffeic acid, ferulic acid, *etc.* that are present in mushrooms exhibit anti-tyrosinase activity and thus inhibit the production of melanin. Kojic acid is present in many commercial skin whitening formulations. It binds with the copper ion at the tyrosinase active site and thus inhibits the activity of tyrosinase. Elagic acid, ascorbyl-3-*p*-coumarate, ascorbyl-2-*p*-coumarate and *p*-coumaric acid have also been reported to exhibit anti-collagenase activity, and thus prevent collagen degradation and maintain healthy skin. Mushroom components are known for their antioxidant potential, which is majorly attributed to the presence of β-glycan. Polysaccharides have the ability to scavenge free radicals and thus exhibit antioxidant potential.[49–53]

## 6.5 Cosmetics Applications of Mushrooms

The skin is known as the biggest organ in the human body, comprising approximately 15 percent of the total weight of the body and serving like a shield between the external and internal environments, as well as providing protection from foreign bodies. Mushrooms are rich in natural bioactive chemicals, which are increasingly being employed as cosmeceutical components.[6] Mushrooms have been extensively researched as medicinal agents for a variety of ailments, as well as to prevent the indications of skin aging. Mushrooms are rich in anti-oxidants, and skin smoothing, hydrating, and brightening ingredients. The high proportion of mushroom polysaccharides has sparked increased interest in their potential for skin benefits by raising the moisture levels of the epidermal and stratum corneum, enhancing the skin's water holding capacity, and reducing moisture loss. When utilized in topical cosmetic products, mushroom polysaccharides work as potent moisturizers similar to hyaluronic acid. Chaiwut *et al.* extracted the polysaccharide from *Volvariella volvacea* mushroom and evaluated its cosmetic potential. This study revealed that its extract subsequently reduces skin roughness, scaliness, and wrinkles and also provides a skin whitening effect.[54]

Phenolic compounds and their derivatives with anti-oxidant potential, such as gallic, caffeic, *p*-hydroxybenzoic, ferulic, homogentisic, myricetin, and protocatechuic acid, have been found in several edible mushrooms species, including *A. bisporus*, *Cantharellus cibarius*, *Boletus edulis*, *Hygrophorus marzuolus*, *Calocybe gambosa*, and *Cratere*.[55] Mushrooms are a rich source of terpenoids, which may be administered topically for a variety of cosmetic applications. Triterpenes have been found in a variety of mushrooms, most notably *Ganoderma* spp. ganoderic acids, ganodermanondiol, ganolucidic acids, ganodermanontriol, lucidumols, ganoluciduones, ganolucidoids, lucidadiols, lucidenic acids, methyl ganolucidates and methyl ganoderates were obtained from the spores or the fruiting body of *G. lingzhi* and exhibited anti-HIV-1 protease activity.[56] Sesquiterpenoids acquired from solid cultured extracts of *Flammulina filiformis*, such as sterpurols, 2,5-cuparadiene1,4-dione, sterpuric acid and enokipodins A and B, demonstrated anti-tumor and anti-bacterial activity against *Bacillus subtilis*, as well as anti-oxidant activity. Canthaxanthin is a reddish-orange keto-carotenoid color found in the edible fungus *Cantharellus cinnabarinus* that acts as an anti-aging rejuvenator and skin strengthener.[8] Mushrooms create a variety of vitamins, including vitB1 (thiamine), vitB2 (riboflavin), vitB3 (niacin), vitB7 (biotin), vitB9 (folic acid or folate), vitB12 (cobalamins), vitC (ascorbic acid), and vitD (calciferols). *L. edodes*, *Agaricus bisporus*, *P. ostreatus* and *P. ferulae* are high in vitD2 (ergocalciferol), which helps the skin's immunological system.[56]

Mushrooms include natural bioactive substances that have been used as components in a variety of cosmeceutical products. Many mushrooms are now used in commercial cosmetic products. The major natural ingredient that promotes anti-aging action in Yves Saint Laurent Temps, Dr Andrew Weil for Origins™ Mega Mushrooms Skin-Relief Micellar Cleanser, Majeur Elixir De Nuit, and Shiseido Ultimune Power Infusing Concentrate is reishi mushroom (*G. lingzhi*) extract. *G. lingzhi* is a common ingredient in anti-aging skin care products such as cream, serum, and micellar cleansers. According to recent pharmacological investigations, extracts of *G. lingzhi* have anti-oxidant, anti-tyrosinase, and anti-inflammatory effects.[57] Moreover, it has been used as a hair stimulant in hair style cream and shampoo to hydrate, protect against UV damage, and prevent color fading. These mushrooms are used in commercial items such as serum, face cream and firming cream. Natural pigments from mushrooms, such as carotenoids, anthraquinones, and melanin, have recently sparked attention for use as natural cosmetic colors, with additional application possibilities as cosmeceutical ingredients.[58] As previously stated, these mushroom species are responsible for a small proportion of the overall number, and also additional species will be found, validated, and furthermore farmed. All these factors indicate the progressive exploitation as well as expansion of the cosmetic business using mushrooms. The molecular basis of the medical properties of mushrooms, called molecular cosmetology, will be revealed through interdisciplinary studies combining metabolomics, genomics, systems pharmacology and proteomics, and more mushrooms can get incorporated into cosmetics *via* various approaches.[59] Table 6.1 presents a few important examples of cosmetic formulations prepared from mushrooms/extracts.

### 6.5.1 Skincare

It is well known that a variety of mushrooms and the components that they contain are good for the hair and skin. Phenolics, terpenoids, polyphenolics, vitamins, polysaccharides, selenium, and volatile organic compounds are the ingredients that stand out. These substances exhibit excellent anti-aging, anti-oxidant, skin whitening, anti-wrinkle, and moisturizing effects, making them excellent cosmetic product candidates.[60]

**Table 6.1** Some examples of mushrooms/extracts with cosmetic applications.

| Mushroom/extract | Cosmetic application |
|---|---|
| *Ganoderma lucidum* | Anti-aging |
| *Ganoderma lucidum* | Sun-protection for hair |
| *Cordyceps sinensis* | Skin whitening |
| *Schizophyllum commune* | Anti-aging; hydrating |
| *Tremella fuciformis* | Hydrating |
| *Tremella polysaccharide* | Hydrating |
| *Cantharellus cinnabarinus* | Skin rejuvenator |
| *G. lingzhi* | Improves skin quality |
| *Ganoderma lucidum* and *Lentinus edodes* | Facial cleanser and makeup remover |
| *Inonotus obliquus* | Facial masks |

### 6.5.1.1 Anti-aging Effect

All body organs gradually lose their reserve capacity and begin to malfunction as we age.[61] Therefore, anti-aging cosmetics are created to maintain and/or repair the body's maintenance and repair systems (MRSs). In the MRSs, antioxidants play a significant role. Numerous varieties of mushrooms are utilized as anti-aging ingredients in topical serums, creams, and facial preparations.[62] *Dictyophora indusiata* (the veiled lady mushroom) fruiting body methanol extract was tested for its anti-cholinesterase, anti-aging, and melanogenesis inhibitory activity.[63]

### 6.5.1.2 Moisturizing Ability/Hydrating Agent

Skin moisture loss is slowed down by moisturizers, and fine wrinkles are reduced. Additionally, they improve the physical and chemical characteristics of the stratum corneum and the skin's surface, making it more hydrated, smooth, and soft.[64] Without the use of a chemical reagent, the polysaccharide isolated from a hot water extract of a *Tremella* mushroom was found to have the novel effect of inhibiting melanin formation and lightening skin spots when applied topically.[65]

### 6.5.1.3 Skin Whitening Effect

Women aspire to have white skin because, particularly in Asia, it denotes wealth and beauty. The main contributor to skin color is melanin synthesis, which is also crucial for UV-induced dermal irritation defense.[66] Tyrosinase controls the biosynthetic pathway for melanin formation in two different oxidation reactions in a variety of bioforms.[67] Thus, the most popular method for skin lightening is tyrosinase inhibition.[68] Numerous mushrooms, such as *Ganoderma* spp., *Agaricus hortensis*, *Agaricus bisporus*, and *Phellinus baumii*, have demonstrated tyrosinase inhibition activity.[69] *Ganoderma lucidum* showed the highest level of inhibition against tyrosinase activity when compared to other basidiomycetes. Ganoderma extracts are currently used in a lot of facial mask cosmetics.[70]

### 6.5.2 Hair Care

Hair cosmetics are made to manipulate the structural characteristics of hair as well as keep the scalp clean. Shampoos, serums, conditioners, hair styling products (gels, waxes, sprays, and mousses), bleaching agents, straightening products, and hair dyes are the main branches of hair cosmetics.[71] Shampoos are designed to clean the hair and scalp, and they have been modified to account for differences in hair type, hair care routine, and specific issues like the treatment of oily hair, dandruff, and androgenic alopecia.[72,73]

Glucuronoxylomannan (GXM), a polysaccharide found in *Tremella* mushrooms, has wound-healing and anti-inflammatory properties and is frequently used in cosmetology.[74] Another formula for promoting hair growth contains *Ganoderma lucidum*, which has the ability to reduce levels of dihydrotestosterone or prostatic hyperplasia.[75] Mushrooms are an excellent source of vitamin D, antioxidants, and minerals, like selenium, iron, and copper, all of which are beneficial for promoting healthy, strong hair and preventing dandruff and hair loss. The cosmetic industry has a lot of potential for incorporating mushrooms and/or extracts into haircare cosmetics.[76]

## 6.6 Conclusion

For centuries, the usage of mushrooms in the food and pharmaceutical industries has been known; however, recently in order to obtain greener cosmetics their bioactive components have been exploited. Various mushroom extracts and their components have been studied, which were found to have valuable effects on the skin and hair mainly due to their antioxidant properties. Mushrooms are rich in phenolics, terpenoids, minerals such as selenium, polysaccharides, vitamins (mainly vitamin D), *etc.*, which offer desired properties such as antioxidant, anti-aging, skin whitening, hydrating and anti-wrinkle effects in a cosmetic preparation. This chapter discusses the types of mushroom and several examples of mushroom extracts and their ingredients presently used in cosmeceutical formulation. However, only a small number of mushrooms have been explored for their cosmeceutical potential and hence the identification and discovery of new mushroom species should be increased.

## References

1. S. Ismail, A. R. Zainal and A. Mustapha, *ISCAIE 2018 - 2018 IEEE Symposium on Comput. Electr. Eng.*, 2018, p. 412.
2. W. A. Elkhateeb and G. M. Daba, *Int. J. Pharm. Chem. Anal.*, 2022, **9**, 1.
3. G. Ma, W. Yang, L. Zhao, F. Pei, D. Fang and Q. Hu, *J. Food Sci. Nutr., Wellness*, 2018, **7**, 125.
4. H. Rahman, M. O. Faruq, T. A. Hai, W. B. Rahman, M. M. Hossain, M. Hasan, S. Islam, M. Moinuddin, M. T. Islam and M. M. Azad, *J. Agric. Res.*, 2022, 100267.

5. C. Sánchez, *Appl. Microbiol. Biotechnol.*, 2004, **64**, 756.
6. O. Taofiq, F. Rodrigues, L. Barros, M. F. Barreiro, I. C. F. R. Ferreira and M. B. P. P. Oliveira, *Food Chem. Toxicol.*, 2019, **127**, 228.
7. S. M. Badalyan, A. Barkhudaryan and S. Rapior, *Int. J. Med. Mushrooms*, 2022, **24**, 1.
8. M. Usman, G. Murtaza and A. Ditta, Nutritional, Medicinal, and Cosmetic Value of Bioactive Compounds in Button Mushroom (Agaricus bisporus): A review, *Appl. Sci.*, 2021, **11**, 5943.
9. O. Tarawneh, M. Tarawneh and Y. Sharrab, Mushroom classification using machine-learning techniques, 2022, 1.
10. A. F. El Sheikha and D. M. Hu, *Trends Food Sci. Technol.*, 2018, **78**, 292.
11. F. Wu, L. W. Zhou, Z. L. Yang, T. Bau, T. H. Li and Y. C. Dai, *Fungal Diversity*, 2019, **98**, 1.
12. M. Dowlati, H. R. Sobhi, A. Esrafili, M. FarzadKia and M. Yeganeh, *Trends Food Sci. Technol.*, 2021, **109**, 527.
13. J. Patocka, R. Wu, E. Nepovimova, M. Valis, W. Wu and K. Kuca, *Int. J. Mol. Sci.*, 2021, **22**, 1.
14. J. White, S. A. Weinstein, L. De Haro, R. Bédry, A. Schaper, B. H. Rumack and T. Zilker, *Toxicon*, 2019, **157**, 53.
15. M. Q. He, M. Q. Wang, Z. H. Chen, W. Q. Deng, T. H. Li, V. Alfredo, R. Jeewon, K. D. Hyde and R. L. Zhao, *Fungal Biol. Rev.*, 2022, **42**, DOI: 10.1016/j.fbr.2022.06.002.
16. A. D. L. Lima, R. C. Fortes, M. R. C. G. Novaes and S. Percário, *Nutr. Hosp.*, 2012, **27**, 402.
17. W. S. Jo, M. A. Hossain and S. C. Park, *Mycobiology*, 2014, **42**, 215.
18. W. A. Elkhateeb, G. M. Daba, P. W. Thomas and T.-C. Wen, *Egypt. Pharm. J.*, 2019, **18**, 88.
19. L. W. Zhou, M. Ghobad-Nejhad, X. M. Tian, Y. F. Wang and F. Wu, *Food Rev. Int.*, 2022, **38**, 589.
20. A. Ganeshpurkar, G. Rai and A. Jain, *Pharmacogn. Rev.*, 2010, **4**, 127.
21. M. L. Gargano, L. J. L. D. van Griensven, O. S. Isikhuemhen, U. Lindequist, G. Venturella, S. P. Wasser and G. I. Zervakis, *Plant Biosys.*, 2017, **151**, 548.
22. G. Venturella, V. Ferraro, F. Cirlincione and M. L. Gargano, *Int. J. Mol. Sci.*, 2021, **22**, 1.
23. A. Podkowa, A. Kryczyk-Poprawa, W. Opoka and B. Muszyńska, *Eur. Food Res. Technol.*, 2021, **247**, 513.
24. A. González, M. Cruz, C. Losoya, C. Nobre, A. Loredo, R. Rodríguez, J. Contreras and R. Belmares, *Food Funct.*, 2020, **11**, 7400.
25. A. M. Abdelshafy, T. Belwal, Z. Liang, L. Wang, D. Li, Z. Luo and L. Li, *Crit. Rev. Food Sci. Nutr.*, 2022, **62**, 6204.
26. Z. M. Thu, K. Ko Myo, H. T. Aung, M. Clericuzio, C. Armijos and G. Vidari, *Molecules*, 2020, **25**, 1972.
27. H. Kumar, K. Bhardwaj, R. Sharma, E. Nepovimova, N. Cruz-Martins, D. S. Dhanjal, R. Singh, C. Chopra, R. Verma, K. A. Abd-Elsalam, A. Tapwal, K. Musilek, D. Kumar and K. Kuča, *J. Fungi*, 2021, 7, 427.

28. H. Dilfy, M. J. Hanawi, A. W. Al-Bideri and A. T. Jalil, *J. Green Eng.*, 2020, **10**, 6200.
29. G. Waktola and T. Temesgen, *Adv. Biotechnol. Microbiol.*, 2018, **113**, 1.
30. U. T. Dudekula, K. Doriya and S. K. Devarai, *Biotechnology*, 2020, **10**, 1.
31. I. A. Kutaiba, M. S. Mustafa and A. Ugis, *World J. Pharm. Res.*, 2020, **8**, 31.
32. S. H. Dilfy, M. J. Hanawi, A. W. Al-Bideri and A. T. Jalil, *J. Green Eng.*, 2020, **10**, 6200.
33. V. Rezvani, H. R. Pourianfar, S. Mohammadnejad, A. M. Ansari and L. Farahmand, *Appl. Microbiol. Biotechnol.*, 2020, **104**, 6855.
34. A. González, M. Cruz, C. Losoya, C. Nobre, A. Loredo, R. Rodríguez, J. Contreras and R. Belmares, *Food Funct.*, 2020, **11**, 7400.
35. D. J. Seo and C. Choi, *Viruses*, 2021, **13**, 1.
36. N. Landi, A. Clemente, P. V. Pedone, S. Ragucci and A. Di Maro, *Toxins*, 2022, **14**, 84.
37. X. Xu, H. Yan, J. Chen and X. Zhang, *Biotechnol. Adv.*, 2011, **29**, 667.
38. D. Yadav and P. S. Negi, *Int. Food Res. J.*, 2021, **148**, 110599.
39. D. K. Rahi and D. Malik, *J. Mycol.*, 2016, **1**.
40. D. W. A. Abate, *Int. J. Food Process. Technol.*, 2015, **06**(03).
41. K. Orywal, K. Socha, P. Nowakowski, W. Zoń, B. Mroczko and M. Perkowski, *Nutrients*, 2022, **14**, 1.
42. K. Orywal, K. Socha, P. Nowakowski, W. Zoń, P. Kaczyński, B. Mroczko, B. Łozowicka and M. Perkowski, *PLoS One*, 2021, **16**, e0252834.
43. M. Kozarski, A. Klaus, D. Jakovljevic, N. Todorovic, J. Vunduk, P. Petrović, M. Niksic, M. M. Vrvic and L. Van Griensven, *Molecules*, 2015, **20**, 19489.
44. P. Diamantopoulou, S. Papanikolaou, E. Katsarou, M. Komaitis, G. Aggelis and A. Philippoussis, *Appl. Biochem. Biotechnol.*, 2012, **167**, 1890.
45. E. Egwim and R. Elem, *J. Food Nutr. Res.*, 2011, **1**, 89.
46. V. Loria-Kohen, T. Lourenço-Nogueira, I. Espinosa-Salinas, F. R. Marín, C. Soler-Rivas and A. R. de Molina, *J. Pharm. Nutr. Sci.*, 2014, **4**, 187.
47. J. J. Zhang, Y. Li, T. Zhou, D. P. Xu, P. Zhang, S. Li and H. Li Bin, *Molecules*, 2016, **21**, 1.
48. P. Bobek, E. Ginter, M. Jurčovičová and L. Kuniak, *Ann. Nutr. Metab.*, 1991, **35**, 191.
49. O. Taofiq, S. A. Heleno, R. C. Calhelha, M. J. Alves, L. Barros, M. F. Barreiro, A. M. González-Paramás and I. C. Ferreira, *Molecules*, 2016, **21**, 1372.
50. N. Alam, K. N. Yoon, K. R. Lee, P. G. Shin, J. C. Cheong, Y. B. Yoo, J. M. Shim, M. W. Lee, U. Y. Lee and T. S. Lee, *Mycobiology*, 2010, **38**, 295.
51. N. Alam, K. N. Yoon and T. S. Lee, *Afr. J. Biotechnol.*, 2011, **10**, 2978.
52. N. Alam, K. N. Yoon, J. S. Lee, H. J. Cho and T. S. Lee, *Saudi J. Biol. Sci.*, 2012, **19**, 111.
53. I. Mirończuk-Chodakowska, K. Kujawowicz and A. M. Witkowska, *Nutrients*, 2021, **13**, 3960.
54. S. Sangthong, P. Pintathong, P. Pongsua, A. Jirarat and P. Chaiwut, *J. Fungi*, 2022, **8**, 572.

55. G. Anusiya, U. G. Prabu, N. V. Yamini, N. Sivarajasekar, K. Rambabu, G. Bharath and F. Banat, *Bioengineered*, 2021, **12**, 11239.
56. S. M. Badalyan, A. L. Barkhudaryan and N. G. Gharibyan, *Proc. YSU B: Chem. Biol. Sci.*, 2022, **56**, 74.
57. K. Sujarit, N. Suwannarach, J. Kumla and T. Lomthong, *Chiang Mai J. Sci.*, 2021, **48**, 699.
58. H. A. El Enshasy and R. Hatti-Kaul, *Trends Biotechnol.*, 2013, **31**, 668.
59. Y. Wu, M. H. Choi, J. Li, H. Yang and H. J. Shin, *Cosmetics*, 2016, **3**, 1.
60. S. Visvanathan, R. Krishnamoorthy and G. S. Sabesan, Fungal Cosmetics: Mushrooms in Beauty Care and the New Age of Natural Cosmetics, in *Applied Mycology*, Springer, Cham, 2022, pp. 1–37.
61. A. Luchowska, M. Sroczyńska and A. Żaczek, *J. Educ. Health Sport*, 2023, **13**, 172.
62. M. P. Lupo and A. L. Cole, Cosmeceutical peptides, *Dermatol. Ther.*, 2007, **20**, 343.
63. T. K. Nguyen, D. B. Shin, K. R. Lee, P. G. Shin, J. C. Cheong, Y. B. Yoo and T. S. Lee, *J. Mushroom*, 2013, **11**, 269.
64. L. N. Fitria, M. A. Jabar, N. Dzulfiana, S. A. W. Cahyati and N. Yuniarsih, *Arch. Clin. Med. Case Rep.*, 2022, **3**, 293.
65. K. Zhang, X. Y. Meng, Y. Sun and P. Y. Guo, *Deterg. Cosmet.*, 2013, **36**, 28.
66. W. Zhao, A. Yang, J. Wang, D. Huang, Y. Deng, X. Zhang and C. Huang, *J. Cosmet. Dermatol.*, 2022, **21**, 6669.
67. S. Parvez, M. Kang, H. S. Chung and H. Bae, *Phytother. Res.*, 2007, **21**, 805.
68. T. S. Chang, *Int. J. Mol. Sci.*, 2009, **10**, 2440.
69. J. S. Lee, D. B. Shin, S. M. Lee, S. H. Kim, T. S. Lee and D. C. Jung, *Korean J. Mycol.*, 2013, **41**, 104.
70. C. C. Chien, M. L. Tsai, C. C. Chen, S. J. Chang and C. H. Tseng, *Mycopathologia*, 2008, **166**, 117.
71. U. S. Abelan, A. C. de Oliveira, É. S. P. Cacoci, T. E. A. Martins, V. M. Giacon, M. V. R. Velasco and C. R. R. D. C. Lima, *J. Cosmet. Dermatol.*, 2022, **21**, 1407.
72. Y. Wu, M. H. Choi, J. Li, H. Yang and H. J. Shin, *Cosmetics*, 2016, **3**, 1.
73. B. R. Shete and P. N. Phule, *ECS Trans.*, 2022, **107**, 5459.
74. S. De Baets and E. J. Vandamme, *Biotechnol. Lett.*, 2001, **23**, 1361.
75. K. U. S. Meehan, *US pat. and Trademark Office*, 9144542, Washington, DC, 2015.
76. S. M. Badalyan, A. Barkhudaryan and S. Rapior, Medicinal Macrofungi as Cosmeceuticals: a Review, *Int. J. Med. Mushrooms*, 2022, **24**(4).

CHAPTER 7

# *Algal and Microalgal Compounds in Cosmeceuticals*

TRIDIP BORUAH[a*], BARSHA DEVI[b], NIBEDITA GOGOI[c] AND CHIRANJIB MILI[d]

[a]P.G. Department of Botany, Madhab Choudhury College, Barpeta, Assam 781301, India; [b]Department of Botany, PDUAM, Tulungia, Assam 783383, India; [c]Department of Chemistry, Madhab Choudhury College, Barpeta, Assam 781301, India; [d]Department of Botany, B.P. Chaliha College, Nagarbera, Assam 781127, India
*E-mail: tridip_boruah@gauhati.ac.in

## 7.1 Introduction

Beauty that pleasures the senses is possibly the desire of every human being on Earth. To achieve their desire for beauty, various types of cosmetics of synthetic origin are being extensively used by people to shield their skin from exterior stimuli. However, synthetic cosmetics have limitations such as ineffectiveness, minimal absorption rates, and toxic side effects.[1] For example, esters of hydroxybenzoic acids (parabens) are extensively used in cosmetic products and are reported to imitate estrogen, leading to the risk of malignant melanoma as well as breast cancer.[2] Moreover, undesired irritation of synthetic cosmetics on the skin is due to the presence of large size molecular compounds. In 2000, Bos and Marcus reported that compounds with a molecular weight of no more than 500 Dalton could infiltrate through the skin.[3] A topical immunosuppressant – cyclosporin, with a molecular weight of 1202 Da – was found to be ineffective against allergic contact dermatitis and psoriasis due to its high molecular weight that suppresses penetration

Bioprospecting of Natural Sources for Cosmeceuticals
Edited by Deepika Kathuria, Ajay Sharma, Meenakshi Verma and Gulzar Ahmad Nayik

Published by the Royal Society of Chemistry, www.rsc.org

of the skin. At the same time, it was efficient against psoriasis when directly injected into the skin, but with a few side effects, such as allergic reaction and irritation. Apart from skin cosmetics, nail care products are also hazardous to the human body. Nail polish remover can cause nail dryness, distal onychoschizia and irritation in the perionychial tissues, and produce onycholysis.[4] Moreover, formaldehyde-containing nail hardeners can cause paronychia, onycholysis, subungual hyperkeratosis, and dryness of the fingertips.[5] In addition, phthalate, a constituent of plastic that is present in cosmetics, can cause DNA damage to human sperm.[6] Therefore, people have changed their preferences and moved towards natural products, for which the demand has increased in recent times.

Algae possess a wide range of bioactive compounds that contribute to products such as pharmaceuticals, cosmetics, and nutritious feeds, which are beneficial for human health.[7] For instance, astaxanthin, a high-value carotenoid extracted from the biomass of microalga (*Haematococcus pluvialis* and *Chlorella zofingiensis*), is regarded as a potential antioxidant candidate due to its free radical scavenging properies.[8] Some bioactive compounds, such as fucoidan, phlorotannin and fucoxanthin from brown algae, possess several activities, such as antioxidant, anti-inflammation, anti-melanogenic, anti-cancer, and anti-aging.[9–12] Likewise, mycosporine-like amino acids or MAAs are a common constituent recorded in green and red algae and also have anti-inflammation, antioxidant, and anti-aging activities.[13] Apart from microalgae, the significance of marine macroalgae in cosmeceutical formulations is also non-negligible. They contain various bioactive compounds, including amino acids, proteins, carbohydrate derivatives, polysaccharides, vitamins, phenolic compounds, fatty acids, minerals, and pigments.[14] For example, the polysaccharide fucoidan from *Chnoospora minima* has been recorded to have anti-inflammation activity by a group of researchers led by Fernando *et al.*[15] Moreover, many polysaccharides such as alginate, fucoidan, ulvan, carrageenan, laminarin, agarose, sulfated polysaccharides, and agar derived from different macroalgae are also reported as antioxidant and anti-inflammation, anti-aging, anti-wrinkle, and moisturizing agents.[16–19] Furthermore, macroalgae act as a source of amino acids (valine, alanine, glycine, proline, arginine, leucine, histidine, serine, tyrosine), phenolic compounds (bromophenols, phlorotannins, flavonoids, terpenoids, *etc.*),[14] fatty acids (glycolipids, triglycerides, sterols, and phospholipids) and minerals, including chlorine, potassium, phosphorous, zinc, silica, copper, selenium, iodine, manganese, and iron, which are essential for the making of cosmetic products.[20,21] Applications of algae-derived compounds in the cosmeceutical industry have been growing in recent times. Several industrial outlets, such as Fuji Chemical Seambiotic, Cyanotech, and Mera Pharmaceuticals, have been producing biomass of microalgal origin contributing to products in nutritious feeds, cosmetics, and pharmaceuticals.[7] Notably, phycocyanin and phycoerythrin found in cyanobacteria and red algae are recognized color additives for cosmetics and food products by the Food and Drug Administration (FDA).[22] Moreover, many leading cosmeceutical companies, such as Beiersdorf, L'Oreal, Unilever, and Henkel, are projected to pick up and expand the market value of carotenoids in the European market and they are likely to accomplish about $1.53 billion market value by 2025.[23] In addition, *Gracilaria* is a species of red algae extensively used in the cosmetics industry due to its gelling, thickening and stabilizing characteristics. Some products from red algae *Gracilaria*, including a hydrating cream by Thalasso®, a facial mask by Balinique Cosmetic, and a hydrogel soap by Sealaria™, have been widely marketed in recent times.[24] Furthermore, various skin products of algal origin, such as Algenist, which is a popular anti-aging moisturizer, Helionori® by Gelyma, Helioguard™ 365 (a sunscreen product), Monaco (an anti-aging agent) and Protulines® by Exsymol S.A.M., Monaco and Dermochlorella D by Codif Technologie Naturelle (an anti-wrinkling agent) have also been marketed.[25] This chapter is an attempt to organize all the authenticated information regarding algal-based cosmeceuticals in terms of their secondary metabolites, bioactive compounds, pharmacological potential, and market products in a single platform to providing necessary direction to future researchers in this emerging field of study.

## 7.2 Cultural and Historical Background

There have been numerous alarms raised recently over the use of cosmetics. The fact that many substances used in these products are absorbed through the skin by the body makes them hazardous to the environment, as well as to the human body. The majority of cosmetics and beauty goods are made from complex chemical mixtures harming our skin and the environment. Synthetic chemicals, microbeads and plastic packaging used by the beauty and wellness industry are endangering marine and aquatic life. According to recent studies, some sunscreen ingredients have been found to damage coral reefs, causing them to bleach.[26] The reasons why switching to a natural cosmetic line is something we should think about are: (a) the base of the natural cosmetics is "water-based" compared to "oils" in synthetic cosmetics; (b) many cosmetics companies test their products on lab animals to ensure the safety of the products before releasing them to customers.[27] It is common for beauty product manufacturers to use animals for testing their ingredients and products for possible reactions and irritation. This animal testing is painful, torturous, inhumane, and usually results in death. Synthetic cosmetics contain petrochemicals and parabens, which extend their shelf lives. It has been reported that parabens can cause a number of diseases, such as cancer, a weakened immune system, and allergies.[28] Plankton populations are decreased by chemicals such as parabens, butylated hydroxytoluene, and butylated hydroxyanisole used in cosmetics. Deforestation and the loss of wildlife habitats are consequences of producing cosmetic ingredients. The ability of resources to be replenished, whether naturally or through recycling, is a key component of sustainability. Sustainable beauty brands take that renewability into account. They use organic components that are sustainably obtained and sustainable materials to create the most sustainable cosmetics.

## 7.3 Use of Algal Secondary Metabolites in Cosmeceuticals

Macroalgae, also known as seaweed, are eukaryotic, macroscopic multicellular algae. They are a rich source of vital nutrients such as essential amino acids, proteins, vitamins, pigments, minerals, and long-chain polyunsaturated fatty acids. These ingredients are utilized in the formulations of cosmetics either as additives, excipients, or active substances. Among all the cultivable macroalgae, a total of close to 59% are brown algae, approximately 40% of all algae are red algae, and less than 1% are green algae.[29] Some species of brown macroalgae used in cosmetics are *Fucus vesiculosus*, *Postelsia palmaeformis*, *Laminaria digitata*, *Pelvetia canaliculate*, *Ascophyllum nodosum*, *Undaria pinnatifida*, *Sargassum muticum*, *Cystoseira nodicaulis*, *Costaria costata*, *Laminaria* sp., *Sargassum* sp., *Pelvetia wrightii*, *Halopteris scoparia*, *Ishige foliacea*, and *Padina* sp. Fucoxanthin, a pigment derived from brown macroalgae, has an anti-tyrosinase activity that lessens or controls skin pigmentation and promotes the production of collagen, which slows down the natural aging process of the skin.[9] Due to its affluence in mannitol, amino acids, and alginate, *Laminaria digitata* has numerous skin benefits (see Table 7.1). Frequently used red macroalgae in the cosmetics industry are *Palmaria palmata*, *Chondrus crispus*, *Gelidium* sp., *Laurencia pacifica*, *Porphyra umbilicalis*, *Rhodella* spp., *Porphyra dentata*, *Porphyra haitanensis*, *Acanthophora muscoides*, *Chondria armata*, *Asparagopsis armata*, *Corallina officinalis*, and *Bangia fuscopurpurea*. *C. crispus*, which is rich in omega-6 fatty acids, phenols, carrageenan, and omega-3 fatty acids, acts as a protective colloid to the skin while preserving the skin's natural defensive film. 3-Sitosterol, ergosterol, fucosterol and brassicasterol from *P. dentata* act as scavengers of reactive oxygen species (ROS).[30] Green seaweeds in cosmetics are Ulva sp., *Haematococcus pluvialis*, *Cladophora glomerata*, *Codium tomentosum*, *Ulva rigida*, *Ulva compressa*, *Enteromorpha linza*, and *Rhizoclonium hieroglyphicum*. Polysaccharides from *C. tomentosum* prevent water loss from the skin.[30] *U. compressa* protects delicate skin and has moisturizing characteristics.[30] Table 7.1 summarizes the major classes of macroalgae and their secondary metabolites along with their bioactive compounds used in cosmetics.[30–48]

## 7.4 Use of Microalgal Secondary Metabolites in Cosmeceuticals

Only a small number of microalgal species have been investigated and employed in the cosmeceutical industry, including *Dunaliella salina*, *Nannochloropsis salina*, *Spirulina platensis*, *Chlorella* sp., *Chlamydomonas hedleyi*, *Tetraselmiss uecica*, *Botryococcus braunii*, *Isochrysis*, *Nannochloropsis oculate*, *Nannochloropsis gaditana*, *Porphyridium cruentum*, *Rhodella violacea*, and *Rhodosorus marinus*. *Chlorella* sp. is rich in polysaccharides, palmitic acid, palmitoleic acid, lutein, and zeaxanthin and act as an antioxidant,

**Table 7.1** Major classes of algae with their key secondary metabolites and bioactive compounds used in cosmetics.

| Class | Algae name | Common use or application of extract | Key bioactive compounds | Reference |
|---|---|---|---|---|
| Brown algae | *Fucus vesiculosus* | Emollient, skin conditioning, smoothing, anti-aging, skin brightening | Fucoidan, chlorophyll c, fucoxanthin | 30 |
| | *Postelsia palmaeformis* | Skin conditioning, antiwrinkle, skin moisturizing, skin softening, anti-inflammatory | Fucoxanthin, chlorophyll c | 30 |
| | *Laminaria digitata* | Protective, skin moisturizing, exfoliant, shower gel, lipolytic | Mannitol, alginate, amino acid, iodine | 30 |
| | *Pelvetia canaliculata* | Antioxidant, soothing, hydrating | Polyphenols, carotenoids, amino acids | 30 |
| | *Ascophyllum nodosum* | Skin conditioning, anti-aging, smoothing, antiwrinkle, skin moisturization, anticellulite, face creams | Fucoxanthin, alginates, acid hydrolyzed fucoidan, phlorotannins | 31 |
| | *Undaria pinnatifida* | Anti-pollution protection, skin conditioning, enhancing skin immunity | Fucoidan | 31 |
| | *Sargassum muticum* | Skin hydration, detoxifying recovery creams, oxygen boosters | Tetraprenyltoluquinol chromane meroterpenoid | 30 |
| | *Cystoseira nodic aulis* | Anti-wrinkle, antioxidant, inhibition of lipid peroxidation, anti-inflammatory | Phlorotannins | 32 |
| | *Laminara japonica* | Skin moisturization, antioxidant, antiphotoaging agents, anti-tyrosinase activity, anti-aging, anti-inflammatory | Fucoxanthin, carotenoids, polysaccharides | 31 and 33 |
| | *Halopteris scoparia* | Anti-wrinkle, anti-aging | Phytohormones | 34 |
| | *Ishige foliacea* | Skin whitening, tyrosinase inhibition | Octaphlorethol A (phlorotannin) | 35 |

| | | | | |
|---|---|---|---|---|
| Red algae | *Palmaria palmata* | Stimulates cutaneous microcirculation, reduces varicose veins by inhibiting the development of fresh blood vessels, anti-UV, natural sunscreens | Vitamins A and B12, MAAs, amino acids | 36 and 37 |
| | *Chondrus crispus* | Protective colloids, protect the skin's innate defensive film, emollient, thickeners, gelling agents, bath salt, cleansing gel, moisturizing cream, body mask, shampoo, body exfoliating soaps, make-up removers | Fatty acids (both omega-6 and omega-3), phenol, carrageenan | 33 and 38 |
| | *Gelidium* sp. | Gelling agents, emulsion stabilizers, thickeners | Agar | 39 |
| | *Porphyra umbilicalis* | Anti-UVA, skin-conditioning agent, sunscreens | Fatty acids, proteins, α-linolenic acid | 38 and 40 |
| | *Porphyra dentata* | Radical scavengers, anti-inflammatory agents, antioxidant, antiallergic | 3-Sitosterol, ergosterol, fucosterol, brassicasterol | 40 |
| | *Acanthophora muscoides* | Gelling agents, anticoagulants, anti-inflammatory | Carrageenan | 41 and 42 |
| | *Corallina officinalis* | Sun protection, facial cleansing, baby care products | Sulfated polysaccharides | 43 |
| Green algae | *Ulva lactuca* | Skin moisturization, collagen synthesis, anti-inflammatory, gelling, antioxidant, anti-wrinkle, emollient, anti-elastase | Fatty acids (linoleic, linolenic, oleic acid), carotenoids (lutein, carotene, fucoxanthin, astaxanthin) Sulfated polysaccharides, chlorophylls | 44 and 45 |
| | *Haematococcus pluvialis* | Tyrosinase inhibitors, radical scavengers, antioxidant, anti-aging, colorant | Carotenoids (astaxanthin) | 46 |
| | *Cladophora glomerata* | Anti-acne effect, anti-allergic preparations, purifying gels, make-up cosmetics, anti-dandruff shampoos | Linoleic acid, β-sitosterol, β-carotene, alginic acid, ascorbic acid, gallic acid | 47 |
| | *Codium tomentosum* | Moisturizing, face masks, hydric regulator | Glucuronic acid, β-hydroxy acid, sulfated polysaccharides, proteins, minerals | 33 |
| | *Ulva rigida* | Chelating agent, gelling, moisturizing | Ulvan (sulfated polysaccharide) | 44 |
| | *Ulva compressa* | Reducing hyperreactivity, hydration, exfoliating gels, cleansing creams | Magnesium | 30 |
| | *Enteromorpha linza* | Radical scavenging effects | Sulfated polysaccharides | 48 |

anti-inflammatory, de-pigmenting, and collagen repair agent.[49] *Spirulina*, a freshwater alga, is utilized in cosmetics as a refreshing grain exfoliant.[30] Extracts prepared from microalgae are a rich source of vitamins (A, B12, C, E, biotin, and folic acid), iron, calcium, phosphorous, pantothenic acid, beta-carotene, astaxanthin, and polyunsaturated fatty acids.[50] Astaxanthin from *Haematococcus pluvialis*, and beta-carotenes from *Dunaliella salina* are potent antioxidants.[51] Table 7.2 summarizes a list of the major microalgae and their bioactive compounds used in cosmetics.[51–57]

## 7.5 Traditional Uses of Bioactive Cosmeceuticals

Traditionally, it has been found that most of the natural bioactive substances used in contemporary cosmetics production are extracted from algae. Algae, both macroalgae and microalgae, produce two types of metabolites, *viz.*, primary metabolites such as polysaccharides, amino acids, proteins and fatty acids, and secondary metabolites, including pigments, phenols, vitamins, bioactive compounds and sterols.[58] Polysaccharides of algal origin help to regulate the distribution of water in the skin. In cosmetics, these are mainly used as antioxidant, stabilizing, gelling and moisturizing agents, and also as an alternative source for lightweight oils, such as acetyl alcohol, or silicone-derived ingredients. Polysaccharides such as chitin, agar, laminarin and alginates, and sulphated polysaccharides such as fucoidan, ulvans and carrageenan are mainly obtained from diverse building blocks, forming elongated carbohydrate molecules of monosaccharide units united together by glycosidic bonds. These are non-toxic, water soluble and hydrophilic, and they have a somewhat standard structure.[59,60] Algae are also sources of various phenolic compounds, including simple phenols with a single aromatic ring, such as gallic acid, pyrocatechol *etc.*, and polyphenols with two or more aromatic rings, like usnic acid, hypericin *etc.* The main characteristic features in these bioactive compounds are the occurrence of two or more hydroxyl groups in the aromatic ring. These phenolic complexes exhibit various activities, including antiradical, anti-inflammatory, anti-allergic, antibiotic, antioxidant, antifungal, antiviral, and ultraviolet (UV) protection functions, as they have scavenging activities for reactive oxygen species and prevent the generation of free radicals.[61] Phlorotannins are an essential group of polyphenols among the other polyphenols like bromophenols, terpenoids, and phenolic pigments for the cosmetics industry. Fucophlorethol, tetrafucols, fuhalols, eckols, phlorethols, carmalols *etc.* are a group of phlorotannins that are oligomers of phloroglucinol and are mainly extracted from brown algae. Bromophenols are obtained from red algae, while phenolic terpenoids such as plastoquinones, chromanols and chromenes occur mainly in brown algae.[62,63] Algae may produce glyco- and phospholipids and a broad range of phytosterols, such as stigmasterol, brassicasterol, fucosterol, sitosterol, and ergosterol. They also contain either unsaturated fatty acids or saturated fatty acids such as omega-9, omega-7, omega-6 (*e.g.*, γ-linoleic acid, linoleic acid, arachidonic acid), omega-3 (*e.g.*, eicosapentaenoic acid, docosahexaenoic acid, α-linoleic acid) *etc.* They exhibit properties

**Table 7.2** Major classes of microalgae with key secondary metabolites and bioactive compounds.

| Class | Name of algae | Common use or application of extract | Key bioactive compounds | Reference |
|---|---|---|---|---|
| Green microalgae | *Dunaliella salina* | Radical scavenger, antiphotoaging, smoothing agent | Palmitic acid, β-cryptoxanthin, linolenic acid, β-carotene, fucoxanthin, astaxanthin, lutein | 51 |
| | *Nannochloropsis salina* | Tanning cosmetics | Canthaxanthin | 49 |
| | *Spirulina platensis* | Nourishing agents, coloring pigments, shining agents | Phycocyanin | 52 |
| | *Chlorella vulgaris* | De-pigmentation, collagen repair | Polysaccharides, palmitic acid, palmitoleic acid | 49 |
| | *Chlorella sorokiniana* | Photoprotection | Lutein | 52 |
| | *Chlorella saccharophila* | Antioxidant, anti-inflammatory | Zeaxanthin | 52 |
| | *Chlamydomonas hedleyi* | Photoprotection | MAAs | 52 |
| | *Tetraselmiss uecica* | Skin protection, antioxidant | α-Tocopherol, stearic acid, palmitic acid | 54 |
| | *Botryococcus braunii* | Antioxidant | Palmitic acid, stearic acid | 54 |
| Brown microalgae | *Isochrysis* | Anti-irritant, sunscreen, soothing agent | Fatty acids (oleic acid, mistiric acid), pigments (fucoxanthin, canthaxanthin) | 30 |
| | *Nannochloropsis oculata* | Skin tightening properties, moisturizing agent | Polyunsaturated fatty acids (linolenic acid or omega 6), vitamin B12, and vitamin C | 55 |
| | *Nannochloropsis gaditana* | Prevents early skin aging, protects skin against oxidative stress | Astaxanthin, violaxanthin, canthaxanthin, zeaxanthin, β-carotene | 56 |
| Red microalgae | *Porphyridium cruentum* | Radical scavenger, prevents premature aging, hydrates skin | Sulfated polysaccharides, phycoerythrin, arachidonic acid, carotenoids, tocopherol | 57 |
| | *Rhodella violacea* | Moisturizes and softens skin | Detoxondria | 57 |
| | *Rhodosorus marinus* | Reduces skin sensitivity to external aggravation | Carotenoids | 57 |

like antiallergic, anti-aging, anti-wrinkle, anti-inflammatory, and antioxidant activities. Furthermore, they also act as emollients, *i.e.*, they can soften and smoothen the skin and also protect them from excessive loss of water and exfoliation of the epidermis.[64–68] Furthermore, algae may also contain some protein and derivatives of protein, including peptides, preferably enzymes, and open amino acids like serine, alanine, proline, histidine, tryptophan, glutamic acid, taurine, tyrosine and MAAs (see Figure 7.1). These bioactive components also possess antioxidant, anticancer, anti-aging and anti-inflammatory activities. Algal proteins are also capable of acting as moisturizers of the hair and skin, which makes them more attractive in the cosmetics industry. Moreover, the MAAs absorb UV radiation in the range 310–365 nm, and therefore they are utilized as innate sunscreens in cosmeceuticals.[69] In addition to the above bioactive metabolites, algae are also a resource of different pigments, such as beta-carotene, carotenoids, chlorophyll a, b, c and d, allophycocyanin, phycoerythrin, phycocyanin, phycobiliproteins, fucoxanthin, lutein, and astaxanthin, and vitamins, like vitamin C, vitamin A, B vitamins (B12, B6, B5, B3, B2, and B1), vitamin E ($\alpha$-tocopherol) *etc.*[70–72]

## 7.6 Methods of Instrumental Extraction of Algal-based Raw Materials (Biometabolites) for Cosmeceuticals

The nature of the extraction technology varies depending upon the temperature, pressure and nature of the solvents used. Since different bioactive compounds have different chemical stability, solubility and polarity, the temperature, pressure and solvent used for their extraction will also be different.[73,74] For example, extraction of proteins, polysaccharides, and other water-soluble bioactive compounds can be carried out using water and other polar solvents. On the other hand, fatty acids, phospholipids, steroids and other lipophilic compounds can be extracted using nonpolar solvents. In recent years, various extraction methods were employed for the extraction of active constituents from algae and this is emerging as an important and popular area of research.[75–77] The methods of extraction of biometabolites can be categorized into two major types. These are: (i) conventional methods such as maceration, classical solvent extraction, Soxhlet extraction, saponification, hydrodistillation *etc.* and (ii) novel techniques such as microwave-assisted extraction (MAE), enzyme-assisted extraction (EAE), subcritical water extraction (SWE), supercritical fluid extraction (SFE), ultrasound-assisted extraction (UAE), pressurized liquid extraction (PLE), *etc.*[65] These methods are briefly described below.

### 7.6.1 Conventional Methods

Among various conventional methods, the maceration method is one of the most popular techniques.[78] It is a multistep process for miniature scale extraction, in which the sample is first ground into minute pieces to enlarge

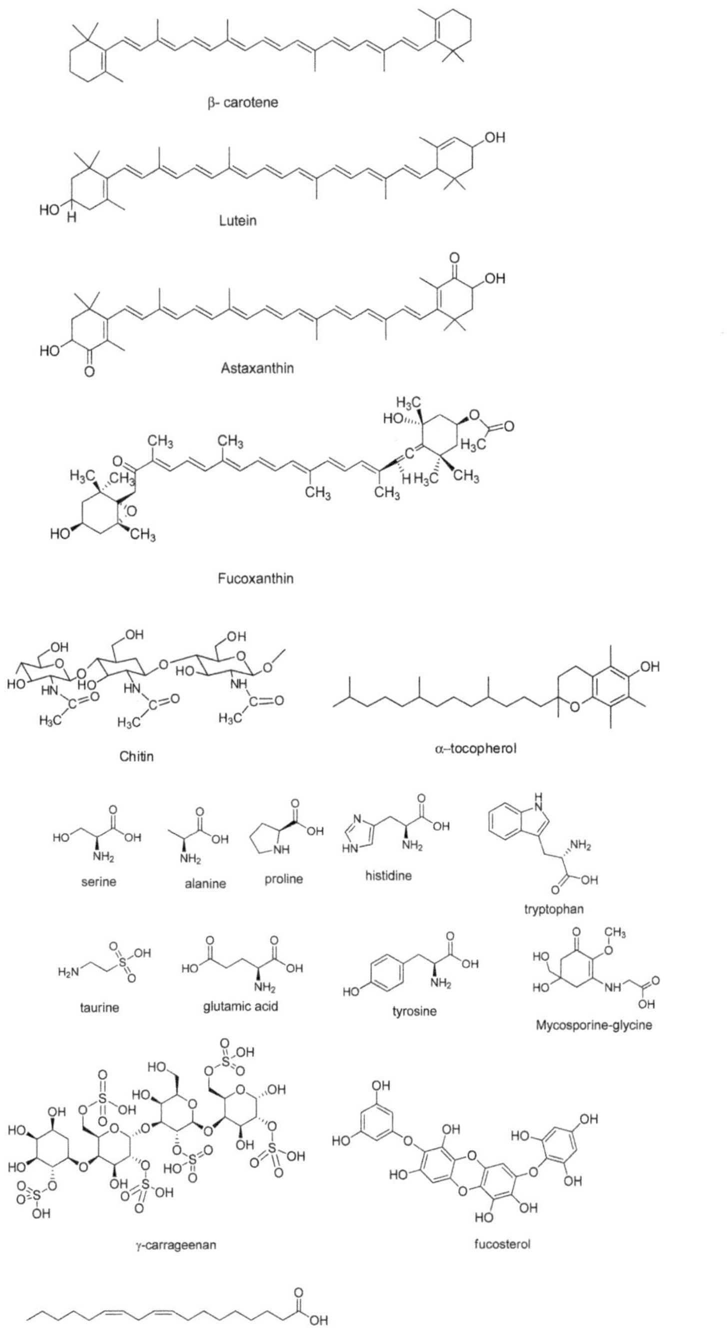


**Figure 7.1** Chemical structures of bioactive compounds extracted from algae.

the surface area for suitable amalgamation with the solvent. Then a suitable solvent (menstruum) is supplemented in a closed container followed by the straining off of the liquid. The next step of this process is carried out to eliminate all the solid and liquid impurities through a rapid filtration process. In the final step, intermittent shaking during maceration smoots the progress of the extraction process by escalating diffusion and removing the concentrated solution from the surface of the sample to bring fresh solvent to the menstruum for improved extraction yield.[79]

Solid–liquid extraction, liquid–liquid extraction, or extraction employing a Soxhlet apparatus are some other widely used extraction techniques. The extraction method by employing Soxhlet has widely been utilized to extort a wide range of bioactive compounds available from natural sources. In this process, a thimble is used to contain a small amount of dry sample, followed by placing it in the distillation unit with a particular solvent. When the solution in the thimble holder is near the overflow level, it is aspirated by a siphon. Then, the solution is unloaded back into the distillation unit, which helps it to dissolve into the bulk liquid. Finally, the required solute is left in the distillation flask and the solvent leaves the flask for the solid bed of the plant. This process is repeated several times to complete the extraction.[79]

Another important conventional technique for the drawing out of essential oils and bioactive compounds from plants is hydrodistillation, of which there is three types *viz.*, direct steam distillation, water and steam distillation, and water distillation. In this method, a sufficient amount of water is added to the plant materials already packed in a still compartment and boiled. This process may also be carried out by injecting steam directly into the sample containing the plant material. This initiates the condensation stage of the vapor mixture containing oil and water and allows it to move out to a separator from the condenser, where the separation process of the bioactive compounds and oils from water will be initiated.[80] Since high extraction temperature may be the reason for the destruction of a few volatile compounds, this method cannot be used for the extraction of thermo-labile compounds.[79] In addition, this conservative extraction technique has some shortcomings, such as low efficiency, large consumption of solvents, and long processing times, meaning these methods are less practical to attain valuable commodities from algae.

### 7.6.2 Novel Extraction Methods

There are a lot of novel extraction methods used for the extraction of algal-based biometabolites, some of which are briefly discussed in the following sections.

#### *7.6.2.1 Enzyme-assisted Extraction (EAE)*

In this method, various hydrolytic enzymes, including xylanase, proteases, α-amylase, cellulase, pectinase, glycosidases, carbohydrases, peptidases, *etc.*, have been used. These enzymes are capable of degrading the cell walls of algae with the assistance of water that acts as a solvent. The EAE method concerns the mining of antioxidants and other polyphenol ingredients,[81] along with bioactive polysaccharides.[82] However, the algal cell walls of some species are composed of glycoproteins and a range of polysaccharides, such as carrageenan, mannans, xylans, starch, cellulose, and pectin that can influence the mining of algal-based active phytochemicals.[83] Therefore, selection of proper enzymes with hydrolytic capabilities for extracting intracellular bioactive compounds from algae is very important, such that the selected enzyme can degrade the cell walls and thereby facilitate the discharge of the compounds with bioactivity. The EAE method has several advantages, such as a shorter processing time, higher extraction yield, and less contamination[81] and energy consumption in comparison to the other methods.[82] However, the suitable optimization of the conditions prevailing in the extraction period, including enzyme concentration and composition, hydrolysis time, pH, temperature, ratio of algal material to water, and specific particle size of the algal material, is very important in order to obtain a good extraction yield in the EAE method.[84] There are also some limitations of this method such as the substrate specificity and cost of enzymes, and the complexity in maintaining the ambient conditions of the bioreactor,[85] which can be overcome by adopting hyphenated extraction technologies.[65]

#### *7.6.2.2 Microwave-assisted Extraction (MAE)*

Since microwave energy is absorbed by polar bioactive molecules, it can be used in the extraction process. Polar solvents such as ethyl acetate, acetonitrile, ethanol, water and methanol are generally used in this method.[86] Although both open and closed systems of bioactive compound extraction techniques are conducted to get the best out of the procedure, the closed system technique is found to be most popular in the cosmetics industry due to its high efficiency. In the case of the closed system, Teflon vessels are used to place the samples with solvents in some special microwave ovens. The MAE method provides rapid and efficient extraction even with small amounts of sample and solvent, which makes it a green technology in comparison to the others. MAE is widely employed for the extraction of diverse active components from algae, both the nonpolar ones and polar ones, including fucoidans such as phytol and phytosterols.[87] Several reports have also been found for the adoption of MAE for extraction of polysaccharides from *Fucus vesiculosus* and polyphenols from *Enteromorpha prolifera*.[87,88] It has been found that the use of pressure, temperature and high power may lead to the degradation and reduction of the acquiesce of compounds with phenolic properties. Similarly, the use of low pressure may minimize the obliteration of the algal cell structure, but it results in a lower yield of carbohydrates, especially polysaccharides, from the cells, and the use of excessive time can augment the yield, but it can cause the degradation of polysaccharides. Since the utilization of toxic chemicals can be minimized through this method, it is a preferable technique for the extraction of ulvan.[65]

### 7.6.2.3 Supercritical Fluid Extraction (SFE)

This method is very useful for the extraction of compounds of non-polar nature, including carotenoids, tocopherols, triglycerides, phytosterols, phenolic, and fatty acids. Various reports have been found about the employment of this method for the extraction of carotenoids and phenolic compounds from *Gracilaria mammillaris*, *Chara fragilis*, *Ulva flexuosa*, and *Cladophora glomerata*.[89] This technique uses supercritical state fluids since such fluid exhibit the explicit properties of liquids and/or gasses like solvent power, density, surface tension, viscosity, and diffusion, which supply an appropriate environment for the extraction of bioactive complexes with elevated yields in a small amount of time. Due to low cost and toxicity, supercritical $CO_2$ is used as a solvent. Furthermore, it is inexpensive, eco-friendly, and effortless to isolate from the extract. Due to its ability to function in nonoxidant conditions and at low temperature, it is very helpful in the extraction of easily oxidized or thermally labile compounds. This extraction methodology is also effective against the extraction of bioactive metabolites of non-polar origin due to its very low polarity values. However, ethanol is effective as a co-solvent to boost the extraction effectiveness of carotenoids and phenolic compounds as compared to the solitary utilization of supercritical $CO_2$.[65] This method has many advantages, such as it is faster, more efficient and eco-friendly in comparison to the other methods, and it is also a discerning method appropriate for getting bioactive compounds that are thermally labile, along with straightforward severance of the solvent from the solute, and low consumption of organic solvents. This method has a wide range of applicability, such as in the extraction of chlorophylls and carotenoids from various classes of algae compared to the UAE method, fatty acids and carotenoids from *Chlorella vulgaris*, $\gamma$-linolenic acid and vitamin E from *Spirulina platensis*, volatiles from *Dictyopteris membranacea*, and lutein from *Chlorella pyrenoidosa*, as well as for an extensive array of other phytochemicals.[89,90]

### 7.6.2.4 Ultrasound-assisted Extraction (UAE)

UAE is a new technique employed for the extraction of algal bioactive metabolites, mainly phenolic compounds, and is considered a simple and inexpensive technique. Firstly, the sample coming from an algal thallus portion is allowed to mix with a solvent in the UAE method, followed by treatment in an ultrasonic cleaning bath that uses ultrasound waves. Various factors, such as the power of an ultrasonic bath, time, and temperature, affect the efficiency of the UAE method. Although the yield increases with increasing temperature, the extraction time must be optimized in order to evade the deformation of the phenolic compounds. It has been observed that when two extraction techniques, like ultrasonication and maceration, are amalgamated, maximum acquiesce is obtained as well as increasing the effectiveness of the extraction procedure of phycobiliproteins from *Gelidium pusillum*.[91] The UAE method has several advantages, such as higher extraction yield, faster energy transfer and also a decline in the time required for extraction, temperature, energy and process steps, because of which it is extremely useful for obtaining plant-based bioactive compounds with a heat labile nature. Furthermore, UAE may be integrated with a variety of extraction methodologies, such as Soxhlet extraction or maceration, using diverse solvents and thereby the efficiency of conventional systems can be increased. As a result, the UAE technique is very relevant and practical in the cosmetics industry in the extraction of algae extracts. For example, the UAE technique can be used for the isolation of β-carotene from the bacterial species *Streptomyces platensis* by using *n*-heptane and for lipid extraction from *Chlorella vulgaris* employing solvents like ionic liquids, which exhibited advanced effectiveness compared to techniques performed using organic solvents.[78]

### *7.6.2.5 Pressurized Liquid Extraction (PLE)*

This technique is extensively used as an environmentally friendly, fast, and efficient means for extracting bioactive compounds, such as antioxidants, carotenoids, fatty acids, and lipids of algal origin. This method is also known by some other terms, like enhanced solvent extraction (ESE), high pressure solvent extraction (HSPE), pressurized fluid extraction (PFE), and accelerated fluid extraction (ASE). Various solvents, including hexane, water, petroleum ether, and ethanol, are used in this extraction method and it is carried out at high temperature and pressure so that the solvent stays in the liquid state even after reaching their normal boiling point.[79] Furthermore, the combination of high pressure and temperature in this method facilitates a faster extraction process. The increase of the mass transfer rate and solubility, as well as decrease of the surface tension and viscosity of the solvents with increasing temperature make this method more efficient than the others. In addition, due to the requirement of a very small amount of solvent in the process, this method has become a very attractive and popular extraction method for researchers in recent years.[92]

### *7.6.2.6 Subcritical Water Extraction (SWE)*

The SWE method is considered the most promising and eco-friendly method in bioactive metabolite extraction from algae. In this case, water is added as a solvent instead of other organic solvents. It operates at pressures between 50 and 300 psi for 5 to 10 min and at high temperatures in the range 50–200 °C. SWE techniques also require a small amount of solvent kept in a liquid state in their designated critical region by applying suitable pressure and temperature.[93] During the entire extraction, the chemical and physical characteristics of $H_2O$ are altered by applying high pressure and temperature.[94] For example, a report suggested that the phenolic content amplified as the temperature started to rise from 100 °C to approximately 225 °C, but once it reached 250 °C, the overall phenolic content started to decrease.[26]

## 7.7 Cosmeceutical Activities

### 7.7.1 Anti-aging

A cosmetic product is suggested to improve the manifestation of aging skin caused by several intrinsic, extrinsic, and genetically programmed factors. Several classes of macroalgae and compounds extracted from microalgae have recently shown promise as anti-aging agents.[95,96] Bioactive compounds from algae used for the preparation of anti-aging formulations are phlorotannins, sulfated galactan, sulfated galactofucans, polyunsaturated fatty acids, polysaccharides, MAAs, fucoxanthin, and astaxanthin (see Table 7.3).[30] A variety of products have been created from algae extracts that are intended for anti-aging, skin whitening, and pigmentation reduction. Extracts from the three genera of algae, *Laminaria*, *Fucus*, and *Chondrus*, are popular skincare ingredients due to their capability to hydrate and nourish the skin.[39] Fucoidan, a polysaccharide extracted exclusively from *Fucus vesiculosus*, has been shown to improve the hydration and elasticity of the skin when consolidated into creams and lotions.[97] It is known that fucoidan helps to produce heparin growth factor (HGF), which encourages the development of several types of tissues and cells, Takara Bio Inc. are currently exploiting this extract commercially.[98] An extract of *F. vesiculosus* induced collagen contraction in fibroblasts by increasing integrin molecule expression.[99] Furthermore, fucoidan inhibited the senescence-induced senescence-associated beta-galactosidase (SA-β-Gal) activity in long-term cultured endothelium colony-forming cells.[100] Fucoidan extracted from *Costaria costata* exhibited skin anti-aging actions in human foreskin fibroblast cells.[101] A series of studies have reported the antioxidant and anti-aging properties of the protein hydrolysates and residual biomass of several green microalgae species, including *Dunaliella tertiolecta*, *Nannochloropsis* sp., and *Tetraselmis suecica*.[102,103] In different cell lines, crude polysaccharide and polyphenolic fractions from

**Table 7.3** Functional applications of algae in cosmeceuticals.

| Property | Phyto-constituents |
|---|---|
| Moisturizing | Amino acids, carotenoids, fatty acids, oligosaccharide |
| Anti-cellulite | Quercetin, flavonoids, phlorotannins |
| Antioxidant | UV-absorbing MAAs, phenolic compounds, pigments, polyunsaturated fatty acids |
| Thickening agents | Alginates, carrageenan and agar |
| Whitening | Tannin, saponin, flavonoids, amines, glutathione, phenols, sugars, terpenoids |
| Water-binding agents | Polysaccharides |
| Emollient | Calcium alginate, galactofucans |
| Healing agents | Proteins, vitamins A and B1, polysaccharides |
| Anti-inflammatory | Sulfated polysaccharides, phytosterols, abscisic acid, lectin, phycobiliproteins, phenolic compounds |
| Anti-aging | Sulfated polysaccharide, polyunsaturated fatty acids, MAAs, fucoxanthin, astaxanthin |

*Ecklonia cava* served as suppressors of tumor cell growth, as well as antioxidants and radical scavengers with low toxicity.[104] Proteins, amino acids, and their derivatives from macroalgae activate collagen production in the skin.[105] The anti-elastase activity of a polypeptide from *Ulva lactuca* has been demonstrated, which stimulates collagen production in human fibroblasts.[106] Many essential components from macroalgae, such as minerals, vitamins, omega-6 and omega-3 fatty acids, promote cell rebirth and healthy skin. Extracts from *Chlorella vulgaris* boost the production of collagen in the skin, which aids in wrinkle reduction, and tissue regeneration.[58,107]

Anti-aging products can incorporate extracts from Thalassiosira sp., Chaeloceros sp., *Chlorococcum* sp., and *Monodus* sp. to boost collagen stimulation.[108] Green and red algae contain β-carotene, an antioxidant that protects against skin aging.[109] Previous reports have suggested the possibility of the brown algae species *Fucus vesiculosus* and the red alga species *Furcellaria lumbricalis* in skin anti-aging cosmetic ingredients.[110] Similarly, the skin anti-aging property of *Lola implexa* (green algae), *Ulva lactuca* (green algae), and *Spirulina maxima* (blue algae) have been reported.[111] In numerous cell-based models, the *Botryococcus braunii* microalgal extract showed skin anti-aging properties.[112] Algal species used for making anti-aging products include *Polysiphonia*, *Laurencia*, *Colpomenia*, *Gracilaria*, *Hydroclathrus*, *Halymenia*, *Padina*, *etc.*[113] Mycosporine-like amino acids present in the cold-water seaweed *Porphyra umbilicalis* protect the skin against harmful UV radiation.[114]

### 7.7.2 Skin Whitening

Mammalian skin produces melanin, which is crucial for regulating skin tone, shielding skin from UV radiation's damaging effects, and scavenging noxious chemicals. It is, however, important to note that melanin production that is abnormal can produce significant cosmetic problems and dermatological complications. Antimelanogenesis, or skin whitening *via* reduction of tyrosinase activity, is a crucial component in skin whitening. Increasingly, natural products that reduce, regulate, or stop melanogenesis, thereby reducing skin pigmentation, are in demand.[115] The subsequent types of chemical complexes derived from algae that can carry out this function include phenols, terpenoids, tannins, saponins, flavonoids, sugars, and amino acids.[116] The pigment fucoxanthin, obtained from the brown algae *Laminaria japonica*, *Macrocystis*, *Alaria*, and *Chorda*, is a solitary instance of an algal pigment that decreases the action of melanogenesis and tyrosinase.[9] Diphlorethohydroxycarmalol (DPHC), a carmalol derivative present in the methanolic extract of *Ishige okamurae*, has been shown to be an effective tyrosinase inhibitor, which had greater levels of activity than the widely used whitening agent arbutin.[117] Silab's product WHITONYL® works as a skin-whitening and skin-lightening product. It is made from *Palmaria palmata*, an alga with anti-pigmenting properties. It reduces melanogenesis, melanosome transport, and photoinduced pigmentation. It results in a radiant and extremely polished complexion. Zeaxanthin, an antityrosinase compound found in the pure extract of *Nannochloropsis oculata*, was patented for cream formulations.[118]

### 7.7.3 Moisturizing

For healthy, hydrated skin, moisturization and hydration are essential. Meanwhile, strengthening its protection against harmful environmental factors will also improve its appearance and elasticity.[105] As a moisturizing agent, algae extracts contain the following biologically active compounds: natural moisturizing factors, which include amino acids (alanine, serine, and threonine) and derivatives of amino acids (pyrrolidone carboxylic acid and citrulline), polysaccharides, fatty acids, and carotenoids.[46,105] Polysaccharides are essential humectants and moisturizers in cosmetic compositions. Carrageenan, fucoidans, alginate, and agar are polysaccharides from specific algae species that assist in controlling the distribution of water in the skin *Saccharina japonica* polysaccharides have superior moisturizing characteristics compared those of to hyaluronic acid, raising the possibility that these polysaccharides could be a useful cosmetic ingredient.[33] Additionally, it was discovered that the principal active site for retention and absorption of moisture was the sulfated group. *Spirulina* sp., *Chlorella* sp., and certain proteins from the *Porphyra* genus have a significant affinity for skin and hair, as they retain moisture and are an appropriate viscosity.[119] According to Tominaga *et al.* carotenoids also have moisturizing qualities.[46] Astaxanthin extracted from the microalgae *Haematococcus pluvialis* improved the condition of the all skin layers, including the elasticity, texture, moisture content, and condition of the corneocyte layer.[46] The effect of a mixture of dead sea minerals, *Dunaliella salina* algal extract, and dessert plants was investigated. Following four weeks of use, the cream with 5% enhancement reduced roughness by 43%, showing its potential as an ideal moisturizing agent.[120–122]

### 7.7.4 Antioxidant

For some time, a wide range of bioactive substances, including polyphenols (*e.g.*, phlorotannins), pigments (*e.g.*, the carotenoid astaxanthin), zinc sulfate, vitamin E, ascorbic acid, superoxide dismutase, coenzyme Q10, UV-absorbing MAAs, and others, have been employed successfully in cosmetic products as free-radical-scavenging molecules.[105,112] Retinoic acid from cyanobacteria improves skin elasticity while minimizing wrinkles, black spots, and dark under-eye circles.[113,123] α-Tocopherol, the body's primary source of vitamin E, has potent antioxidant properties that are helpful in avoiding light-induced diseases of the human skin.[124] The ability of carotenoids to react with peroxyl radicals makes them ideal for protecting membranes and lipoproteins from oxidative damage.[125,126] MAAs from *Porphyra umbilicalis* extract can lessen skin damage from UVA radiation.[127] In human skin cells exposed to UVA radiation, astaxanthin demonstrated a shielding outcome with a decrease of damage in nuclear DNA and upholding the natural properties of cellular antioxidant capacity.[128] Sulfated polysaccharides extracted from *Enteromorpha linza*, *Ulva pertusa*, *Bryopsis plumose* (green algae), *Porphyra haitanensis* (red algae), and *Laminaria japonica* (brown algae) demonstrated antioxidant activities.[48] Naturally occurring antioxidants like vitamins A and C that are present in *Chlorella vulgaris* and *Spirulina maxima* help to tone and cleanse the skin and repair dark circles.[129]

### 7.7.5 Healing and Therapeutic Agent

There are a number of advantageous elements (Ca, Cd, Ag, Bi, Co, Cr, Na, Ni, Fe, K, and Zn) contained in seaweeds, which can be easily absorbed by the skin.[130] For a range of skin conditions, including warts, psoriasis, scabies, rashes, acne, itching, eczema, burns, and boils, such elements are employed in therapeutic applications.[130] As an inhibitor of scar development, alginic acid might prevent scarring by inhibiting the invasion of fibroblasts and speeding up healing.[131] Ethanolic extract from *Kappaphycus alvarezii* (red algae) has been shown to accelerate hair growth and wound healing (primarily because of the carrageenans contained in the extract).[132] Phlorotannins, such as eckol and dieckol extracted from brown alga *Ecklonia stolonifera,* have been verified to be effective in treating skin diseases.[133] Premarathna *et al.* indicated that *Sargassum illicifolium*, through encouraging fibroblast proliferation and migration, may be a viable remedial agent for skin wound healing.[134] Furthermore, they compared the wound healing efficacy of *Sargassum illicifolium* with *Ulva lactuca* and found that the former had stronger wound healing properties *in vivo* and *in vitro* than the latter.[135] Fucoidan (very low molecular weight) isolated from algae can be used to improve wound healing.[136] Recent studies suggested that *Padina gymnospora*, a brown seaweed, has strong anti-inflammatory, antibacterial, and antinociceptive properties; therefore, this methanolic extract can be used as a natural wound-care solution.[137]

## 7.8 Cosmeceutical-based Formulations and Marketed Products

There has been an apparent increase in the use of algal metabolites in cosmetic products among both small- and large-scale businesses. Numerous reputable companies include algal extracts in their products. As of today, various algal products are being used as cosmetic additives. The "Red algae mask" marketed by OSEA Malibu is one such example. They use an extract of a red alga *Gigartina skottsbergii*. They claim that the product removes impurities from the skin surface, eliminates excess oil, and keeps the skin hydrated at the same time. Another product from OSEA Malibu, "Undaria Algae Body Butter", which includes *Butyrospermum parkii*, claims to transform dry skin into soft, smooth, and healthy-looking skin. Algenist, a renowned algal cosmetic brand, produces several skincare products containing a blend of

*Spirulina*, *Chlorella*, *Protothecoides*, *Parachlorella beijerinckii*, *Butyrospermum parkii*, *Corallina officinalis*, *Dunaliella salina*, and *Undaria pinnatifida* that enhance skin tone and minimize the appearance of fine lines and wrinkles (see Table 7.4). The "Sea Buckthorn + Algae Instant Glow Face Oil" produced by Björk & Berries uses Kalpariane™ – a natural ingredient from the brown algae *Alaria esculenta* that hydrates and smoothens the skin, maintaining the skin's firmness and elasticity. There are various cosmetic products with algal extracts on the market. Some of which are shown in Table 7.4.

## 7.9 Conclusions and Future Prospects

Whenever we think of cosmetics, what first comes to our mind is "full of chemicals", but times have now changed. People are becoming aware and looking towards natural and organic cosmetic products. India is one of the largest markets for natural and organic cosmetics that have been highlighted by the global certification bodies – COSMOS and NATRUE. In recent decades, Asian markets have provided 2417 products with COSMOS certification from 246 different companies, which is a far better ratio than the 24 984 certified products from 1722 companies in the European market. Furthermore, more than 700 raw materials and 12 000 ingredients are either termed as COSMOS-approved or COSMOS-certified, while the signature COSMOS NATURAL or COSMOS ORGANIC has been accredited to more than 29 000 products from 71 countries. At the same time, 'NATRUE' is also a top representative body that covers 7000 certified products across 325 brands in 70 member countries in the international cosmetics market. In 2021, a close to 4.2% increase in the valuation of the global organic and natural product market compared to previous records was recorded, reaching a value of USD 12.4 billion. Moreover, recent reports suggest a whopping 1.8% rise in the organic and natural cosmetics industry sector in the Indian financial year 2021–22. It is also noteworthy that in Germany, there was an increase of 1.6 million customers of nature-based cosmetic commodities in only the past year, which is an enormous transformation in the adoption of natural cosmetics over synthetic ones. In addition, according to a research report, the emerging Indian market is expected to reach USD 1239.04 million by 2026 with a compound annual growth rate of close to 14.69%.

Products are not tested for safety, but regulations exist, which results in a grey area for consumers. A large number of chemical ingredients are used in cosmetic products which are harmful to human beings. For instance, the average woman in the US utilizes 12 cosmetics consisting of 168 chemical compounds on a daily basis. Some makeup products such as lip products, mascaras, and foundations contain high concentrations of presumably noxious chemicals called per and polyfluoroalkyl substances. Diethanolamine is used as an emulsifier and foaming agent, which is correlated to cancer in animals and creates a risk to humans also. As another example, parabens used as a preservative in cosmetics are associated with disturbance to the reproductive system, disruption in the endocrine system, and skin irritation.

**Table 7.4** List of algae-based cosmetics on the market.

| Product name | Brand name | Product function | Algal extract/algal pigments | References (Webpage last accessed, 14th September, 2022) |
|---|---|---|---|---|
| Red algae hyaluronic acid face cream | Spaveda | Replenishes skin and restore moisture levels | Astaxanthin | 138 |
| Skin repair red algae face serum | W2 Why Wait | Provides antioxidant protection to the skin, is anti-inflammatory, improves moisture-holding capacity of the skin, neutralizes free-radicals | *Chondrus crispus* extract | 139 |
| Rose sea algae peel-off mask | FABEYA | Protects skin against sun damage, ageing and inflammation | Diatomites | 140 |
| Pro-collagen marine cream | ELEMIS | Anti-aging, improves skin firmness, tone and hydration | *Chlorella vulgaris* extract, *Porphyridium cruentum* extract | 141 |
| Red algae revitalizing serum | MISSHA Time Revolution | Antioxidant | *Chondrus crispus* extract (irish moss extract) | 142 |
| Sea buckthorn + algae instant glow face oil | Björk & Berries | Hydrates and smooths the skin, maintains skin firmness and elasticity | Kalpariane™ – a natural ingredient from the brown algae *Alaria esculenta* | 143 |
| Marine algae skin firming and reawakening sace masque | Oriental Beauty Cosmetics | Tightens pores, deeply hydrates skin | *Spirulina* powder, diatomaceous earth, algin | 144 |
| Red algae detoxifying toner | USTAWI | Hydrates and detoxifies skin | *Chondrus crispus* (carrageenan) extract | 145 |
| Undaria algae body butter | OSEA Malibu | Transforms dry skin to soft, smooth and healthy-looking skin | *Butyrospermum parkii* (shea) butter | 146 |
| Red algae mask | OSEA Malibu | Removes surface impurities, eliminates excess oil, hydrates skin | *Gigartina skottsbergii* extract | 147 |
| Brightening solution triple care serum | Spawake | Protects against damaging effects of UV radiation | *Laminaria japonica* extract | 148 |

(*continued*)

**Table 7.4** (*Continued*)

| Product name | Brand name | Product function | Algal extract/algal pigments | References (Webpage last accessed, 14th September, 2022) |
|---|---|---|---|---|
| HA4 plus serum | FCL Skincare | Protects the dermal matrix | Osmogeline (red algae extract) | 149 |
| Nocturnal signature anti age cream | BYNACHT | Restores and replenishes the skin, minimizes wrinkles and signs of aging | *Butyrospermum parkii* (shea) butter, *Haematococcus pluvialis* extract | 150 |
| Triple algae eye renewal balm | Algenist | Brightens the overall look of the eye area | *Butyrospermum parkii* (shea) butter, *Parachlorella beijerinckii* exopolysaccharides, *Corallina officinalis*, *Dunaliella salina* and *Undaria pinnatifida* extracts | 151 |
| Blue algae vitamin C™ skinclarity brightening serum | Algenist | Enhances brightness, evens skin tone, smooths texture, improves skin clarity | Ascorbic acid from *Spirulina* | 152 |
| Mara algae enzyme cleansing oil | Allure_Skin | Cleanses, brightens and hydrates the skin | *Laminaria digitata* extract, *Spirulina maxima* extract | 153 |
| Detox effect mask | lavera NATURKOSMETIK | Removes skin impurities, protects against atmospheric pollution | *Fucus serratus* extract, *Butyrospermum parkii* (shea) butter | 154 |

Moreover, polyethylene glycol compounds are used in cosmetic care provisions that have also been connected to a range of different types of cancer. Besides, petroleum jelly or petrolatum, siloxanes or silicones, sodium laureth sulfate, and triclosan are also widely used chemical ingredients in cosmetic products that cause a variety of side effects. Due to their adverse effect, chemical ingredients such as zirconium-containing complexes, vinyl chloride, methylene chloride, mercury compounds, hexachlorophene, halogenated salicylanilides, chloroform, propellants, chlorofluorocarbon, and bithionol are prohibited by the FDA of the US. Moreover, in 2019, the FDA advised stopping the use of several cosmetics, including Johnson's Baby Powder (Lot #22318RB), due to it containing carcinogenic asbestos.

The 21st century will be the century for biotechnology if the world survives the ever growing threats of geopolitical problems. Discovery of new technologies and biological equipment has opened many doors of opportunity for the cosmetics industry as well. In the last century, when people talked about "cosmetics", they were often referring to chemicals that we can wear as makeup, but now the game has changed for good. Researchers are looking towards nature to find an alternative for synthetic chemical-based cosmetics. Algae are considered as seaweeds and they play a key role in providing oxygen to the atmosphere of the Earth. These algae are now being used for the extraction of many biologically important compounds with therapeutic and cosmeceutical properties. Integration of modern technologies such as bioinformatics, tissue culture, biostatistics, biological modelling, bioprospecting and ecological engineering in the field of cosmeceutical science can definitely improve the quantity and quality of algae-based cosmeceuticals. The market for cosmeceuticals was always there; the increasing footsteps of digital media marketing and e-commerce platforms have provided the distributors with a whole new dimension for selling these products. The new generation seems to be very aware of the conservation of nature and natural product-based industries, and with proper strategies these algae-based cosmeceuticals will be a product to look forward to in the near future.

## References

1. K. Thiyagarasaiyar, B. H. Goh, Y. J. Jeon and Y. Y. Yow, *Mar. Drugs*, 2020, **18**, 323.
2. A. Kerdudo, P. Burger, F. Merck, A. Dingas, Y. Rolland, T. Michel and X. Fernandez, *C. R. Chim.*, 2016, **19**, 1077.
3. J. D. Bos and M. M. Meinardi, *Exp. Dermatol.*, 2000, **9**, 165.
4. R. Baran and J. Andre, *J. Cosmet., Dermatol.*, 2005, **4**, 204.
5. J. C. Mitchell, *Contact Dermatitis*, 1981, **7**, 173.
6. J. Barrett, *Environ. Health Perspect.*, 2005, **113**, A24.
7. S. Khanra, M. Mondal, G. Halder, O. N. Tiwari, K. Gayen and T. K. Bhowmick, *Food Bioprod.*, 2018, **110**, 60.
8. Y. Ren, J. Deng, J. Huang, Z. Wu, Y. Bi and F. Chen, *Bioresour. Technol.*, 2021, **1**, 340.

9. H. Shimoda, J. Tanaka, S. J. Shan and T. Maoka, *J. Pharm. Pharmacol.*, 2010, **62**, 1137.
10. S. Anastyuk, N. Shervchenko, S. Ermakova, O. Vishchuk, E. Nazarenko, P. Dmitrenok and T. Zvyagintseva, *Carbohydr. Polym.*, 2012, **87**, 186.
11. N. V. Thomas and S. K. Kim, *Mar. Drugs*, 2013, **11**, 146.
12. Z. J. Wang, W. Xu, J. W. Liang, C. S. Wang and Y. Kang, *Afr. J. Tradit., Complementary Altern. Med.*, 2017, **14**, 149.
13. S. S. Suh, J. Hwang, M. Park, H. Seo, H. S. Kim, J. Lee, S. Moh and T. K. Lee, *Mar. Drugs*, 2014, **12**, 5174.
14. H. S. Kalasariya, L. Pereira and N. B. Patel, *Phycology*, 2022, **2**, 172.
15. I. S. Fernando, K. A. Sanjeewa, K. W. Samarakoon, W. W. Lee, H. S. Kim, N. Kang and Y. J. Jeon, *Int. J. Biol. Macromol.*, 2017, **104**, 1185.
16. A. Robic, C. Rondeau-Mouro, J. F. Sassi, Y. Lerat and M. Lahaye, *Carbohydr. Polym.*, 2009, **77**, 206.
17. M. F. Raposo, R. M. De Morais and A. M. B. de Morais, *Mar. Drugs*, 2013, **11**, 233.
18. S. K. Kim, Y. D. Ravichandran, S. B. Khan and Y. T. Kim, *Biotechnol. Bioprocess Eng.*, 2008, **13**, 511.
19. D. B. Stengel and S. Connan, in *Natural Products from Marine Algae, Methods in Molecular Biology*, ed. D. B. Stengel and S. Connan, Humana Press, USA, 2015, p. 1.
20. J. Cotas, A. Leandro, P. Monteiro, D. Pacheco, A. Figueirinha, A. M. Gonçalves, G. J. da Silva and L. Pereira, *Mar. Drugs*, 2020, **18**, 384.
21. T. Kuda and T. Ikemori, *Jpn. J. Food Chem.*, 2009, **112**, 575.
22. F. Pagels, A. C. Guedes, H. M. Amaro, A. Kijjoa and V. Vasconcelos, *Biotechnol. Adv.*, 2019, **37**, 422.
23. R. R. Ambati, D. Gogisetty, R. G. Aswathanarayana, S. Ravi, P. N. Bikkina, L. Bo and S. Yuepeng, *Crit. Rev. Food Sci. Nutr.*, 2019, **59**, 1880.
24. P. Torres, J. P. Santos, F. Chow and D. Y. A. dos Santos, *Algal Res.*, 2019, **37**, 288.
25. K. Thiyagarasaiyar, B. H. Goh, Y. J. Jeon and Y. Y. Yow, *Mar. Drugs*, 2020, **18**, 323.
26. I. B. Miller, S. Pawlowski, M. Y. Kellermann, M. Petersen-Thiery, M. Moeller, S. Nietzer and P. J. Schupp, *Environ. Sci. Eur.*, 2021, **33**, 1.
27. L. M. Csorba and V. A. Boglea, *Reg. Bus. Stud.*, 2011, **3**, 167.
28. P. D. Darbre, *J. Appl. Toxicol.*, 2003, **23**, 89.
29. K. Miyashita, S. Nishikawa, F. Beppu, T. Tsukui, M. Abe and M. Hosokawa, *J. Sci. Food Agric.*, 2011, **91**, 1166.
30. I. Michalak, A. Dmytryk and K. Chojnacka, *Encyclopedia of Marine Biotechnology*, 2020, vol. 1, p. 65.
31. W. A. J. P. Wijesinghe and Y. J. Jeon, *Carbohydr. Polym.*, 2012, **88**, 13.
32. F. Ferreres, G. Lopes, A. Gil-Izquierdo, P. B. Andrade, C. Sousa, T. Mouga and P. Valentão, *Mar. Drugs*, 2012, **10**, 2766.
33. J. Wang, W. Jin, Y. Hou, X. Niu, H. Zhang and Q. Zhang, *Int. J. Biol. Macromol.*, 2013, **57**, 26.

34. G. Andre, M. Pellegrini and L. Pellegrini, Cosmetic or dermatological compositions especially useful as antiaging and antiwrinkle products comprise extracts of Ascophyllum and Helopteris seaweeds, *French Pat.*, FR2837386-A1, 2003.
35. K. N. Kim, H. M. Yang, S. M. Kang, D. Kim, G. Ahn and Y. J. Jeon, *Food Chem. Toxicol.*, 2013, **59**, 521.
36. M. C. Houston, *Prog. Cardiovasc. Dis.*, 2005, **47**, 396.
37. Y. V. Yuan, N. D. Westcott, C. Hu and D. D. Kitts, *Food Chem.*, 2009, **112**, 321.
38. A. M Sanghvi and Y. Martin Lo, *Recent Pat. Food, Nutr. Agric.*, 2010, **2**, 187.
39. L. Pereira, *Cosmetics*, 2018, **5**, 68.
40. D. B. Stengel, S. Connan and Z. A. Popper, *Biotechnol. Adv.*, 2011, **29**, 483.
41. A. L. Quinderé, B. P. Fontes, E. de SO Vanderlei, I. N. de Queiroz, J. A. Rodrigues, I. W. de Araújo and N. Benevides, *Pharmacol. Rep.*, 2013, **65**, 600.
42. J. A. G. Rodrigues, I. N. L. de Queiroz, A. L. G. Quinderé, N. M. B. Benevides, A. M. F. Tovar and P. A. de Souza Mourão, *Acta Sci., Technol.*, 2016, **38**, 273.
43. Y. Yang, D. Liu, J. Wu, Y. Chen and S. Wang, *Int. J. Biol. Macromol.*, 2011, **49**, 1031.
44. M. Lahaye and A. Robic, *Biomacromolecules*, 2007, **8**, 1765.
45. R. Wang, V. J. Paul and H. Luesch, *Free Radicals Biol. Med.*, 2013, **57**, 141.
46. K. Tominaga, N. Hongo, M. Karato and E. Yamashita, *Acta Biochim. Pol.*, 2012, **59**, 47.
47. J. Fabrowska, B. Leska and G. Schroeder, *Chemik*, 2015, **69**, 495.
48. Z. Zhang, F. Wang, X. Wang, X. Liu, Y. Hou and Q. Zhang, *Carbohydr. Polym.*, 2010, **82**, 118.
49. M. Koller, A. Muhr and G. Braunegg, *Algal Res.*, 2012, **6**, 52.
50. I. Priyadarshani and B. Rath, *J. Algal Biomass Util.*, 2012, **3**, 89.
51. I. Hamed, *Compr. Rev. Food Sci. Food Saf.*, 2016, **15**, 1104.
52. S. M. Arad and A. Yaron, *Trends Food Sci. Technol.*, 1992, **3**, 92.
53. J. Y. Berthon, R. Nachat-Kappes, M. Bey, J. P. Cadoret, I. Renimel and E. Filaire, *Free Radical Res.*, 2017, **51**, 555.
54. C. Couteau and L. Coiffard, in *Microalgae in health and disease prevention*, ed. I. Levine, and J. Fleurence, Academic Press, 2018, p. 317.
55. S. Letsiou, K. Kalliampakou, K. Gardikis, L. Mantecon, C. Infante, M. Chatzikonstantinou and E. Flemetakis, *Front. Mar. Sci.*, 2017, **4**, 221.
56. R. Martínez, A. García-Beltrán, G. Kapravelou, C. Mesas, L. Cabeza, G. Perazzoli and J. M. Porres, *Mar. Drugs*, 2022, **20**, 318.
57. R. Winter, in *A Consumer's Dictionary of Cosmetic Ingredients: Complete Information about the Harmful and Desirable Ingredients Found in Cosmetics and Cosmeceuticals*, Harmony/Rodale, United States, 2009, p. 576.
58. M. A. Borowitzka, *J. Appl. Phycol.*, 2013, **25**, 743.
59. P. Malinowska, *Univ. Econ.*, 2011, **212**, 123.

60. J. Surabhi, K. Roshani and N. U. Vivek, *Int. J. Innov. Res. Sci. Eng. Technol.*, 2018, **7**, 1269.
61. N. V. Thomas and S. K. Kim, *Mar. Drugs*, 2013, **11**, 146.
62. M. Liu, P. E. Hansen and X. Lin, *Mar. Drugs*, 2011, **9**, 1273.
63. W. A. J. P. Wijesinghe and J. You-Jin, *Phytochem. Rev.*, 2011, **10**, 431.
64. S. S. Suh, J. Hwang, M. Park, H. Seo, H. S. Kim, J. Lee, S. Moh and T. K. Lee, *Mar. Drugs*, 2014, **12**, 5174.
65. T. Krishnapriya, G. Bey-Hing, J. You-Jin and Y. Yoon-Yen, *Mar. Drugs*, 2020, **18**, 323.
66. S. Saka, Y. Komura, Y. Nishimura, T. Sugawara and T. Hirata, *Food Sci. Technol. Res.*, 2011, **17**, 171.
67. A. Hartmann, J. Gostner, J. E. Fuchs, E. Chaita, N. Aligiannis and L. Skaltsounis, *Planta Med.*, 2015, **81**, 813.
68. D. L. Maria, P. Ilaria, R. L. Antonina, V. Emanuele, P. R. Rosa, T. Simona, M. Giuseppe, V. Infantino and V. Antonio, *Cosmetics*, 2021, **8**, 52.
69. C. Ana-Marija, J. Stela, S. Drago and J. Igor, *Mar. Drugs*, 2018, **16**, 348.
70. J. Afroz, Z. A. Iffat, F. Nida, A. A. Vaseem and A. Juber, *Phycologia*, 2017, **56**, 410.
71. O. Pulz and W. Gross, *Appl. Microbiol. Biotechnol.*, 2004, **65**, 635.
72. A. M. O'sullivan, Y. C. O'callaghan, M. N. O'grady, B. Queguineur, D. Hanniffy, D. J. Troy, J. P. Kerry and N. M. O'brien, *Food Chem.*, 2011, **126**, 1064.
73. A. Molino, S. Mehariya, G. Di Sanzo, V. Larocca, M. Martino and G. P. Leone, *J. $CO_2$ Util.*, 2020, **36**, 196.
74. A. Aslam, A. Bahadar, R. Liaquat, M. Saleem, A. Waqas and M. Zwawi, *Sci. Total Environ.*, 2021, **772**, 144905.
75. S. Dey and V. K. Rathod, *Ultrason. Sonochem.*, 2013, **20**, 271.
76. Y. H. Kim, S. Park, M. H. Kim, Y. K. Choi, Y. H. Yang, H. J. Kim, H. Kim, H. S. Kim, K. G. Song and S. H. Lee, *Biomass Bioenergy*, 2013, **56**, 99.
77. A. P. Martins, N. S. Yokoya and P. Colepicolo, *Braz. J. Pharmacogn.*, 2012, **4**, 854.
78. S. K. Kim and K. Chojnacka, in *Marine Algae Extracts: Processes, Products, and Applications*, John Wiley & Sons, 2015, p. 784.
79. J. Azmir, I. S. M. Zaidul, M. M. Rahman, K. M. Sharif, A. Mohamed, F. Sahena, M. H. A. Jahurul, K. Ghafoor, N. A. N. Norulaini and A. K. M. Omar, *J. Food Eng.*, 2013, **117**, 426.
80. L. V. Silva, D. L. Nelson, M. F. B. Drummond, L. Dufossé and M. B. A. Glória, *Food Res. Int.*, 2005, **38**, 1087.
81. T. Wang, R. Jónsdóttir, H. G. Kristinsson, G. O. Hreggvidsson, J. O. Jónsson, G. Thorkelsson and G. Ólafsdóttir, *Food Sci. Technol.*, 2010, **43**, 1387.
82. S. Li, H. Zhang, D. Han and K. H. Row, *Korean J. Chem. Eng.*, 2012, **29**, 650.
83. B. Łęska, B. Messyasz and G. Schroeder, in *Algae Biomass: Characteristics and Applications*, Springer, Cham, 2018, p. 89.

84. J. Azmir, I. S. M. Zaidul, M. M. Rahman, K. M. Sharif, A. Mohamed, F. Sahena, M. H. A. Jahurul, K. Ghafoor, N. A. N. Norulaini and A. K. M. Omar, *J. Food Eng.*, 2013, **117**, 426.
85. S. U. Kadam, B. K. Tiwari and C. P. O'Donnell, *J. Agric. Food Chem.*, 2013, **61**, 4667.
86. I. Nowak and S. Kowalska, *Analityka: nauka i praktyka*, 2007, p. 4.
87. R. M. Rodriguez-Jassoa, S. I. Mussattoa, L. Pastranab, C. N. Aguilarc and J. A. Teixeira, *Carbohydr. Polym.*, 2011, **86**, 1137.
88. H. Y. Lou, B. Wang, C. G. Yu and Y. F. Xu, *Chin. Herb. Med.*, 2010, **2**, 321.
89. J. Fabrowska, E. Ibanez, B. Leska and M. Herrero, *Algal Res.*, 2016, **19**, 237.
90. S. P. Sivagnanam, S. Yin, J. H. Choi, Y. B. Park, H. C. Woo and B. S. Chun, *Mar. Drugs*, 2015, **13**, 3422.
91. R. Mittal, H. A. Tavanandi, V. A. Mantri and K. S. M. S. Raghavarao, *Ultrason. Sonochem.*, 2017, **38**, 92.
92. B. Kaufmann and P. Christen, *Phytochem. Anal.*, 2002, **13**, 105.
93. K. Duarte, C. I. L. Justino, A. M. Gomes, T. A. P. Rocha-Santos and A. C. Duarte, *Compr. Anal. Chem.*, 2014, **65**, 59.
94. C. Turner and E. Ibanez, in *Enhancing Extraction Processes in the Food Industry*, ed. N. Lebovka, E. Vorobiev and F. Chemat, Taylor & Francis Group, LLC, Abingdon, UK, 2011, p. 223.
95. T. Vo Dinh, P. S. Saravana, H. C. Woo and B. S. Chun, *Sep. Purif. Technol.*, 2018, **196**, 287.
96. C. Ana-Marija, J. Stela, S. Drag and J. Igor, *Mar. Drugs*, 2008, **16**, 348.
97. T. Fujimura, K. Takahara, S. Moriwaki, T. Kitahara and Y. Takema, *Biol. Pharm. Bull.*, 2000, **23**, 291.
98. J. H. Fitton, G. Dell'Acqua, V. A. Gardiner, S. S. Karpiniec, D. N. Stringer and E. Davis, *Cosmetics*, 2015, **2**, 66.
99. T. Fujimura and K. Tsukahara, *J. Cosmet. Sci.*, 2002, **53**, 1.
100. J. H. Lee, S. H. Lee, S. H. Choi, T. Asahara and S. M. Kwon, *Stem Cells*, 2015, **33**, 1939.
101. H. J. Moon, K. S. Park, M. J. Ku, M. S. Lee, S. H. Jeong, T. I. Imbs and Y. H. Lee, *J. Nat. Prod.*, 2009, **72**, 1731.
102. C. D. Norzagaray-Valenzuela, A. Valdez-Ortiz, L. M. Shelton, M. Jiménez-Edeza, J. Rivera-López, M. A. Valdez-Flores and L. J. Germán-Báez, *J. Appl. Phycol.*, 2017, **29**, 189.
103. L. J. German-Báez, M. A. Valdez-Flores, J. V. Félix-Medina, C. D. Norzagaray-Valenzuela, D. U. Santos-Ballardo, C. Reyes-Moreno and A. Valdez-Ortiz, *Food Sci. Technol. Int.*, 2017, **23**, 681.
104. Y. Athukorala, K. N. Kim and Y. J. Jeon, *Food Chem. Toxicol.*, 2006, **44**, 1065.
105. G. Bedoux, K. Hardouin, A. S. Burlot and N. Bourgougnon, *Adv. Bot. Res.*, 2014, **71**, 345.
106. M. Guglielmo and D. Montanari, *WO Pat.*, 2008146116A2, 2008.
107. H. M. D. Wang, C. C. Chen, P. Huynh and J. S. Chang, *Bioresour. Technol.*, 2015, **184**, 355.

108. L. Zanella and M. Alam, in *Microalgae Biotechnology for Food, Health and High Value Products*, ed. M. Alam, J. L. Xu and Z. Wang, Springer, Singapore, 2020, p. 295.
109. S. K. Schagen, V. A. Zampeli, E. Makrantonaki and C. C. Zouboulis, *Dermato-Endocrinology*, 2012, **4**, 298.
110. T. Al-Bader, A. Byrne, J. Gillbro, A. Mitarotonda, A. Metois, F. Vial and A. Laloeuf, *J. Cosmet., Dermatol.*, 2012, **11**, 17–26.
111. E. Xhauflaire-Uhoda, K. Fontaine and G. E. Piérard, *Int. J. Cosmet. Sci.*, 2008, **30**, 131.
112. S. Buono, A. L. Langellotti, A. Martello, M. Bimonte, A. Tito, A. Carola and V. Fogliano, *Arch. Dermatol. Res.*, 2012, **304**, 755.
113. D. Kelman, E. K. Posner, K. J. McDermid, N. K. Tabandera, P. R. Wright and A. D. Wright, *Mar. Drugs*, 2012, **10**, 403.
114. S. Daniel, S. Cornelia and Z. Fred, *Cosmet Toiletries Manufacture Worldwide*, 2004, p. 139.
115. C. H. A. Seon-Heui, K. O. Seok-Chun, K. I. M. Daekyung and J. E. O. N. You-Jin, *J. Dermatol.*, 2011, **38**, 354.
116. Y. Y. Chan, K. H. Kim and S. H. Cheah, *J. Ethnopharmacol.*, 2011, **137**, 1183.
117. S. J. Heo, S. C. Ko, S. M. Kang, S. H. Cha, S. H. Lee, D. H. Kang and Y. J. Jeon, *Food Chem. Toxicol.*, 2010, **48**, 1355.
118. S. Babitha and E. K. Kim, in *Effect of Marine Cosmeceuticals on the Pigmentation of Skin*, CRC Press, Boca Raton, FL, USA, 2011, p. 63.
119. H. Hagino and M. Saito, *European Pat.*, EP1433463B1, 2010, p. 18.
120. Z. Ma'Or, G. Meshulam-Simon, S. Yehuda, J. A. Gavrieli and D. Sea, *J. Cosmet. Sci.*, 2000, **51**, 27.
121. C. Zappelli, A. Barbulova, F. Apone and G. Colucci, *Cosmetics*, 2016, **3**, 39.
122. A. Maadane, N. Merghoub, T. Ainane, H. El Arroussi, R. Benhima, S. Amzazi and I. Wahby, *J. Biotechnol.*, 2015, **215**, 13.
123. X. Wu, J. Jiang, Y. Wan, J. P. Giesy and J. Hu, *Proc. Natl. Acad. Sci. U. S. A.*, 2012, **109**, 9477.
124. A. A. Hamid, O. O. Aiyelaagbe, L. A. Usman, O. M. Ameen and A. Lawal, *Afr. J. Pure Appl. Chem.*, 2010, **4**, 142.
125. W. Stahl and H. Sies, *Mol. Asp. Med.*, 2003, **24**, 345.
126. W. Stahl and H. Sies, *Biochim. Biophys. Acta, Mol. Basis Dis.*, 2005, **1740**, 101.
127. D. Schmid, C. Schürch and F. Zülli, *Eur. Cosmet.*, 2006, **9**, 1.
128. N. M. Lyons and N. M. O'Brien, *J. Dermatol. Sci.*, 2002, **30**, 73.
129. S. Joshi, R. Kumari and V. N. Upasani, *Int. J. Innov. Res. Technol. Sci. Eng.*, 2018, **7**, 1269.
130. M. A. Rizvi and E. E. Valeem, *Int. J. Phycol. Phycochem.*, 2012, **8**, 95.
131. D. H. Park, W. S. Choi, S. H. Yoon, J. S. Shim and C. H. Song, in *Key Engineering Materials*, Trans Tech Publications Ltd, 2007, vol. 342, p. 125.
132. G. F. Samaneh, T. S. Fatemeh, E. Mozhdeh, G. Meng, M. Kharidah and M. Suhaila, *J. Anim. Vet. Adv.*, 2011, **10**, 601.

133. M. J. Joe, S. N. Kim, H. Y. Choi, W. S. Shin, G. M. Park, D. W. Kang and Y. K. Kim, *Biol. Pharm. Bull.*, 2006, **29**, 1735.
134. A. D. Premarathna, T. H. Ranahewa, S. K. Wijesekera, R. R. M. K. K. Wijesundara, A. P. Jayasooriya, V. Wijewardana and R. P. V. J. Rajapakse, *Wound Med.*, 2019, **24**, 1.
135. A. D. Premarathna, S. K. Wijesekera, A. P. Jayasooriya, R. N. Waduge, R. R. M. K. K. Wijesundara, R. Tuvikene and R. P. V. J. Rajapakse, *Biochem. Biophys. Rep.*, 2021, **26**, 100986.
136. J. H. Park, S. H. Choi, S. J. Park, Y. J. Lee, J. H. Park, P. H. Song and C. H. Song, *Mar. Drugs*, 2017, **15**, 112.
137. A. P. Baliano, E. F. Pimentel, A. R. Buzin, T. Z. Vieira, W. Romão, L. V. Tose and D. C. Endringer, *Rev. Bras. Farmacogn.*, 2016, **26**, 714.
138. https://spaveda.in/products/light-non-greasy-deep-hydrating-moisturiser.
139. https://www.w2whywait.com/product/w2-skin-repair-face-serum/.
140. https://www.desertcart.in/products/85941413-fabeya-rose-sea-algae-peel-off-mask.
141. https://www.cosme-de.com/product-page?pdid=20444&src=googlefeed.
142. https://misshaus.com/products/time-revolution-red-algae-revitalizing-serum.
143. https://eu.bjorkandberries.com/products/sea-buckthorn-algae-instant-glow-face-oil.
144. https://www.youcarelifestyle.com/product-details/timeless-beauty-secrets-oriental-marine-algae-skin-firming-and-reawakening--face-masque-15-gm?.
145. https://www.ssense.com/en-us/everything-else/product/ustawi/red-algae-detoxifying-toner-200-ml/9225651.
146. https://oseamalibu.com/products/undaria-algae-body-butter.
147. https://oseamalibu.com/products/red-algae-mask.
148. https://www.spawake.in/products/brightening-solution-triple-care-serum/.
149. https://fclskincare.com/products/fcl-ha4-plus-serum.
150. https://www.bynacht.com/en/shop/creams/nocturnal-signature-anti-age-cream.
151. https://www.algenist.com/products/triple-algae-eye-renewal-balm.
152. https://www.algenist.com/products/blue-algae-vitamin-c%E2%84%A2-skinclarity-brightening-serum.
153. https://www.allure.com/shop/product/chia-moringa-algae-enzyme-cleansing-oil.
154. https://cosmeticobs.com/en/products/lavera-163/anti-pollution-3202/detox-effect-mask-11683.

CHAPTER 8

# Cosmetic Applications of Lichens

YINGSHU ZHAO[a], BAOJUN XU[b] AND JESSICA PANDOHEE[c*]

[a]China Food Publishing Co., Beijing Academy of Food Sciences, Beijing, 100068, China; [b]Food Science and Technology, Program, BNU-HKBU United International College, Zhuhai, Guangdong 519087, China; [c]Sydney Mass Spectrometry, University of Sydney, Camperdown, New South Wales 2006, Australia
*E-mail: jessica.pandohee@gmail.com

## 8.1 Introduction

Lichens are symbiotic organisms consisting of algae (phytobiont) and fungi (mycobiont).[1] The mycobionts are predominantly ascomycetes, with a few basidiomycetes, while the phytobionts are cyanobacteria or green algae.[2] The unique symbiotic structure of lichens allows them to produce lichen-specific bioactive substances, so they are considered to be one of the most important sources of bioactive compounds other than plants.[3] Lichens are highly viable and widely distributed, with 1002 genera and 19 409 species known worldwide.[4]

Lichens and their metabolites have been used since ancient times around the world as raw materials for a variety of products, such as food, medicine, cosmetics, fragrances and dyes.[5] *Dermatocarpon miniatum*, *Lobaria pulmonaria*, *Umbilicaria esculenta*, *Alectoria asiatica* and other edible lichens have been used as dishes in China after simple processing over the centuries.[6] *Roccella* spp. were used as a dye lichen in Europe.[7] *Pseudevernia furfuracea* and *Evernia prunastri* have been extensively processed in France and used in

Bioprospecting of Natural Sources for Cosmeceuticals
Edited by Deepika Kathuria, Ajay Sharma, Meenakshi Verma and Gulzar Ahmad Nayik

Published by the Royal Society of Chemistry, www.rsc.org

the perfume industry.[8] Atranorin, pannarin, gyrophoric acid and usnic acid from lichens have been used in sunscreen preparations, with usnic acid also being used as a preservative in cosmetic creams.[8] *Ramalina thrausta* was used in Finland to treat wounds, athlete's foot or other skin diseases, and orally for sore throats and toothaches.[9] In addition, lichens also have important applications in the evolution of ecosystems, environmental monitoring, geological exploration and glacier age estimation.[10]

Even though lichens have been used for a variety of purposes since ancient times, they are considered a difficult biological material to obtain because of their particularly slow growth and difficulty in artificial cultivation. Numerous constraints have resulted in extensive research on lichens and their metabolites being neglected for a long time in the past.[8] Since the end of the 20th century, research relating to the culture, production, biochemical analysis, and application of metabolites of lichens has gradually increased and evidence for the biological activity of lichens and their metabolites has been presented.[8] However, current knowledge about the uses of lichens for food, medicine, and cosmetics is still scattered, and they are still generally disregarded by industries like modern pharmaceutical and cosmetic manufacture.[11]

There is no doubt that the natural product resources contained in lichens are of great research value to mankind in exploring new sources of medicines, new foods, new cosmetic ingredients, and many other aspects. This chapter therefore provides an overview of lichens and their metabolites, and summarizes and analyzes the potential health benefits of lichen extracts and their bioactive substances in the field of cosmetics with a view to providing some support for the application of lichens in cosmetic development, and the selection of ingredients.

## 8.2 Lichen Morphology, Symbiosis, and Diversity

In the early days, there was a lack of research on lichens, which were considered to be separate organisms.[3] It was not until the 1860s that the Swiss botanist Simon Schwendener first realized that lichens were not single green autotrophs but intimate symbioses of specialized fungi and phototrophic unicellular green algae, when he observed from lichens that fungal hyphae were closely intertwined with individual algal cells, but he likened the relationship between mycobiont and phytobiont to that of master and slave.[1,12] Some scholars believe that the symbiotic nature of lichens is commensalism because the mycobiont seems to gain more benefit while the algae grow much more slowly in the symbiotic state with the fungi than in the non-symbiotic state. In fact, there has been a debate about the symbiosis of fungi and algae in lichen. The idea of mutualism between mycobiont and phytobiont in lichens was first proposed by the famous children's author Beatrix Potter.[1] Hawksworth and Honegger proposed a definition of lichens as "*ecologically obligate, stable mutualisms between an exhabitant fungal partner (the mycobiont) and an inhabitant population of extracellularly located unicellular or filamentous algal and/or cyanobacterial cells (the photobiont)*".[13] This is also the widely accepted concept of lichen. Many biology textbooks also use lichens as a typical example of mutualism.

The algal cells or cyanobacteria involved in the symbiosis are located inside, and fungal hyphae wrap around the algal cells and surround the algae from the outside, which together form a thallus with a unique structure and a set of unique biological characteristics.[1] In this type of symbiotic complex, the algae contain photosynthetic pigments for photosynthesis and provide the fungi with most of the organic nutrients produced, while the fungi provide the algae with the raw material for photosynthesis by absorbing water and inorganic salts from the external environment through hyphae, and entangle and surround algal cells to prevent them from drying out and dying in direct light, as well as maintaining their form and moisture content.[11] In addition, cyanobacteria are not only able to provide nutrients through photosynthesis, but they are also able to fix atmospheric nitrogen, converting it into a form that is available for the growth of the symbiont.[14] The mycobiont and phytobiont are interdependent and cannot be separated under natural conditions.

However, not any fungus can form a lichen with any algae or cyanobacteria. Only fungi that have survived in a symbiotic state with the corresponding algae or cyanobacteria after a long-term survival competition and evolution can form lichens. Approximately 20 000 known mycobionts have been identified in lichens, accounting for 20% of all known fungi.[12] Most of them are ascomycetes, but a few are basidiomycetes and deuteromycetes.[2] Phytobionts are usually cyanobacteria of the genus *Nostocs* and unicellular green algae of the genera *Trebouxia* and *Trentepohlia*.[2] Only approximately 120 photobiont species have been found, far fewer than the number of mycobiont species.[12] In addition, as the same phytobiont species can be found in various lichens, and mycobionts are conspicuous and specific among lichens, lichens are systematically classified and named according to their mycobionts.[12]

The unique and successful symbiosis of lichens allows them to colonize different substrates such as trees, rocks and soil, and to tolerate harsh environments such as heat, extreme cold, oligotrophy, drought and intense radiation. Lichens cover 8% of the Earth's surface and are found in mountains, plains, forests, deserts, and equatorial and polar regions, as well as on cliffs.[3] They also come in a variety of colors, sizes, shapes, and lifespans.

Lichens are divided into three main categories based on the morphology of the lichenized thalli: crustose, foliose and fruticose lichens.[15] Crustose lichens make up approximately 80% of all lichens, including the genus *Lecanora* growing on rocks and the genus *Graphis* growing on bark. The thalli of crustose lichens are flat and form crusts. They adhere firmly to the substrate with filaments of their fungal partners, and some even extend into the substrate, making them difficult to peel off.[16,17] Foliose lichens include the genus *Peltigera* growing on grasslands and the genus *Parmelia* growing on rocks or bark. The thalli of foliose lichens are leaf-like with distinct upper and lower surfaces and consist of flat sheets of tissue.[16,17] They are usually loosely attached to the substrate with rhizines on the lower surface and are easily detached from the substrate. Fruticose lichens include the genus *Cladonia* that grows erect from the ground and the genus *Usnea* that hangs from branches. The thalli of fruticose lichens are well developed, branched, shrub-like and generally erect and pendulous.[16,17] Typically, they are only affixed to the substrate at the base. In addition to these three types, lichenized thalli have intermediate types that are between crustose and foliose lichens, such as filamentose and leprose lichens.[11]

According to the distribution of algal cells in fungal tissues, lichenized thalli can also be divided into heteromerous lichens and homolomerous lichens.[17] A distinct layer of algae cells presents between the cortex and medullary layers in heteromerous lichens, whereas algae cells in homolomerous lichens are evenly dispersed throughout the cortex and medullary layers without a distinct layer. Most lichens are heteromerous lichens.

## 8.3 Overview of Biochemical Metabolites in Lichens

The unique structure of lichens in mutualism allows them to produce a range of unique metabolites not found in higher plants. Chemical substances in lichens were reported by William Nylander as early as the mid-19th century.[18] Zopf systematically reported on the chemical and pharmacological properties of lichen substances and the analytical techniques used for their detection in 1907.[8] However, knowledge of lichen substances was still very limited until the 1960s, when the introduction of analytical techniques such as thin layer chromatography, high performance liquid chromatography, spectroscopy and mass spectrometry enabled a large number of lichen substances to be isolated and identified.[3,8]

Lichen substances are derived from four main biosynthetic pathways.[8] Goga *et al.* gave an overview of the metabolites produced by the acetyl-malonate pathway, the shikimate pathway, and the mevalonate pathway.[19] The acetyl-malonate pathway synthesizes depsides (evernic acid, lecanoric acid, gyrophoric acid, atranorin, thamnolic acid, and umbilicaric acid), depsidones (protocetraric acid, fumarprotocetraric acid, physodic acid, and stictic acid), dibenzofurans (usnic acid and alectosarmentin), chromones (lepraric acid), xanthones (norlichexanthone and thiophanic acid) and anthraquinones (emodin and parietin). Polyketide synthases (PKSs) involved in this pathway have been reported to be responsible for regulating the assembly of the carbon spine in many secondary metabolites for the synthesis of basic structures.[6,18] Metabolites synthesized *via* the mevalonate pathway include terpenes (limonene, phytol, and zeorin), steroids (brassicasterol, ergosterol, and lichesterol), and carotenoids (β-carotene, lutein, and zeaxanthin). The shikimate pathway involves the synthesis of terphenylquinones (polyporic acid and thelephoric acid) and pulvinic acid derivatives (vulpinic acid and pulvinic acid). In addition, another is the synthesis of products of photosynthesis in phycobionts, including sugar alcohols, monosaccharides and polysaccharides.[8]

Lichen metabolites are divided into primary and secondary metabolites depending on their role in lichens. Primary metabolites, such as polysaccharides, amino acids, carotenoids, polyols, and vitamins, are bound in the cell wall and protoplasm and provide energy and raw materials to sustain the life activities of lichens.[19] They are also closely associated with the growth, development, and reproduction of lichens. The main component of the lichen cell wall is polysaccharide, which is abundant in the lichen body. More than 100 lichen polysaccharides have been examined to date, with lichenan, isolichenan, and galactomannans being the major structural types.[20] Lichenan is a β-glucan, a lichen polysaccharide with linear or almost no substituents. Isolichenan was the first α-glucan isolated from lichen, and its structure is mostly straight-chain (1→3)-(1→4)-α-D-glucan. The main chain of galactomannans consists of (1→6)-α-D-galactopyranose, and its side chains usually contain galactose, mannose and a small amount of glucose. These lichen polysaccharides are generally water soluble, can be extracted using hot water,[19] and have been reported to have immunoactivating, antitumor, cytotoxic, antioxidant, antiviral and other biological activities.[21,22]

The secondary metabolites of lichens are specific to the lichen and are produced extracellularly by mycobionts and deposited as amorphous or crystalline forms on the surface of the hyphae cells. They are maintained at very high levels in the lichen body, usually at 0.1–10% of the dry weight of the lichen, but sometimes up to 30%.[19] Over 1000 secondary metabolites have been reported and can be classified as xanthone derivatives, dibenzofurans, depsides, depsidones, depsones, lactones, quinones, pulvinic acid derivatives, terpenes, sterols, nitrogenous compounds and other phenolic compounds,[6] of which lactones, phenolic compounds, depsides, pulvinic acid derivatives and dibenzofurans are common and well-studied metabolites in lichens.[11] These secondary metabolites are insoluble in water and can be isolated from the lichens using organic solvents.[19] The role of lichen secondary metabolites in nature is unclear, but some researchers believe that environmental factors may affect the production of secondary metabolites,[19] which have been proposed to have anti-microbial, insecticidal, antiherbivore, allelopathic, photoprotective and water retention properties, and protect against heavy metal toxicity.[23] Additionally, it has been demonstrated that the secondary metabolites of lichens have a wide range of biological activities, including antibacterial, antifungal, antiviral, antioxidant, antitumor, anti-inflammatory, analgesic, enzyme inhibitor, photoprotection and cytotoxic activities.[6] These biological activities make them extremely valuable for research and have potential applications in many fields.

## 8.4 Pharmacological and Cosmeceutical Applications of Lichens

### 8.4.1 Antimicrobial Agents

The high content of nutrients and water in the raw materials and additives of cosmetic products provides favorable conditions for the growth of microorganisms. It is important to include preservatives or ingredients with antimicrobial activity to maintain the effectiveness of the product and protect it from contamination.

Lichen extracts and their metabolites have the potential to be used as cosmetic preservatives because they have the capacity to kill harmful bacteria and fungi, promote the microbiological stability of cosmetic formulations, and keep the skin in a stable state of floral balance. In traditional medicine, *Cladina arbuscula*, *Cladonia fruticulosa*, *Rhizoplaca chrysoleuca* and *Vulpicida pinastri* have been used to treat skin infections.[24] In modern studies, usnic acid is the most common and extensively studied lichen metabolite with strong anti-microbial activity. Cocchietto *et al.*, Ingólfsdóttir, Luzina and Salakhutdinov provided systematic reviews of the antibacterial and other activities of usnic acid.[3,9,25] Multi-resistant strains of *Staphylococcus aureus*, *Mycobacteria* spp., and *Enterococcus* spp. are all susceptible to usnic acid.[11] Because of its broad-spectrum antimicrobial activity, usnic acid has long been shown to be used as a preservative in cosmetic creams.[26] It was once used as an ingredient in antibacterial creams throughout Europe and marketed under the names "Usno®" and "Evosin®".[27] Higuchi *et al.* also developed facial acne control creams containing usnic acids or lichesterinic acids.[28] However, the potential allergic reactions of some people to this natural compound have led to these products not being widely used.[27] The application of usnic acid in cosmetics is restricted by its yellow color and the difficulties it has in dissolving in water and some organic solvents.[29] In addition to this, researchers have identified alternative substances to the preservatives commonly used in cosmetics from some lichen species. The methyl β-orsellinate and orsellinic acid esters isolated from Icelandic lichen species (*Stereocaulon alpinum* and *Peltigera aphthosa*) have a broad antimicrobial spectrum and higher potency compared to preservatives (methyl and propyl *p*-hydroxybenzoates, and chlorocresol) commonly used in medicine and cosmetics.[30] In terms of inhibitory activity against *S. aureus* by the broth microdilution assay, Dieu *et al.* discovered that condidymic acid, didymic acid, and (−)-usnic acid were significantly more effective than Phenonip® commonly used in the cosmetic industry.[29] In contrast to the widely studied usnic acid, condidymic acid and didymic acid can be good candidates to replace parabens in cosmetics and skin care products because of their colorlessness and higher solubility in polar solvents.[29] Using a similar method with Phenonip® as a positive control, Dieu *et al.* found that acetone extracts of *Usnea florida* and *Flavoparmelia caperata*, (+)-usnic acid, protocetraric acid and caperatic acid showed better anti-*S. aureus* activity. These may be good candidates for replacing parabens in cosmetics and skin care products.[31]

These studies have stimulated interest in the further development of lichen extracts and their metabolites as potentially useful preservatives as well as in the broader study of other lichens. The following is an overview of recent studies on the antibacterial and antifungal activities of lichens to inform the development of additional, more effective, and hypoallergenic natural preservatives.

### 8.4.2 Antibacterial Properties

Scientists have been particularly interested in the antibacterial activities of lichen substances. 50% of lichens have been found to exhibit antibiotic activity, and the indicators most frequently used to assess the antibiotic potential of lichens include minimum inhibitory concentration (MIC) and inhibition zone (IZ).[11,32] In earlier studies,[11] lichen extracts and their metabolites from the genera *Cetraria*, *Cladonia*, *Parmelia*, *Ochrolechia*, *Parmeliopsis*, *Lecanora*, *Usnea*, *Anaptychia*, *Peltigera*, *Umbilicaria*, *Xanthoria*, *Xanthoparmelia*, *Everniastrum*, *Ramalina*, and others were found to inhibit Gram-negative bacteria, primarily *Aeromonas*, *Enterobacter*, *Escherichia*, *Klebsiella*, *Pseudomonas*, *Proteus* and *Helicobacter*. Some of the lichens from these genera also have Gram-positive inhibitory activity, such as *Cladonia foliacea*, *C. furcata*, *Parmelia caperata*, *Parmelia conspresa*, *Ochrolechia androgyna*, *Parmeliopsis hyperopta*, *L. frustulosa*, *U. barbata*, *Anaptychia ciliaris*, *Peltigera polydactyla*, *P. praetextata*, *Umbilicaria vellea*, *Xanthoria elegans*, *Xanthoparmelia tinctina*, *Everniastrum cirrhatum* and *Ramalina farinacea*.[11] Besides, anti-Gram-positive activity was also observed in lichens from the genera *Everniastrum*, *Lasallia*, *Stereocaulon*, *Flavoparmelia*, *Heterodermia*, *Leptogium*, *Lobaria*, *Phaeophyscia*, *Rimelia*, *Stereocaulon*, *Rhizoplaca* and others.[11] The most studied Gram-positive genera are mainly *Bacillus*, *Staphylococcus*, *Listeria*, *Mybobacterium*, *Streptococcus*, and *Micrococcus*.[11] The main metabolites contributing to this activity are usnic acid, atranorin, fumarprotocetraric acid, protocetraric acid, stictic acid, lecanoric acid, divaricatic acid, zeorin, protolichesterinic acid, salazinic acid, and norstictic acid.[11,33]

In recent years, studies on antibacterial activity have focused on lichens of the genera *Usnea*, *Cladonia* and *Ramalina*. Bazarnova *et al.* observed that the 1,4-dioxane extract (0.24–0.6 mg $mL^{-1}$) of *U. barbata* lichen had an inhibitory effect on *Bacillus subtilis* and *Pseudomonas fluorescens* (IZ 27–29 mm).[34] Popovici *et al.* and Popovici *et al.* found that ethanol and acetone extracts of this lichen also had inhibitory activity against *S. aureus* and *Streptococcus pneumoniae*, but no inhibition was observed against *Klebsiella pneumoniae*.[35,36] Effective activity against bacteria of the genus *Staphylococcus* (mainly against *S. aureus*) was observed in organic extracts of lichens of the genera *Cladonia* (*C. foliacea*, *Cladonia uncialis*, *Cladonia substellata*, *Cladonia rangiformis*, *Cladonia crispatula* and *C. furcata*) and *Ramalina* (*Ramalina fastigiata*, *Ramalina sinensis* and *R. farinacea*), but no or only weak antibacterial activity was found against *Escherichia coli*.[37–41] Lysozyme potential was found in methanol and acetone extracts of *C. foliacea*.[42] Acetone extracts of *C. uncialis* showed an inhibition effect against *Staphylococcus epidermidis* and *Enterococcus faecium* and this was comparable to (+)-usnic acid activity.[41] The acetone extract of *R. farinacea* was effective against Gram-positive bacteria (methicillin-resistant *S. aureus*, *B. subtilis*, and *Listeria innocua*), but not against Gram-negative bacteria (*E. coli*, *Pseudomonas aeruginosa*, and *Proteus mirabilis*).[43]

Moreover, the antibacterial activity of the genera *Parmelia*, *Pseudevernia*, *Physcia*, *Dermatocarpon*, *Roccella*, *Evernia*, *Hypogymnia*, *Flavoparmelia*,

*Punctelia*, *Peltigera*, *Cryptothecia*, *Umbilicaria* and *Bulbothrix* lichens has also been reported in recent years. Rehman *et al.* reported significant inhibition of *S. epidermidis* by methanolic extracts of *Pseudevernia furfruracea*, *Physcia* sp., and *Dermatocarpon vellerum* (IZ 11.3–14.3 mm) and *Parmellia* spp. extracts had a moderate inhibitory effect on *S. epidermidis* (IZ 8.0 mm).[42] Tannins, phenols, saponins, and terpenoids isolated from these extracts showed effective inhibition of *E. coli*.[42] Sargsyan *et al.* studied eight lichens from the Armenia region (*R. sinensis*, *R. farinacea*, *F. caperata*, *E. prunastri*, *Punctelia subrudecta*, *P. furfuracea*, *P. praetextata* and *P. sulcata*), whose methanol, ethanol and acetone extracts (500 μg $mL^{-1}$) showed antibacterial activity against *B. subtilis* and *S. aureus* (IZ 3–25 mm).[44] Acetone extracts of *Hypogymnia tubulosa*, *Umbilicaria muhlenbergii* and *Bulbothrix setschwanensis* and methyl β-orcinolcarboxylate of *Cryptothecia faveomaculata* also demonstrated antibacterial activity against *S. aureus*.[45–48] Maurya *et al.* observed the disruption of bacterial cell walls by lichen extracts.[46] A wide range of antibacterial activity was found from the extract of *Roccella phycopsis*, with inhibition against *S. aureus*, *Enterococcus faecalis*, *E. coli*, *Acinetobacter baumannii* and *K. pneumoniae*.[37,49]

Therefore, lichen extracts and their metabolites have diverse antibacterial activities and may be used as potential natural preservatives in cosmetic products. The inhibitory activity against bacteria causing skin infections, such as *S. epidermidis* and *E. faecium*, makes lichens potentially valuable in cosmeceutical products as well. Recent studies on the synthesis of lichen metabolite derivatives and their antimicrobial activity (usnic acid, monoaromatic compounds and butyrolactone derivatives) have had a positive effect on optimizing the biological activity and toxicity of the metabolites and reducing concentrations and/or reaction times,[50–56] which has increased the possibility of lichen substances being used in cosmetics.

### 8.4.3 Antifungal Properties

Previous studies have demonstrated the antifungal effect of lichen substances on *Aspergillus flavus*, *Aspergillus niger*, *Aspergillus fumigatus*, *Candida albicans*, *Penicillium notatum*, *Cladosporium sphaerospermum*, *Fusarium moniliforme*, dermatophytes and a number of phytopathogenic fungi.[6,11] The lichen metabolites with antifungal activity that have been studied more are depsides (atranorin, lecanoric acid and divaricatic acid), monocyclic aromatic derivatives, depsidones (fumarprotocetraric acid, protocetraric acid, stictic acid and salazinic acid), aliphatic acids (protolichesterinic acid), usnic acid and zeorin.[11,32]

Pandey *et al.* found that *Cladia aggregata* extract was an effective inhibitor of *Malassezia furfur*, *Malassezia globosa* and *Malassezia sympodialis*, which cause pityriasis versicolor, seborrheic dermatitis and dandruff (MIC 0.63–2.72 mg $mL^{-1}$), and it was considered to have cosmeceutical potential.[57] Furmanek *et al.* reviewed the inhibitory effects of lichen extracts and their secondary metabolites on dermatophytic fungi, focusing on genera *Microsporum* (*Microsporum canis*, *Microsporum audouinii*, *Microsporum nanum* and *Microsporum gypseum*) and *Trichophyton* (*Trichophyton rubrum*, *Trichophyton mentagrophytes*,

*Trichophyton tonsurans* and *Trichophyton violaceum*) and *Epidermophyton floccosum* fungi, which cause fungal infections of the skin, hair and nails, usually by degrading keratin. *T. mentagrophytes* are among the most studied antidermatophytic fungi and inhibitory activity against them has also been observed from a large number of lichen species (more than 40).[32] Lichen extracts and metabolites from the genera *Hypotrachyna*, *Heterodermia*, *Variospora*, *Caloplaca*, *Flavoplaca*, *Protousnea*, *Usnea*, *Myelochroa*, *Ramalina*, *Cladia*, *Pseudocyphellaria*, *Lecanora*. *Cladonia*, *Bunodophoron*, *Cetraria*, *Leifidium*, *Nephroma*, *Peltigera*, *Psoroma*, *Siphula*, *Sphaerophorus*, *Stereocaulon*, *Hypogymnia*, *Notoparmelia*, *Parmelia*, *Xanthoparmelia* and *Evernia* have varying degrees of antidermatophytic activity.[32] Therefore, lichen substances can be used not only as potential antifungal agents in cosmetics, but also as potential active ingredients in cosmeceutical products against skin fungi.

Studies in recent years have focused on the antibacterial activity of lichen extracts against pathogenic fungi of the genus *Candida* and *A. fumigatus*. Sargsyan *et al.* observed that the methanolic extract of *P. furfuracea* exhibited antibacterial activity against *C. albicans* (MIC 3.75 mg $mL^{-1}$).[44] Mendili *et al.* studied the antifungal activities of methanol, acetone and quench extracts from *Diploschistes ocellatus*, *F. caperata*, *Squamarina cartilaginea* and *Xanthoria parietina* against *C. albicans*, *Candida sake* and *Candida parapsilosis*.[58] All the extracts were found to show anti-*Candida* activity and inhibited spore germination of the tested fungi.[58] Popovici *et al.* reported that acetone, ethanol, and methanol extracts of *U. barbata* showed inhibitory activity against *C. albicans* and its ethyl acetate extract showed inhibitory activity against *C. parapsilosis*, while usnic acid produced inhibitory effects against both fungi.[35] Girardot *et al.* evaluated the inhibitory activity of 11 polyphenols from lichens against *C. albicans* biofilms and their biofilm maturation.[59] Three depsides (evernic acid, squamatic acid and thamnolic acid) were able to retard biofilm maturation (MMIC50 $\leq$ 12.5 μg $mL^{-1}$) after 48 h of contact with *C. albicans* cells.[59] Evernic acid also showed a reduction in the formed biofilms (MBIC50 $\leq$ 12.5 μg $mL^{-1}$) after 48 h of contact with the pre-formed biofilm for 24 h.[59] Brakni *et al.* found that cyclohexane extracts of *C. rangiformis*, *R. farinaceae*, *R. fastigiata* and *R. phycopsis* all showed antifungal activity against *C. albicans* and *Candida glabrata* (MIC 8–62.5 μg $mL^{-1}$), but not against *A. fumigatus*.[37] Gandhi *et al.* studied three fractions of ethyl acetate extracts of *P. sulcata*, showing an inhibitory capacity against *A. fumigatus* (MIC 7.25–18.25 mg $mL^{-1}$), but still weaker than the inhibitory activity against *C. albicans* (MIC 5.18–8.25 mg $mL^{-1}$).[60] Maurya *et al.* not only observed effective antifungal activity of an acetone extract from *B. setschwanensis* against *Cryptococcus neoformans* (MIC 6.25 mg $mL^{-1}$), but also clearly saw cell wall damage of *C. neoformans* caused by the extract.[46]

### 8.4.4 Antioxidants

Both the external environment and the internal factors of the body contribute to the production of free radicals. Free radicals have extremely strong oxidation capacity and play a role in many chemical processes involved in life activities. The body's antioxidant enzyme system scavenges excess free radicals and keeps them at a normal physiological level under normal physiological conditions. When homeostasis is disrupted, oxidative stress occurs, leading to cell damage and aging, and promoting the development of many chronic diseases.[6] Antioxidants are effective by providing electrons to inhibit oxidation, absorbing free radicals to protect biological macromolecules from attack and scavenging free radicals. The application of antioxidant ingredients in cosmetics not only acts to slow down skin aging, but also maintains the sensory properties of cosmetics by inhibiting lipid oxidation. Additionally, antioxidants added to sunscreens can enhance photoprotection to better protect against UV radiation damage.

Kohlhardt-Floehr *et al.* investigated the antioxidant and pro-oxidant activities of usnic acid isolated from *Xanthoparmelia farinosa* on human lymphocytes under UVB irradiation.[61] The results showed that usnic acid exhibited a dual antioxidant-promotional behavior under UVB irradiation, but its pro-oxidant activity was weak.[61] At UVB doses up to 0.1 J $cm^{-2}$, low concentrations of usnic acid ($1 \times 10^{-8}$ and $1 \times 10^{-6}$ mol $L^{-1}$) exhibited high protective and antioxidant activities on cells.[61] It is advantageous for application in sunscreen products. Recently, Sim *et al.* observed that *U. esculenta* extract inhibited reactive oxygen species (ROS) production and matrix metalloprotein-1 (MMP-1) expression in human keratinocyte lines (HaCaT) and human skin tissues.[62] It was also revealed that *U. esculenta* extracts could enhance UV-inhibited phosphatase activity by inhibiting ROS and enhanced dephosphorylation of ErbB2 to suppress UV-induced MMP-1 expression and inhibit skin wrinkle formation.[62] Although the use of lichen extracts and metabolites as antioxidants in cosmetics is rare, antioxidant activities of various lichen extracts and their metabolites have been widely demonstrated. The following is an overview of recent studies on the antioxidant activity of lichens, which will provide a reference for the subsequent development of natural antioxidants in cosmetic products.

The lichens of the family Parmeliaceae are the most widely distributed and studied, including the genera *Bryoria*, *Cetraria*, *Evernia*, *Flavocetraria*, *Flavoparmelia*, *Hypogymnia*, *Lethariella*, *Parmelia*, *Parmotrema*, *Platismatia*, *Pseudevernia*, *Pseudophebe*, and *Usnea*.[63] In addition to this family, lichen extracts and metabolites from the genera *Arthothelium*, *Toninia*, *Cladonia*, *Graphis*, *Thamnolia*, *Lecanora*, *Lobaria*, *Nephroma*, *Ochrolechia*, *Peltigera*, *Anaptychia*, *Heterodermia*, *Physcia*, *Ramalina*, *Sphaerophorus*, *Stereocaulon*, *Calopaca*, *Fulgensia*, *Xanthoria*, *Laurera*, *Lasallia*, *Umbilicaria*, *Dermatocarpon* and *Sticta* have also been found to have antioxidant activity.[63]

In recent years, *in vitro* chemical studies have been carried out on the antioxidant effects of lichen substances mainly from the following lichens: *P. furfuracea*, *P. sulcata*, *Parmelia saxatilis*, *U. barbata*, *Usnea cornuta*, *H. tubulosa*, *Parmotrema tinctorum*, *R. farinacea*, *R. sinensis*, *Cladina mitis*, *Hypogymnia physodes*, *C. furcata*, *C. rangiformis*, *Cladonia chlorophaea*, *C. uncialis*, *F. caperata*, *P. subrudecta*, *P. praetextata*, *Cetraria islandica*, *S. cartilaginea*, *X. parietina*, *D. miniatum*, *Acarospora socialis*, *Xanthoparmelia mexicana*, *Xanthoparmelia stenophylla*, *Lobothallia alphoplaca*, *Hypotrachyna cirrhata*, *Hypotrachyna caraccensis*, *Dirinaria consimilis*, *V. pinastri*, *Himantormia lugubris*, *Stereocaulon montagneanum*, *D. ocellatus*, *Candelariella vitellina* and *Umbilicaria aprina*. Different levels of antioxidant activity were observed in the extracts of these lichens, mainly reflected in the scavenging activity of free radicals (ATBS+, DPPH, superoxide anions and hydroxyl radicals), oxidative radical absorption capacity, iron ion reduction ability and ferrous ion chelation activity.

Most researchers believe that the antioxidant activity of lichen extracts is related to the presence of phenolic compounds. Shcherbakova *et al.* found that the hexane extract of *E. prunastri* had the highest antioxidant capacity of all the extracts, corresponding to the highest total phenolic content (73 mg gallic acid per g extract), suggesting that the deoxidant capacity of the extracts was directly related to their total phenolic content.[64] Similar results were observed in a study by Mendili *et al.*[65] The methanolic extract of *D. ocellatus* had the highest DPPH scavenging capacity ($IC_{50}$ 0.029 mg $mL^{-1}$) and iron chelating capacity ($IC_{50}$ 0.425 mg $mL^{-1}$), obtaining the highest tannin and proanthocyanidin content (5.5 µg TAE per g DW and 35.12 µg CE per g DW).[65] Hawryl *et al.* studied the antioxidant properties of methanolic extracts of 20 lichens and found that the methanolic extracts of *H. physodes*, *Cladonia digitata* and *P. furfuracea* with the highest antioxidant activity had the highest phenolic content.[66] Aoussar *et al.* observed that the methanolic extract of *P. furfuracea* exhibited the strongest DPPH radical scavenging ability and iron reduction capacity compared to *E. prunastri* and *R. farinacea*.[67] They also found that the total phenolic content of the extracts showed a highly significant positive correlation with the antioxidant activity ($P < 0.01$).[67] Kalra *et al.* also investigated the correlation between the antioxidant potential (DPPH, ABTS+ and hydroxyl scavenging) of *P. furfuracea* extracts and the total polyphenol content and total flavonoid content.[68] The correlations between the total polyphenol content and antioxidant potential (0.979, 0.979, 0.962) were stronger than those between the total flavonoid content and antioxidant potential (0.920, 0.842, 0.936), but all were highly significant ($P < 0.01$).[68]

Phenolic compounds that contribute to the antioxidant properties of lichens include dibenzofurans, depsides, depsidones and pulvinic acid.[63,67] In a recent study, El-Garawani *et al.* identified gallic, chlorogenic, syringic, pyrocatechol, protocatechuic, caffeic and ellagic acids from the dromethanolic extract of *C. vitellina* with good DPPH radical scavenging activity.[69] Kalra *et al.* isolated atraric acid and olivetoric acid from *P. furfuracea* extracts and found a direct correlation between the concentrations of these two phenolics and the free radical scavenging potential of the extracts.[68] Elecko *et al.* investigated the antioxidant activity of gyrophoric acid isolated from *Umbilicaria hirsuta*, evernic acid isolated from *E. prunastri*, physodic acid isolated from *H. physodes*, 3-hydroxyphysodic acid, physodalic acid, atranorin and usnic acid.[70] The strongest scavenging of superoxide anion radicals was observed for 3-hydroxyphysodic acid (80%) and was higher than that of ascorbic acid (74%), while the scavenging of superoxide anion radicals for the rest of the substances ranged from 51% to 63%.[70] Legouin *et al.* found that vulpinic acid and pinastric acid from *V. pinastri* were moderate antioxidant compounds in the DPPH test ($IC_{50}$ 80.0 and 12.5 μg $mL^{-1}$), and usnic acids below 500 μg $mL^{-1}$ did not show DPPH scavenging activity, but vulpinic, pinastric and usnic acids all had good superoxide anion scavenging activity ($IC_{50}$ 30, 70 and 24 μg $mL^{-1}$).[71]

The polysaccharides in lichens have also been shown to have antioxidant properties.[6] In a recent study, Yang *et al.* used water extraction and graded ethanol precipitation to obtain four polysaccharides from *Usnea longissima*, ULP15, ULP30, ULP50 and ULP70 (different ethanol concentrations), all of which were heteropolysaccharides and contained small amounts of protein.[72] The four polysaccharides showed different degrees of antioxidant activity, with ULP15 showing the highest DPPH radical scavenging activity ($IC_{50}$ 3.38 mg $mL^{-1}$), hydroxyl radical scavenging activity (65.33% at 5 mg $mL^{-1}$) and the strongest iron-reducing ability, and ULP70 having the best ABTS+ radical scavenging activity ($IC_{50}$ 4.17 mg $mL^{-1}$).[72] Shang *et al.* obtained two purified grades (UTP-1 and UTP-2) from crude water-soluble polysaccharides of *Umbilicaria tornata*, both consisting of glucose and xylose.[73] The DPPH radical scavenging activity of these polysaccharides was $IC_{50}$ 0.886–1.003 mg $mL^{-1}$ and the hydroxyl radical scavenging activity was $IC_{50}$ 1.821–2.258 mg $mL^{-1}$, and they also possessed slight reducing activity.[73]

Moreover, mycosporine-like amino acids of lichen have been shown to have antioxidant effects. de la Coba *et al.* isolated mycosporine-glycine from aqueous extracts of marine lichen *Lichina pygmaea* and found that the mycosporine-glycine showed good antioxidant activity (ABTS+ radical scavenging and superoxide radical scavenging) at all pH levels.[74]

Studies on cellular substrates and animal models have contributed to further understanding of the antioxidant mechanisms of lichens,[6] but relatively little research has been done on this aspect. Lichen extracts or metabolites have been observed to exert their antioxidant effects by enhancing the antioxidant defence system of cells or tissues (increasing the levels of antioxidant enzymes SOD, CAT, and GPx) and by inhibiting lipid peroxidation. Most studies have focused on the antioxidant effects and cytotoxicity of lichen substances in metabolic diseases, aging-related diseases, cancer and other diseases. However, such studies investigating *in vitro* cellular and *in vivo* antioxidant studies of lichen substances in relation to beauty are still rare and require extensive research.

### 8.4.5 Sunscreen Agents

The skin is the first line of defence against UV radiation. Due to its short wavelength and high energy, UV radiation is able to pass through the epidermis and penetrate the basement membrane zone of the epidermis. Long-term exposure to UV radiation causes skin damage, accelerates skin aging, and may even lead to skin cancer. UV radiation is classified as UVA (315–400 nm), UVB (280–315 nm) and UVC (170–280 nm), depending on the wavelength.[75] The regular use of cosmetics containing various UV filters is recommended to prevent skin damage in addition to physical protection.[76] Some commercial synthetic UV filters may have adverse effects on humans and are also toxic to marine life,[76] so it may be safer to use natural products as sunscreens. Based on the UV absorbing region and antioxidant capacity of lichen substances, lichens are considered to be an important biological resource with photoprotective potential.

Nguyen *et al.* reviewed the metabolites with UV protective properties in lichens.[75] Depsides, depsidones, dibenzofuranes, and anthraquinones are the most abundant UVB screening compounds in lichens, and xanthones and pulvinic acid derivatives are used for UVA filtration.[75] Atranorin, panarin, usnic acid, parietin and calycin are among the most studied compounds.[75] Millot *et al.* screened *Diploicia canescens* from 12 lichen species as potential sources of UVB and/or UVA absorbing compounds by high performance thin layer chromatography.[77] A theoretical analysis of the *in vitro* photoprotective capacity and UV/visible light absorption functions of the compounds isolated from *D. canescens* revealed that xanthone derivatives and no chlorinated compounds should be the focus of the selection of photoprotection compounds in lichens.[77] After a preliminary assessment of photostability and cytotoxicity (human epidermal keratin-forming cells), the authors concluded that secalonic acid B showing a predominant band at 338 nm (UVA range) was less toxic compared to other compounds, suggesting that further study of this compound as a UV screening agent would be valuable.[77] Lohezic-Le Devehat *et al.* performed *in vitro* sun protection factor (SPF) and protection factor-UVA (PF-UVA) assays and a radiation exposure experiment on three lichen extracts (*C. islandica*, *Lasallia pustulata* and *Usnea hirta*), isolated metabolites (fumarprotocetraric acid, gyrophoric acid and (+)-usnic acid) and seven purchased lichen metabolites.[78] They found that *Lasallia pustulata* extract and gyrophoric acid were the most effective protectors and UV filters (SPF > 5, PF-UVA > 2), while variolaric acid, evernic acid, vulpinic acid and salazinic acid could be used as UV filters even if the SPF values were not high.[78] They also suggest that the additivity or synergy of lichen compounds with known sunscreens may also be of interest.[78] Legouin *et al.* proposed a photoprotection index calculation strategy and tested the *in vitro* photoprotective activity of vulpinic acid, pinastric acid and usnic acid from *V. pinastri* on the human keratinocyte cell line.[71] They found that vulpinic acid had the properties of a UVA filter due to its high spectral uniformity index value (7.6), low ideal spectral profile value (19%) and high PF-UVA value, pinastric acid had a high critical wavelength and low PF-UVA considered as a UVA booster, and usnic acid was similar to the sunscreen ethylhexyl methoxycinnamate and had the effect of a UVB filter.[71] Besides, these compounds were noncytotoxic and had better photostability in the UVA and UVB range compared to the sunscreen ethylhexyl methoxycinnamate.[71] Varol *et al.* preliminarily revealed the potential mechanism of the photoprotective activity of lichen substances on the human keratinocyte cell line and found that vulpinic and gyrophoric acids from *Letharia vulpina* and *Xanthoparmelia pokornyi* lichens could exert their photoprotective effects by attenuating UVB-induced cytotoxicity, apoptosis and cytoskeletal alterations.[79] All doses of vulpinic acid tested were non-cytotoxic, whereas high doses (≥400 μmol $L^{-1}$) of gyrophoric acid were cytotoxic.[79] Rancan *et al.* investigated the photoprotective effects of 1′-chloropannarine, usnic acid, epiphorelic acids I and II and calicin isolated from lichen extracts of *Erioderma leylandii*, *X. farinosa*, *Coelopogon epiphorellus* and *Pseudochellaria bereberina* using *in vivo* combined with *in vitro* cellular assays.[80] In *in vitro* experiments, the membrane protection factor (MPF) of lichen substances on human T lymphocytes under UVA and UVB radiation was calculated and the results indicated that usnic acid (4.5) and epiphorelic acids I and II (3.41, 3.38) were effective UVB filters and their mean MPF values were higher than those of reference octylmethoxycinnamate (3.22), while calicin showed better UVA filter properties and its mean MPF value (3.14) was only slightly lower than the reference 4-*tert*-butyl-4′-methoxy dibenzoylmethane (3.52).[80] In *in vivo* experiments, lichen substances were applied to the skin of volunteers to determine the UV protection factor (UVPF) and it was found that the mean UVPF of usnic acid as a UVB filter was 4.1, which was comparable to the reference Nivea Sun Spray LSF 5 (4.2), and the mean UVPF values of epiphorelic acids I and II and 1′-chloropannarine were 2.1–2.5.[80]

Studies on the skin penetrability of lichen substances are helpful in assessing their potential as UV filters in cosmetics, but research in this area is still limited. Rojas *et al.* used partition coefficients ($P$) expressed as $\log P$ to predict the degree of skin penetration of 3-methoxycarbonyl-2-hydroxy-6-methoxy-4-metylbenzoic acid, (+)-(9b-*R*)-usnic acid and decarboxythamnolic acid isolated from *Usnea roccellina*.[81] 3-Methoxycarbonyl-2-hydroxy-6-methoxy-4-metylbenzoic acid with good UVB protective properties and decarboxythamnolic acid with effective UVA photoprotective properties were observed to have good skin penetration and similarity to commercial sunscreens avobenzone and benzophenone-3.[81] Recently, Galanty *et al.* found that both usnic acid enantiomers as UV filters showed a fairly good ability to penetrate the skin barrier by combining a skin parallel artificial membrane permeability assay with spectrophotometric quantitative analysis.[76] After safety assessment, both usnic acid enantiomers showed a high safety profile against skin cells representing different skin layers (keratinocytes, melanocytes and fibroblasts) even after a longer exposure time of 24 h using a wider concentration range, and (+)-usnic acid was safer against normal skin cells.[76]

In addition to common metabolites of lichens, mycosporines, mycosporine-like amino acids and scytonemin in lichenized fungi are also photoprotective substances of interest.[75] de la Coba *et al.* isolated four mycosporine-like amino acids from three species of red macroalgae (*Porphyra rosengurttii*, *Gelidium corneum* and *Ahnfeltiopsis devoniensis*) and one species of marine lichen (*L. pygmaea*) and used them in sunscreen formulations for photoprotective activity studies.[82] Formulations containing these MAAs (4.1% porphyra-334 (+shinorine) + 2.9% mycosporine-serinol) were found to show similar SPF values to the reference sunscreen and also showed similar UVB-biological effective protection factors mainly in terms of DNA damage, photocarcinogenesis, systemic immunosuppression, and urocanic acid photoisomerization.[82] Torres *et al.* isolated a mycosporine, named collemin A, from the lichenified Ascomycete of *Collema cristatum*.[83] The *in vitro* cellular assays revealed that collemin A prevented the destruction of a UVB-irradiated human keratinocyte line in a dose-dependent manner and significantly reduced the production of pyrimidine dimers.[83] The *in vivo* experiment revealed that collemin A completely prevented UVB-induced skin erythema.[83]

To sum up, lichen extracts and metabolites can be used as potential novel sunscreen agents.

### 8.4.6 Anti-inflammatory Agents

Inflammation is the body's defence response to irritation and is manifested by redness, swelling, itching, pain and increased temperature.[84] Inflammatory reactions are generally beneficial, aiming to repair damage and protect the body from further damage. However, dysregulated inflammatory reactions are harmful, attacking the body's own tissues, causing damage and affecting normal tissue function.[6]

It has been documented that some lichens were used in traditional medicine to relieve inflammatory reactions in the skin.[24] For example, in China, *L. pulmonaria* powder was applied to affected areas to relieve local swelling of the skin and reduce inflammation and itching, and *U. florida* was used to reduce edema.[24] In India, high quality "charila" (consisting mainly of *Parmotrema nilgherrense* and some other lichen mixtures) was used for reducing swelling and soothing irritated skin.[24] *Usnea sikkimensis* was used for skin eruptions and can be placed in shoes to prevent or treat blisters.[24] The decoction of *Usnea* spp. was used by Europeans as an anti-inflammatory agent on affected skin areas.[24] The ancient Greeks used *Ramalina* spp. as a poultice to relieve inflammation for skin diseases.[24] Chewed *Evernia* sp. was used for skin eruptions.[24] Therefore, substances derived from lichens may have potential for development in the relief of skin inflammation. The following is an overview of recent studies on the anti-inflammatory activity of lichens, which will hopefully inform the development of future cosmeceutical products with the potential to alleviate skin inflammation.

Previous *in vivo* and *in vitro* studies have identified anti-inflammatory activity of lichen substances such as depside, depsidone, dibenzofuran and polysaccharides, which are able to exert their anti-inflammatory effects by inhibiting enzymes involved in the inflammatory process (cyclooxygenase, lipoxygenase and prostaglandin synthase), altering cytokine levels and regulating the key signaling pathway.[6] Extracts from lichens such as *P. furfuracea*, *Cetrelia monachorum*, *Parmotrema reticulatum*, *Lepraria incana* and *P. saxatilis* have also been demonstrated to have anti-inflammatory activity.[84]

In recent years, anti-inflammatory activity has been observed in more lichens. The inhibition of pro-inflammatory cytokines, chemokines, nitric oxide (NO) and prostaglandin E2 in lipopolysaccharide (LPS)-stimulated RAW 264.7 macrophages by lichen substances was investigated by *in vitro* cellular assays.[85] Mendili *et al.* found that the extracts of *Parmotrema hypoleucinum*, *R. phycopsis* and *X. parietina* showed good inhibition of NO in LPS-stimulated RAW 264.7 macrophages, ranging from 29.16% at 25 μg $mL^{-1}$ to 86.91% at 100 μg $mL^{-1}$.[86] Atranorin and (+)-iso-usnic acid isolated from the extracts showed higher anti-inflammatory potential, with 75.99% and 57.27% inhibition of NO production at a concentration of 25 μg $mL^{-1}$.[86] Lee *et al.* used a similar approach to demonstrate the ability of stereocalpin A and B from *Ramalina terebrata* to inhibit LPS-induced NO levels in RAW 264.7 macrophages with $IC_{50}$ values of 5–7 μmol $L^{-1}$, demonstrating the effective anti-inflammatory activity of the compounds.[87]

Several experiments have used quantitative reverse transcription-polymerase chain reaction and Western blot to measure mRNA and protein expression of pro-inflammatory mediators in *in vitro* cellular models to analyze the anti-inflammatory mechanisms of lichen substances. Carpentier *et al.* observed that lobaric acid, two pseudodepsidones and methylhaematommate isolated from *Stereocaulon paschale* inhibited nuclear factor kappa-B (NF-κB) activation and secretion of pro-inflammatory cytokines (interleukin (IL)-1β and tumor necrosis factor-α (TNF-α)) in LPS-stimulated RAW 264.7 macrophages.[88] Jin *et al.* found that amycolasporin C and 2-(4-hydroxybenzylacetone)-5-methoxyphenol in *Amycolatopsis hippodromi* were able to reduce NO production in a dose-dependent manner by inhibiting the expression of inducible nitric oxide synthase (iNOS) in LPS-stimulated RAW 264.7 macrophages.[89] Lee *et al.* revealed that lobaric acid from *S. alpinum* significantly reduced the production of TNF-α and IL-6 in LPS-stimulated RAW 264.7 macrophages by inhibiting the activation of mitogen-activated protein kinases (MAPKs) and NF-κB also inhibited the production of IL-1 β and IL-18 and caspase-1 maturation by suppressing NLRP3 inflammasome activation in macrophages.[90] Lim *et al.* suggested that an ethyl acetate extract of *Peltigera neopolydactyla* was able to reduce the production of inflammatory cytokines such as TNF-α, IL-6, IL-1β and granulocyte macrophage-colony stimulating factor (GM-CSF) in LPS-stimulated RAW 264.7 macrophages by inhibiting activator protein 1 (AP-1) signaling and its upstream mitogen-activated protein kinases kinase/mitogen-activated protein kinases (MKK/MAPKs).[85]

Recently, several studies have observed the anti-inflammatory effects of lichen substances using both *in vitro* cellular models and *in vivo* animal models. Kim *et al.* found that the methanolic extract of *Amandinea* sp. reduced the production of NO and pro-inflammatory cytokines (IL-6 and TNF-α) and inhibited the release of inflammatory mediators (iNOS and cyclooxygenase 2 (COX-2)) in LPS-stimulated RAW 264.7 macrophages.[91] Inhibition of mRNA expression of inflammatory cytokines and mediators was also observed in an inflammation-zebrafish model (tail-cutting).[91] Similar results were demonstrated by Hong *et al.*[92] They found that the methanolic extract of *Umbilicaria antarctica* significantly inhibited the production of NO, prostaglandin E2 (PGE2), IL-6 and TNF-α and suppressed mRNA and protein expression of iNOS, COX-2 and NF-κB in LPS-stimulated RAW 264.7 macrophages.[92]

Inhibition of mRNA and protein expression of relevant pro-inflammatory mediators was also demonstrated from an inflammation-zebrafish model (tail-cutting).[92] Mun *et al.* investigated the anti-inflammatory activity of atraric acid from *Heterodermia hypoleuca* in LPS-induced RAW 264.7 macrophages and LPS-induced endotoxic shock mouse models.[93] Atraric acid was found to inhibit the production of inflammatory factors by down-regulating the expression of phosphorylated inhibitor of NF-κB (IκB), extracellular signal regulated kinase (ERK) and NF-κB signaling pathways in macrophages, and was also found to reduce the secretion of cytokines in the peritoneum and serum of mice and to reverse organ damage caused by LPS.[93] Salem *et al.* used a carrageenan-induced paw oedema model in mice and found that both lichen extracts exhibited effective anti-inflammatory activity and the extract of *P. caperata* showed better anti-inflammatory effects and inhibition of paw oedema formation compared to *L. pulmonaria*.[94] During the time-course of carrageenan-induced paw oedema, the extract of *P. caperata* showed a similar profile to that of the reference anti-inflammatory drug diclofenac.[94]

In summary, lichen extracts and their metabolites are rich in anti-inflammatory activity. However, there are few reports of experiments utilizing skin as a test subject, and thus more research is needed to produce cosmeceuticals containing lichen substances that have inflammatory effects on the skin.

### 8.4.7 Skin Lightening and Depigmenting Agents

In Europe and the USA, tanning is fashionable, but Asian women prefer a fairer complexion.[8] Cosmetics with whitening properties are best sellers in Asia. Melanin is a dark brown biopigment, produced and stored by melanocytes in the basal layer, the amount and distribution of which determine the color of the skin.[95] Tyrosinase inhibitors have been frequently used in cosmetics as depigmentation agents for the treatment of hyperpigmentation and are also beneficial in reducing melanin production.[11] Earlier studies have identified that *Graphina glaucorufa*, *Graphina multistriata*, *Graphina salacinilabiata*, *Graphis assamensis*, *Graphis nakanishiana*, *Phaeographosis indica*, *U. esculenta*, *U. longissima*, *Usnea ghattensis*, *Heterodermia podocarpa*, *Arthothelium awasthii*, *P. tinctorum*, *H. physodes*, *L. vulpina* and *Cetraria juniperina* have tyrosinase inhibitory effects in the extracts of lichens.[96–99] Several lichen metabolites have also been shown to have tyrosinase inhibitory activity, including atranoin from *P. furfuracea*, resorcinol from *Protousnea* spp. and ramalin from *R. terebrata*.[11] Among them, ramalin from *R. terebrata* has been studied by Chang *et al.*[95] Ramalin was found to effectively reduce melanin synthesis in melan-a cells in a dose-dependent manner and significantly inhibited tyrosinase activity more effectively than arbutin (a well-known tyrosinase inhibitor).[95] It acted as a melanogenesis inhibitor by significantly reducing the protein levels of the melanogenic enzymes tyrosinase, tyrosinase-related protein 1 (TRP-1) and tyrosinase-related protein 2 (TRP-2). In clinical trials, the use of 0.2% ramalin was found to significantly improve the degree of brightness in human skin after 3 weeks.[95]

More lichen extracts and metabolites have been found to have tyrosinase inhibitory action recently. Lopes *et al.* found that resorcinol, orcinol, lecanoric acid, orsellinic acid, *n*-butyl, iso-propyl, *sec*-butyl, *n*-pentyl, *n*-hexyl and *n*-octyl orsellinates from *P. tinctorumdi* acted as inhibitors with 45.62%–90.59% tyrosinase activity at 0.50 mmol $L^{-1}$ compared to the negative control (100%).[100] *n*-Octyl orsellinates among them had the strongest inhibitory activity and kinetic studies revealed that *n*-octyl orsellinates had a non-competitive inhibition of tyrosinase, with an inhibition constant of 0.99 mmol $L^{-1}$.[100] Areche *et al.* observed that an alcoholic extract of *H. lugubris* and its main compound phenolic atranol also showed good tyrosinase inhibitory activity with $IC_{50}$ values of 22.32 μg $mL^{-1}$ and 7.25 μg $mL^{-1}$, respectively.[101] Malaspina *et al.* conducted *in vitro* and *in vivo* studies on the anti-tyrosinase activity of the chloroform, chloroform–methanol, methanol and water extracts from *C. islandica*, *F. caperata*, *L. vulpina* and *P. perlatum* and concluded that the extracts of *L. vulpina* and *C. islandica* are potential candidates for the development of drugs and cosmetics for skin whitening.[102] The tyrosinase inhibition assay showed that the methanolic extract of *L. vulpina* showed the highest inhibition ($IC_{50}$ 67 μg $mL^{-1}$), followed by the chloroform-methanolic extract of *C. islandica* ($IC_{50}$ 86 μg $mL^{-1}$).[102] The reduction in melanin content of MeWo human melanoma cells induced by 10 and 50 μg $mL^{-1}$ methanolic extracts of *L. vulpina* and 50 μg $mL^{-1}$ chloroform-methanolic extract of *C. islandica* was consistent with the positive control 8 mmol $L^{-1}$ arbutin (the melanin content was reduced by approximately 50%).[102] The *in vivo* experiments showed that methanolic and chloroform-methanolic extracts of *L. vulpina* and *C. islandica* reduced the pigmentation of zebrafish larvae and the methanolic extract of *L. vulpina* ($IC_{50}$ 30 μg $mL^{-1}$) showed higher inhibitory activity than the chloroform-methanolic extract of *C. islandica* ($IC_{50}$ 44 μg $mL^{-1}$).[102] The lichen metabolite 5-pentylresorcinol also showed excellent inhibitory activity against tyrosinase with an $IC_{50}$ value of 34.81 μmol $L^{-1}$.[102]

Acetylcholinesterase (AChE) inhibitors are commonly used for neuroprotection and the treatment of neurodegenerative diseases. AChE inhibitors were added to a cream as one of the skin lightening compositions in a patent,[103] so cholinesterase inhibitors may be valuable in cosmetics. The extracts of *Hypotrachyna formosana*, *C. uncialis*, *H. lugubris* and *Parmotrema perlatum* demonstrated good inhibitory activity against AChE.[13,101,104,105] Lichen secondary metabolites phenolic atranol and usnic acid had good inhibitory activity against AChE.[101,106] Perlatolic acid exhibited similar AChE inhibition ($IC_{50}$ 6.8 mol $L^{-1}$) to the positive compound galanthamine ($IC_{50}$ 2.5 μmol $L^{-1}$).[107]

The above evidence suggests that lichen extracts and their metabolites have the potential to be developed as skin lightening agents.

### 8.4.8 Moisturizers

Moisturizing and hydrating are essential in skin care. This helps to maintain the appearance and elasticity of the skin and enhances the skin's barrier effect against harmful environments. Lichens have been used as moisturizers in traditional medicine. Decoctions of lichens from the genus *Alectoria*, *L.*

*pulmonaria*, *Usnea* sp. and *P. aphthosa* have been used externally on chapped skin of infants or chapped foot skin of adults.[24] In Europe, a decoction of *Usnea* spp. was used as a moisturizer in the treatment of drying skin lesions.[24] In Canada, *L. pulmonaria* was applied to children's faces when their skin peeled.[24] Lichenans in *C. islandica* were used to moisturize the mucous membrane of the throat.[108]

Rahn researchers blended extracts of lichen (*C. islandica*), moss (*Sphagnum magellanicum*) and fern (*Polypodium vulgare*) to develop a new anti-aging moisturizer (Hydractin®-LMF) and found positive effects on skin aging parameters including hydration, elasticity, roughness and firmness.[109] The active ingredients of the lichen extract are polysaccharides, especially lichenan and isolichenan, which hydrate the skin surface by forming a moisturizing physical film. Besic Gyenge *et al.* conducted a double-blind, placebo-controlled, intra-individual study on the same formulation of 3% Hydractin®-LMF and confirmed its ability to increase skin hydration while improving skin firmness and roughness.[108] They concluded that lichenans from lichens have a direct effect on skin hydration and skin feel.[108]

There are limited reports on the moisturizing effect of lichen extracts and metabolites. Some sugar polymers prevent water evaporation by forming a protective film on the surface of the skin through high hydration capacity, such as hyaluronic acid and algal polysaccharides.[108] Therefore, higher levels of lichens with multiple bioactive polysaccharides may be potential moisturizers in cosmetics, and the hygroscopicity and moisture retention of more lichen species as well as other lichen polysaccharides can be investigated in the future.

### 8.4.9 Use in Perfume and Other Products

A large number of lichens have been used as fragrances since the 16th century or earlier, including some species of the families Cladoniaceae, Stictaceae, Parmeliaceae, and Usneaceae.[110] In Africa, *Parmelia andina* was used as a tobacco flavor.[111] In ancient Egypt, lichen was used as a wrapping material for mummies, serving the purpose of embalming and preventing odors.[111] The two main species of lichens used in the modern perfumery industry are *E. prunastri*, commonly known as "oakmoss", collected from oak trees, and *P. furfuracea*, commonly found on pine and cedar trees, known as "treemoss" and "cedarmoss".[112,113] More than 170 components have been identified from oak moss extracts, including 47 depsides or depside-derived compounds and 25 triterpenes or steroids, and more than 90 components from tree moss extracts, including 42 depsides, depsidones or depside-derived compounds, and 42 triterpenes or steroids. Joulaina and Tabacchi have provided a systematic review of these components.[112,113] Other lichen species are also used as fragrances. For example, *Ramalina fraxinea* is sometimes mixed with *Usnea caucasica* for industrial processing.[113] The lichens collected in India are graded. *Parmotrema nilgherrensis* are exported and used in fine perfumery, while lichens containing 90%–95% *P. nilgherrensis* and *Everniastrum*

**Table 8.1** Some lichen substances that have been used in commercial cosmetic products.

| Lichen species | Chemical constituents | Cosmeceutical properties | Application (brand or ingredient name) |
|---|---|---|---|
| Many species | Usnic acid | Anti-microbial, anti-aging, antioxidant | Cosmetic creams (Usno®, Evosin®, Natura Siberica, PhytoCide Lichen, Jurlique) and deodorant (Lavanila and usNeo)[26,27,29,110] |
| *Cetraria islandica* | Polysaccharide | Moisturizing effect | Skin care products (Hydractin®-LMF), hair care products (Aveda Sap Moss, Eudermique, Natural Springs, Natural Balance and Polaar cryo-firming gel)[109,110] |
| | Extract | Perfume | Deodorants (Polaar Men); toothpastes (Kiss my face)[110] |
| *Umbilicaria esculenta* | Polysaccharide | Moisturizing effect | Facial mask (Luxliss)[110] |
| *Evernia prunastri* and *Pseudevernia furfuracea* | Extract (depsides, depside-derived compounds, triterpenes and steroids) | Perfume – used to extend and fix the scent | Perfumery industry (many brands)[110] |
| *Usnea barbata* | Extract | Perfume | Deodorants (Centella Men)[110] |

*nepalense*, mixed with 5%–10% *Ramalina* lichens, are used in the local perfume industry.[110] *R. fastigiata*, *Evernia mesomorpha* and *Cetrariastrum nepalense* are collected in China as "Chinese oak moss" for use in perfumery by ethnic groups.[110]

The fragrance of lichen extract is often categorized as "woody" and "green" and is very popular among perfumers and has been used in many well-known brands.[110,114] In addition to their own aroma, lichen substances are also used in perfumery to fix and prolong the release of fragrances.[27] The applications of lichen substances in other commercial products are shown in Table 8.1.

## 8.5 Conclusion

Lichens have been used since ancient times in various fields, such as food, medicine, cosmetics, spices, and dyes. The unique symbiotic structure of lichens enables them to produce a large number of unique metabolites, such as polysaccharides, dibenzofurans, depsides, depsidones, depsones, lactones, quinones and pulvinic acid derivatives. The diverse biological activities of lichen extracts and their metabolites have been continuously demonstrated in recent decades. At present, the application of lichen substances in modern cosmetics is fragmented and very limited. The biological activities of lichen substances with potential value for cosmetic development and ingredient selection are mainly antibacterial, antifungal, antioxidant, photoprotective, anti-inflammatory, enzymatic inhibitory and moisturizing activities, which may theoretically help to prevent skin aging and sunburn and play a role in skin whitening, antiseptic, anti-inflammatory, moisturizing and hydrating activities. Among them, antibacterial, antifungal, antioxidant and anti-inflammatory activities have been systematically studied, while the photoprotective, enzymatic inhibitory and moisturizing activities have seen relatively little study. Most of these studies are *in vitro* experiments and the mechanism of action of lichen substances needs to be further explored. In addition, the skin penetration ability and skin safety of most lichen bioactive components have not been proven and still require extensive research.

## References

1. M. J. Calcott, D. F. Ackerley, A. Knight, R. A. Keyzers and J. G. Owen, *Chem. Soc. Rev.*, 2018, **47**, 1730–1760.
2. M. Kosanic, B. Rankovic, T. Stanojkovic, P. Vasiljevic and N. Manojlovic, *EXCLI J.*, 2014, **13**, 1226–1238.
3. M. Cocchietto, N. Skert, P. L. Nimis and G. Sava, *Naturwissenschaften*, 2002, **89**, 137–146.
4. R. Lücking, B. P. Hodkinson and S. D. Leavitt, *Bryologist*, 2017, **120**, 58–69.
5. Y. Yamamoto, *Korean J. Mycol.*, 2005, **10a**, 19–23.
6. Y. Zhao, M. Wang and B. Xu, *J. Funct. Foods*, 2021, **80**, 104283.
7. B. Aguirre-Hudson, I. Whitworth and B. M. Spooner, *Bot. J. Linn. Soc.*, 2011, **166**, 185–211.
8. W. Zopf, *Die Flechtenstoffe in chemischer, botanischer, pharmakologischer und technischer Beziehung*, G. Fischer, Jena, 1907, 450 pp.
9. K. Ingólfsdóttir, *Phytochemistry*, 2002, **61**, 729–736.
10. N. T. Manojlovic, P. J. Vasiljevic, P. Z. Maskovic, M. Juskovic and G. Bogdanovic-Dusanovic, *J. Evidence-Based Complementary Altern. Med.*, 2012, **2012**, 452431.
11. V. P. Zambare and L. P. Christopher, *Pharm. Biol.*, 2012, **50**, 778–798.
12. D. Armaleo, O. Muller, F. Lutzoni, O. S. Andresson, G. Blanc, H. B. Bode, F. R. Collart, F. Dal Grande, F. Dietrich, I. V. Grigoriev, S. Joneson, A. Kuo, P. E. Larsen, J. M. Logsdon Jr., D. Lopez, F. Martin, S. P. May, T. R. McDonald, S. S. Merchant, V. Miao, E. Morin, R. Oono, M. Pellegrini, N. Rubinstein, M. V. Sanchez-Puerta, E. Savelkoul, I. Schmitt, J. C. Slot, D. Soanes, P. Szovenyi, N. J. Talbot, C. Veneault-Fourrey and B. B. Xavier, *BMC Genomics*, 2019, **20**, 605.
13. D. L. Hawksworth and R. Honegger, The lichen thallus: a symbiotic phenotype of nutritionally specialized fungi and its response to gall

producers, in *Plant Galls: Organisms, Interactions, Populations*, ed. M. A. J. Williams, Clarendon Press, Oxford, 1994, pp. 77–98.
14. G. Seneviratne and I. K. Indrasena, *J. Biosci.*, 2006, **31**, 639–643.
15. H. Y. Bao and T. Bau, Advance in studies on chemical constituents and pharmacological activity of lichens in Usnea genus, *Zhongguo Zhongyao Zazhi*, 2013, **38**(4), 539–545.
16. H. M. Jahns, *The Lichens*, 1973, pp. 3–58.
17. Y. Zhang and N. Zhou, *Botany*, Huazhong University of Science and Technology Press, Wuhan, 2014.
18. O. Vitikainen, Great Discoveries in Bryology and Lichenology: William Nylander (1822-1899) and Lichen Chemotaxonomy, *The Bryologist*, 2001, **104**, 263–267.
19. M. Goga, J. Elečko, M. Marcinčinová, D. Ručová, M. Bačkorová and M. Bačkor, in *Co-Evolution of Secondary Metabolite*, ed. J. M. Merillon, and K. Ramawat, Springer, Cham, 2018, pp. 175–209.
20. Z. Yang, J. Li, J. Su, H. Li, Z. Zhang and F. Peng, *Chem. Ind. For. Prod.*, 2021, **41**, 112–124.
21. E. S. Olafsdottir and K. Ingolfsdottir, *Planta Med.*, 2001, **67**, 199–208.
22. S. Ullah, A. A. Khalil, F. Shaukat and Y. Song, *Foods*, 2019, **8**, 304.
23. S. Gulijiahan and A. Abbas, *Food Sci.*, 2007, **28**, 624–627.
24. B. Ranković, *Lichen Secondary Metabolites*, Springer, Switzerland, 2015.
25. O. A. Luzina and N. F. Salakhutdinov, *Expert Opin. Ther. Pat.*, 2018, **28**, 477–491.
26. P. Seifert, C. Bertram, S. Ö. F. W. Seifen and F. Öle, *Wachse*, 1995, **121**, 480–485.
27. R. K. Upadhyay, *Advances in Microbial Toxin Research and its Biotechnological Exploitation*, Springer, New York, 2002.
28. M. Higuchi, Y. Miura, Y. Kinoshita, Y. Yamamoto and S. Mayama, *Jpn. Kokai Tokkyo Koho*, 1993, 3.
29. A. Dieu, M. Millot, Y. Champavier, L. Mambu, V. Chaleix, V. Sol and V. Gloaguen, *Planta Med.*, 2014, **80**, 931–935.
30. K. Ingolfsdottir, S. Bloomfield and P. Hylands, *Antimicrob. Agents Chemother.*, 1985, **28**, 289–292.
31. A. Dieu, L. Mambu, Y. Champavier, V. Chaleix, V. Sol, V. Gloaguen and M. Millot, *Nat. Prod. Res.*, 2020, **34**, 3358–3362.
32. L. Furmanek, P. Czarnota and M. R. D. Seaward, *J. Appl. Microbiol.*, 2019, **127**, 308–325.
33. A. Podterob, *Pharm. Chem. J.*, 2008, **42**, 582–588.
34. Y. Bazarnova, N. Politaeva and N. Lyskova, *Z. Naturforsch., C: J. Biosci.*, 2018, **73**, 291–296.
35. V. Popovici, L. Bucur, S. I. Calcan, E. I. Cucolea, T. Costache, D. Rambu, V. Schroder, C. E. Gird, D. Gherghel, G. Vochita, A. Caraiane and V. Badea, *Plants*, 2021, **11**, 32.
36. V. Popovici, L. Bucur, C. E. Gird, A. Popescu, E. Matei, G. C. Cozaru, V. Schroder, E. A. Ozon, A. C. Fita, D. Lupuliasa, M. Aschie, A. Caraiane, M. Botnarciuc and V. Badea, *Pharmaceuticals*, 2022, **15**, 829.

37. R. Brakni, M. A. Ahmed, P. Burger, A. Schwing, G. Michel, C. Pomares, L. Hasseine, L. Boyer, X. Fernandez, A. Landreau and T. Michel, *Chem. Biodiversity*, 2018, **15**, e1800031.
38. M. Mendili, M. R. D. Seaward and A. Khadhri, *Nat. Prod. Res.*, 2022, **36**, 3095–3099.
39. A. C. Micheletti, N. K. Honda, L. M. Ravaglia, T. Matayoshi and A. A. Spielmann, *An. Acad. Bras. Cienc.*, 2021, **93**, e20191194.
40. R. Santiago, M. C. B. Martins, M. D. Vilaca, L. F. B. Barros, T. Nascimento, N. H. D. Silva, E. P. S. Falcao, M. E. Legaz, C. Vicente and E. C. Pereira, *Fitoterapia*, 2018, **131**, 23–34.
41. E. Studzinska-Sroka, T. Hanna, M. Natalia, W. Marta, K. Robert, G. Agnieszka, C.-P. Judyta, L. Dorota and P. Jaroslaw, *Acta Biochim. Pol.*, 2019, **66**, 597–603.
42. S. Rehman, S. Andleeb, A. R. Niazi and S. Ali, *Pak. J. Pharm. Sci.*, 2018, **31**, 1407–1416.
43. N. Aoussar, F. E. Laasri, M. Bourhia, N. Manoljovic, R. A. Mhand, N. Rhallabi, R. Ullah, A. A. Shahat, O. M. Noman, F. A. Nasr, O. M. Almarfadi, M. El Mzibri, P. Vasiljevic, L. Benbacer and F. Mellouki, *J. Evidence-Based Complementary Altern. Med.*, 2020, **2020**, 8104538.
44. R. Sargsyan, A. Gasparyan, G. Tadevosyan and H. Panosyan, *AMB Express*, 2021, **11**, 110.
45. L. Letwin, L. Malek, Z. Suntres and L. Christopher, *Curr. Pharm. Biotechnol.*, 2020, **21**, 1516–1527.
46. I. K. Maurya, S. Singh, R. Tewari, M. Tripathi, S. Upadhyay and Y. Joshi, *Microb. Pathog.*, 2018, **115**, 12–18.
47. N. N. Tuan, P. C. Kuo, T. T. Hieu, L. N. T. Vi, Q. T. Hung, L. T. Dunge, N. D. Trinh, N. Q. Trung, N. C. Khoa, H. V. Hai and T. D. Thang, *Nat. Prod. Res.*, 2021, **35**, 1349–1356.
48. G. Stojanovic, I. Zlatanovic, I. Zrnzevic, M. Stankovic, V. S. Jovanovic and B. Zlatkovic, *Nat. Prod. Res.*, 2018, **32**, 2735–2739.
49. J. G. Zorrilla, T. D'Addabbo, E. Roscetto, C. Varriale, M. R. Catania, M. C. Zonno, C. Altomare, G. Surico, P. L. Nimis and A. Evidente, *Int. J. Mol. Sci.*, 2022, **23**, 8471.
50. H. D. A. Araujo, H. Silva, J. G. D. Silva Junior, M. Albuquerque, L. Coelho and A. L. Aires, *Molecules*, 2021, **26**, 5995.
51. P. K. Bangalore, R. K. Pedapati, A. N. Pranathi, U. R. Batchu, S. Misra, M. Estharala, D. Sriram and S. Kantevari, *Mol. Diversity*, 2023, **27**, 811–836.
52. P. K. Bangalore, S. K. Vagolu, R. K. Bollikanda, D. K. Veeragoni, P. C. Choudante, S. Misra, D. Sriram, B. Sridhar and S. Kantevari, *J. Nat. Prod.*, 2020, **83**, 26–35.
53. D. Cirillo, E. Borroni, I. Festoso, D. Monti, S. Romeo, D. Mazier and L. Verotta, *Arch. Pharm.*, 2018, **351**, e1800177.
54. T. H. Do, T. H. Duong, H. T. Nguyen, T. H. Nguyen, J. Sichaem, C. H. Nguyen, H. H. Nguyen and N. P. Long, *Molecules*, 2022, **27**, 2871.

55. A. Sweidan, I. Smida, M. Chollet-Krugler, A. Sauvager, J. Vallet, N. Gouault, N. Oliviero, Z. Tamanai-Shacoori, A. Burel, P. van de Weghe, A. Chokr, S. Tomasi and L. Bousarghin, *Fitoterapia*, 2019, **137**, 104274.
56. K. Victor, L. Boris, G. Athina, P. Anthi, S. Marija, K. Marina, R. Oliver and S. Marina, *MedChemComm*, 2018, **9**, 870–882.
57. A. Pandey, R. K. Mishra, A. K. Tiwari, A. Kumar, A. K. Bajaj and A. Dikshit, *BioMed Res. Int.*, 2013, **2013**, 169794.
58. M. Mendili, B. Essghaier, M. R. D. Seaward and A. Khadhri, *Arch. Microbiol.*, 2021, **203**, 1461–1469.
59. M. Girardot, M. Millot, G. Hamion, J. L. Billard, C. Juin, G. Ntoutoume, V. Sol, L. Mambu and C. Imbert, *Front. Cell. Infect. Microbiol.*, 2021, **11**, 698883.
60. A. D. Gandhi, K. Umamahesh, S. Sathiyaraj, G. Suriyakala, R. Velmurugan, D. A. Al Farraj, M. R. A. Gawwad, K. Murugan, R. Babujanarthanam and R. Saranya, *J. Infect. Public Health*, 2022, **15**, 491–497.
61. C. Kohlhardt-Floehr, F. Boehm, S. Troppens, J. Lademann and T. G. Truscott, *J. Photochem. Photobiol., B*, 2010, **101**, 97–102.
62. W. J. Sim, Y. Kim, D. U. Jo, J. W. Seo, S. Chung, H. K. Choi, K. T. Kim and T. G. Lim, *J. Med. Food*, 2022, **25**, 770–777.
63. C. Fernandez-Moriano, M. P. Gomez-Serranillos and A. Crespo, *Pharm. Biol.*, 2016, **54**, 1–17.
64. A. Shcherbakova, A. A. Stromstedt, U. Goransson, O. Gnezdilov, A. Turanov, D. Boldbaatar, D. Kochkin, G. Ulrich-Merzenich and A. Koptina, *World J. Microbiol. Biotechnol.*, 2021, **37**, 129.
65. M. Mendili, M. Bannour, M. E. M. Araujo, S. Aschi-Smiti, M. R. D. Seaward and A. Khadhria, *Int. J. Med. Mushrooms*, 2019, **21**, 817–823.
66. A. Hawryl, M. Hawryl, A. Hajnos-Stolarz, J. Abramek, A. Bogucka-Kocka and L. Komsta, *Molecules*, 2020, **25**, 4301.
67. N. Aoussar, M. Achmit, Y. Es-Sadeqy, P. Vasiljevic, N. Rhallabi, R. A. Mhand, K. Zerouali, N. Manojlovic and F. Mellouki, *Arch. Microbiol.*, 2021, **203**, 2887–2894.
68. R. Kalra, X. A. Conlan, C. Areche, R. Dilawari and M. Goel, *Front. Pharmacol.*, 2021, **12**, 629695.
69. I. El-Garawani, M. Emam, W. Elkhateeb, H. El-Seedi, S. Khalifa, S. Oshiba, S. Abou-Ghanima and G. Daba, *Pharmaceutics*, 2020, **12**, 477.
70. J. Elecko, M. Vilkova, R. Frenak, D. Routray, D. Rucova, M. Backor and M. Goga, *Plants*, 2022, **11**, 1077.
71. B. Legouin, F. Lohezic-Le Devehat, S. Ferron, I. Rouaud, P. Le Pogam, L. Cornevin, M. Bertrand and J. Boustie, *Molecules*, 2017, **22**, E1162.
72. Z. Yang, Y. Hu, P. Yue, H. Luo, Q. Li, H. Li, Z. Zhang and F. Peng, *ACS Omega*, 2021, **6**, 25010–25018.
73. J. Shang, M. Fang, L. Zhang, H. Wang, G. Gong, Z. Wang, A. Zhao and H. Yi, *Glycoconjugate J.*, 2018, **35**, 107–117.
74. F. de la Coba, J. Aguilera, F. L. Figueroa, M. V. De Gálvez and E. Herrera, *J. Appl. Psychol.*, 2009, **21**, 161–169.

75. K. H. Nguyen, M. Chollet-Krugler, N. Gouault and S. Tomasi, *Nat. Prod. Rep.*, 2013, **30**, 1490–1508.
76. A. Galanty, J. Popiol, M. Paczkowska-Walendowska, E. Studzinska-Sroka, P. Pasko, J. Cielecka-Piontek, E. Pekala and I. Podolak, *Molecules*, 2021, **26**, 5224.
77. M. Millot, F. Di Meo, S. Tomasi, J. Boustie and P. Trouillas, *J. Photochem. Photobiol., B*, 2012, **111**, 17–26.
78. F. Lohezic-Le Devehat, B. Legouin, C. Couteau, J. Boustie and L. Coiffard, *J. Photochem. Photobiol., B*, 2013, **120**, 17–28.
79. M. Varol, A. Turk, M. Candan, T. Tay and A. T. Koparal, *Phytother. Res.*, 2016, **30**, 9–15.
80. F. Rancan, S. Rosan, K. Boehm, E. Fernández, M. E. Hidalgo, W. Quihot, C. Rubio, F. Boehm, H. Piazena and U. Oltmanns, *J. Photochem. Photobiol., B*, 2002, **68**, 133–139.
81. J. L. Rojas, M. Diaz-Santos and N. A. Valencia-Islas, *Pharm. Biosci. J.*, 2015, **3**, 18.
82. F. de la Coba, J. Aguilera, N. Korbee, M. V. de Galvez, E. Herrera-Ceballos, F. Alvarez-Gomez and F. L. Figueroa, *Mar. Drugs*, 2019, **17**, 55.
83. A. Torres, M. Hochberg, I. Pergament, R. Smoum, V. Niddam, V. M. Dembitsky, M. Temina, I. Dor, O. Lev, M. Srebnik and C. D. Enk, *Eur. J. Biochem.*, 2004, **271**, 780–784.
84. E. Studzińska-Sroka and A. Dubino, *Herba Pol.*, 2018, **64**, 56–64.
85. E. H. Lim, S. K. Mun, J. J. Kim, D. J. Chang and S. T. Yee, *BioMed Res. Int.*, 2022, **2022**, 2717196.
86. M. Mendili, A. Khadhri, J. Mediouni-Ben Jemaa, A. Andolfi, I. Tufano, S. Aschi-Smiti and M. DellaGreca, *Chem. Biodiversity*, 2022, **19**, e202200134.
87. S. Lee, S. Y. Jeong, D. L. Nguyen, J. E. So, K. H. Kim, J. H. Kim, S. J. Han, S. S. Suh, J. H. Lee and U. J. Youn, *Metabolites*, 2022, **12**, 141.
88. C. Carpentier, X. Barbeau, J. Azelmat, K. Vaillancourt, D. Grenier, P. Lague and N. Voyer, *Bioorg. Med. Chem.*, 2018, **26**, 5845–5851.
89. Y. Jin, N. Aobulikasimu, Z. Zhang, C. Liu, B. Cao, B. Lin, P. Guan, Y. Mu, Y. Jiang, L. Han and X. Huang, *J. Nat. Prod.*, 2020, **83**, 3545–3553.
90. H. W. Lee, J. Kim, J. H. Yim, H. K. Lee and S. Pyo, *Planta Med.*, 2019, **85**, 302–311.
91. J. E. Kim, S. K. Min, J. M. Hong, K. H. Kim, S. J. Han, J. H. Yim, H. Park and I. C. Kim, *Fish Shellfish Immunol.*, 2020, **107**, 301–308.
92. J. M. Hong, J. E. Kim, S. K. Min, K. H. Kim, S. J. Han, J. H. Yim, H. Park, J. H. Kim and I. C. Kim, *BioMed Res. Int.*, 2021, **2021**, 8812090.
93. S. K. Mun, K. Y. Kang, H. Y. Jang, Y. H. Hwang, S. G. Hong, S. J. Kim, H. W. Cho, D. J. Chang, J. S. Hur and S. T. Yee, *Int. J. Mol. Sci.*, 2020, **21**, 7070.
94. S. Salem, E. Leghouchi, R. Soulimani and J. Bouayed, *Int. J. Vitam. Nutr. Res.*, 2021, **91**, 143–151.
95. Y. H. Chang, J. S. Ryu, S. H. Lee, S. G. Park, H. D. Bhattarai, J. H. Yim and M. H. Jin, *J. Soc. Cosmet. Sci. Korea*, 2012, **38**, 247–254.

96. B. C. Behera, B. Adawadkar and U. Makhija, *J. Herb. Pharmacother.*, 2006, **6**, 55–69.
97. M. Higuchi, Y. Miura, J. Boohene, Y. Kinoshita, Y. Yamamoto, I. Yoshimura and Y. Yamada, *Planta Med.*, 1993, **59**, 253–255.
98. M. S. Kim and H. B. Cho, *J. Microbiol.*, 2007, **45**, 578–582.
99. N. Smit, J. Vicanova and S. Pavel, *Int. J. Mol. Sci.*, 2009, **10**, 5326–5349.
100. T. I. B. Lopes, R. G. Coelho and N. K. Honda, *Chem. Pharm. Bull.*, 2018, **66**, 61–64.
101. C. Areche, J. R. Parra, B. Sepulveda, O. Garcia-Beltran and M. J. Simirgiotis, *Metabolites*, 2022, **12**, 560.
102. P. Malaspina, E. Catellani, B. Burlando, D. Brignole, L. Cornara, M. Bazzicalupo, S. Candiani, V. Obino, V. De Feo, L. Caputo and P. Giordani, *PeerJ*, 2020, **8**, e9150.
103. S. A. Madison, *US Pat.*, 12/332002, 2008, https://patents.justia.com/patent/8247405.
104. V. Sieteiglesias, E. Gonzalez-Burgos, P. Bermejo-Bescos, P. K. Divakar and M. P. Gomez-Serranillos, *Chem. Res. Toxicol.*, 2019, **32**, 1165–1177.
105. E. Studzinska-Sroka, A. Majchrzak-Celinska, P. Zalewski, D. Szwajgier, E. Baranowska-Wojcik, B. Kapron, T. Plech, M. Zarowski and J. Cielecka-Piontek, *Pharmaceuticals*, 2021, **14**, 1293.
106. K. C. Cakmak and I. Gulcin, *Toxicol. Rep.*, 2019, **6**, 1273–1280.
107. R. G. Reddy, L. Veeraval, S. Maitra, M. Chollet-Krugler, S. Tomasi, F. L. Devehat, J. Boustie and S. Chakravarty, *Phytomedicine*, 2016, **23**, 1527–1534.
108. E. B. Gyenge, S. Hettwer and B. Obermayer, *SOFW J.*, 2021, **147**, 1–5.
109. Anti-Aging Moisturization, https://www.happi.com/issues/2015-05-01/view_features/anti-aging-moisturization/20598, accessed January 2023.
110. C. Illana-Esteban, *Bol. Soc. Micol. Madrid*, 2016, **40**, 217–223.
111. H. D. Sun, *Flavour Fragrance Cosmet.*, 1990, 1–10.
112. D. Joulaina and R. Tabacchi, *Flavour Fragrance J.*, 2009, **24**, 49–61.
113. D. Joulaina and R. Tabacchi, *Flavour Fragrance J.*, 2009, **24**, 105–116.
114. Z. Li, H. Zhang, L. Wang and Z. Du, *Flavour Fragrance Cosmet.*, 2015, 17–20.

CHAPTER 9

# Bacteria as a Source of Bioactive Cosmeceuticals

MD. ANISUR RAHMAN MAZUMDER[a], MD. FAHAD JUBAYER[b], MD. SHAHIDULLAH KAYSHAR[b], DAYANAND PETER[c] AND THOTTIAM VASUDEVAN RANGANATHAN[c*]

[a]Department of Food Technology and Rural Industries, Bangladesh Agricultural University, Mymeningh 2202, Bangladesh; [b]Department of Food Engineering and Technology, Sylhet Agricultural University, Sylhet 3100, Bangladesh; [c]Department of Food Processing Technology, School of Agriculture and Biosciences, Karunya Institute of Technology and Sciences, Coimbatore 641114, Tamil Nadu, India
*E-mail: srivarahe@gmail.com

## 9.1 Introduction

The cosmetics production sector is constantly searching for cutting-edge technologies and greener practices to meet the market demand and reduce environmental issues, which compels the investigation of novel approaches and components. The term "cosmeceuticals" has recently emerged as a trend in the field of cosmetics, and consumers' reliance on these products has grown a lot over the past few years. As part of the modern lifestyle, there has been a worldwide shift in the cosmetics business toward creating "medicinally" active cosmetics and a similar shift in the pharmaceutical sector toward creating "cosmetically" focused medical goods.[1]

Cosmeceuticals are a new category of cosmetics with drug-like properties that improve and preserve the human body's natural beauty and harmony.

Bioprospecting of Natural Sources for Cosmeceuticals
Edited by Deepika Kathuria, Ajay Sharma, Meenakshi Verma and Gulzar Ahmad Nayik

Published by the Royal Society of Chemistry, www.rsc.org

Generally, the concept of cosmeceuticals combines "cosmetics" with "pharmaceuticals." Cosmeceuticals are a cross between cosmetics and pharmaceuticals; they include active substances that improve the skin's health and appearance by influencing its biological structure and function.[2] In other words, cosmeceuticals are cosmetics that claim to have healing or therapeutic effects because of the inclusion of biologically active or bioactive components. Though cosmeceuticals are claimed to behave both as cosmetics and medicines, they fall into neither area of regulation. The market for cosmeceuticals is the most rapidly expanding segment of the natural personal care business. Since this market is not regulated, large pharmaceutical industries and scientists are drawn to operate in this area. Cosmetics producers are using a wide range of methods for identifying items derived from biological products in an effort to maintain their current market share and expand into new niches. Biologically derived cosmetics are becoming more popular to meet the growing demand of health-conscious consumers.[3] Herbal cosmetics and organic elements derived from a variety of biological sources further lessen the potential for adverse reactions and other problems. As a result, more and more people are turning to biological products in the cosmetics industry. Therefore, the preference for beauty products that come from natural sources instead of synthetic ones has helped the cosmetics market grow in a big way.[4] As a trend, the cosmeceutical industry around the world is moving toward natural compounds that are biocompatible, safe, and good for the environment.[5]

Developing cosmeceuticals is a complicated process that involves a lot of different fields of study in order to make a safe, effective, and consumer-friendly product. Cosmeceuticals go through various steps before being used for skin care, hair care, *etc.* Cosmeceuticals have been made in many different forms, such as anti-wrinkle creams, sunscreens, moisturizers, bleaching agents, shampoos that fight dandruff, eye wrinkle creams, and so on. The primary focuses to achieve success by a cosmeceutical manufacturer are safety, affordability and availability. Apart from the above, the manufactured cosmeceutical should possess the ability to maintain the main ingredient, distribute it within a physiologically active condition, and produce a biological effect throughout the process.[6] Thousands of chemical substances derived either naturally or synthetically have been evaluated for their potential usage as active ingredients in cosmeceutical products.[7] Many of the substances are biologically derived and come from bacteria, fungus, and algae. Microorganisms are easy and affordable resources due to the fact that the manufacture of metabolites from microbes is achievable and scalable at a low cost.[8] Bioactive substances in cosmeceuticals are derived from diversified sources, like food, herbs, tissue engineering, and stem cell treatment.[9] A variety of biological components, such as bio-surfactants, vitamins, anti-oxidants, pigments, enzymes, and peptides, offer intriguing qualities and benefits. Furthermore, these goods are easily commercializable.[10]

The current book chapter discusses bacterial sources of bioactive cosmeceuticals and their applications. Although there are a wide variety of bacteria in nature, only a few of them are commercially used in the cosmetics business. Furthermore, bacteria have the potential to adapt and thrive in different environments and provide unique bioactive compounds not present in other species. Despite the fact that cosmeceuticals have traditionally been derived from terrestrial plants, bioactive substances derived from marine bacteria have diverse roles in the development of cosmeceuticals. Several bioactive compounds isolated from marine bacteria are reported to exhibit multifarious activities, including photoprotective, anti-aging, soothing, anti-microbial, anti-oxidant, moisturizing, and rejuvenating properties.[11,12] Bacterial cellulose (BC) is gaining popularity in the cosmetics sector for its extensive array of applications.[13] Additionally, bacterial pigments as a one-of-a-kind and sustainable source of bioactive color molecules are utilized in cosmetics, food, textiles, printing, and pharmaceuticals. According to research, these bioactive pigments may be employed as anti-bacterial, anti-oxidant, and sun-protective components in cosmeceuticals.[14] It is thus necessary to determine the interactions between the complex nature of vessels (skin, hair, *etc.*) and cosmeceuticals. Research developments in skin biology and pharmacotherapy have enabled the development of novel substances. Since there are more active ingredients that are delivered through the skin, it is more important to understand the mechanisms of how these active ingredients are delivered.[15] By creating a strong link between a healthy microbiome and healthy skin, we can explore cutting-edge innovations.

## 9.2 Cultural and Historical Background

When cosmeceuticals were first brought to the attention of dermatologists, they were prescribed items that helped with appearance. Topical tretinoin was the first cosmetic that met these certain criteria. The topical form of minoxidil was the subsequent cosmeceutical that dermatologists were made aware of. After tretinoin and minoxidil were approved, almost everything in the world of prescription cosmetics came to a stop. But the idea of cosmeceuticals is still going strong. Patients are well-informed thanks to television, magazines, and the Internet, so they want sophisticated products when they buy them.[16]

Cosmetics are an essential component of daily grooming. Cosmetics and beauty products do not change or heal the skin; they are only intended to cover and embellish. A cosmeceutical item, on the other hand, will contain active ingredients that work much more actively to protect, recover, and prevent skin damage. They are put on the skin in the same way that cosmetics are, but they contain ingredients that interfere with the skin's natural biological processes.[17] In accordance with the Food, Drugs, and Cosmetics Act of the Food and Drug Administration (FDA), a product can be a medication, a cosmetic, or a combination of the two, but the legal definition of "cosmeceuticals" is unclear. Cosmetics do not even need FDA approval before they can be sold.[18] The term cosmeceuticals, while not too old, is the latest trendy theme in the cosmetics industry, and represents the most rapidly expanding segment of the natural personal care industry. Companies working in medicines, biotechnology, organic products, and cosmetics are continually coming up with new active ingredients for cosmeceuticals. This is made possible by developments in the relevant field as well as advances in our understanding of skin biology and pharmacotherapy. According to some studies, we can conclude that cosmeceuticals are hybrids of pharmaceutical and cosmetic goods that are externally applied to improve the appearance and health of the skin.[19,20] Cosmeceuticals is a field that combines medicine for skin problems and beauty products.[21]

A cosmetic can be purchased from any beauty retail chain without a prescription and is applied to the skin's outer surface. Cosmetic skincare ingredients cannot transfer through the epidermis and into the dermis, where noticeable changes can occur, because the active components are not strong enough (see Figure 9.1). In contrast, cosmeceutical skincare products can penetrate the epidermis because their active constituents allow them to pass through the epidermis and act deep within the dermis. Cosmeceuticals are highly effective, and one can notice visible changes in the skin, once applied.[22] According to Dr Jetske Ultee, a cosmetic dermatologist, a product may only be called a cosmeceutical when it incorporates no rubbish (irritating and valueless ingredients), and only when it contains ingredients that can genuinely have something advantageous for skin, such as anti-oxidants against harmful radiation from the sun, or an increased concentration of niacinamide to decrease the production of sebum in oily skin, *etc.*[23]

Cosmeceutical formulations and developments need to be evaluated for three significant reasons: (i) to make sure they offer an advantage that the consumer can see; (ii) to make sure that they are safe and won't cause any side effects;

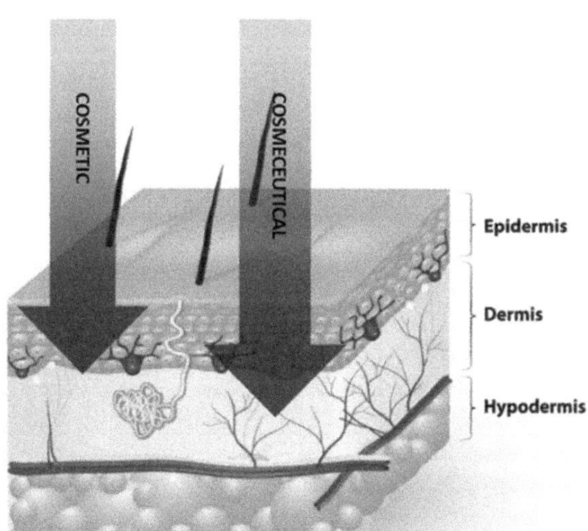


**Figure 9.1** Action modes of cosmeceuticals and cosmetics.

and (iii) to back up marketing claims.[24] The cosmeceutical industry is propelled forward by the introduction of novel ingredients and formulations accompanied by alluring advertising claims. Although it is challenging to generalize due to the continuous release of new products on a monthly basis, there are a few fundamental ideas that can be applied to the development of cosmeceuticals. To begin with, cosmeceuticals are essentially the same as cosmetics, and as such, they are required to be risk-free. This indicates that substances made from phytocompounds or dietary components would be the best source of fresh materials for making formulations. In particular, secondly, cosmeceutical ingredients need to be available as a powder or liquid, since most cosmeceuticals need to be put on the body's surface for an effective application. Third, in order to make a functional claim, cosmeceuticals need to have some benefit that can be readily recognized by consumers. For these reasons, the majority of cosmeceutical compounds may be traced back to either the plant kingdom or the kitchen.[25] A schematic illustration of cosmeceutical ingredient development is given in Figure 9.2.

The development of new active compounds has been made possible by the progress that has been made in the pharmacology and biology of the skin. Optimal delivery of active ingredients is characterized by a high

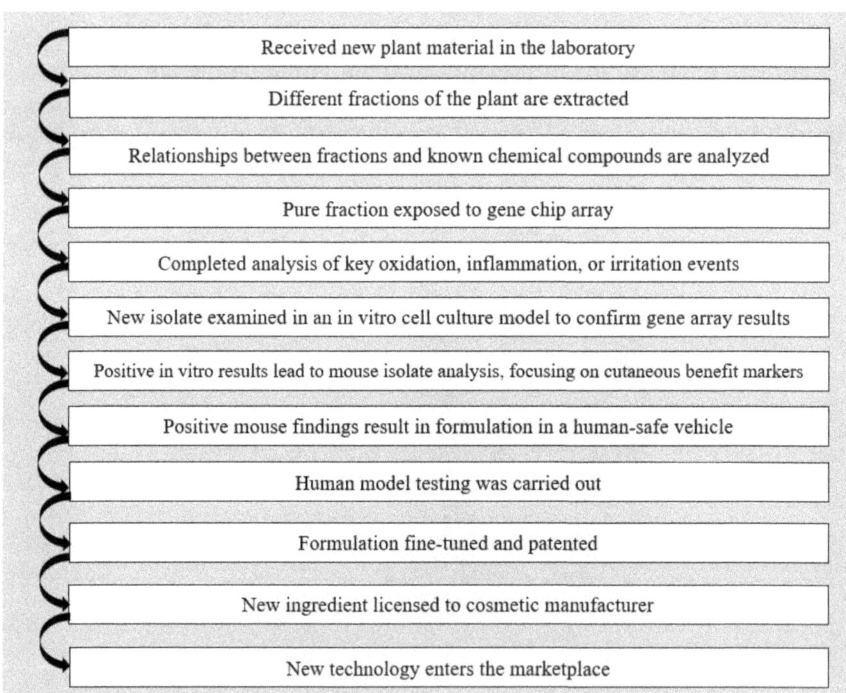


**Figure 9.2** Steps in cosmeceutical ingredient development. Reproduced from ref. 25 with permission from Springer Nature, Copyright 2019 Springer Nature Switzerland AG.

administration capability, formulation integrity, and low adverse effects. Cosmeceuticals can be administered in two ways: trans-epidermal and trans-appendageal. There are two types of delivery systems: active and passive. Active systems make use of different types of physical improvement procedures. Chemical penetration enhancers, emulsions, vesicular lipid-based systems, and lipid particle carrier systems are all examples of passive system components that are used to transport chemicals to their intended destinations. Passive systems also use lipid particulate carrier systems.[15] Corneodesmosomes must disintegrate in order to create an aqueous hole that can then be used for transcellular or intracellular transport. In transappendageal transport, the designed drug is moved through the sweat and follicle cells to the desired point of treatment. The lipid lamellae of the intercellular spaces are involved in intercellular mobility. This is because the majority of solute molecules are able to pass through the intercellular lipid channels.[15]

There are intercellular and intracellular pathways in the transepidermal pathway. Solutes move through the lipid domains between cells as part of the intercellular pathways. In contrast, in the transappendageal pathway, actives enter the body through the openings of hair follicles and sweat glands. Due to the deep penetration of hair follicles into the dermis, they play a crucial part in this pathway.[26] An additional option is to make use of physical augmentation techniques, which involve the use of an external pushing force in order to penetrate the active ingredients. In addition, it has been demonstrated that ablative, nonablative, and fractional lasers are all effective methods for increasing the cutaneous penetration of cosmeceuticals.[26–28] However, both chemical and physical techniques have proven effective for the administration of a wide range of cosmeceuticals.

## 9.3 Use of Secondary Metabolites in Cosmeceuticals

Bacterial secondary metabolites are a source of anti-cholesterol drugs, immune suppressants, antibiotics, anticancer agents, and other medications; secondary metabolite-producing microbes create these bioactive and complex compounds during the late and stationary phases of their growth.[29,30]

### 9.3.1 Bacterial Cellulose

Plants can make cellulose, but it can also be made in a lab with the help of enzyme reactions and chemical synthesis from glucose derivatives. Biosynthesis is the process by which microorganisms like algae and fungi, as well as nonpathogenic aerobic bacteria, can produce cellulose.[32,33] Despite possessing a molecular structure that is comparable to that of plant cellulose (PC), BC possesses capabilities that are distinct and superior in the areas of biology, mechanics, and physics.[32] BC has been manufactured in its most unadulterated form during the biosynthesis process, which is totally free of hemicelluloses, pectin, and lignin.[34] BC is a high molecular weight, water-insoluble β-(1,4) linkage glucose homopolysaccharide.[34,35] It is a biopolymer that is secreted by the bacterium *Gluconacetobacter xylinus*. The *Gluconacetobacter xylinus* strain has received the greatest attention and is being investigated for commercial production.[35] Apart from *Acetobacter* species, for example *G. hansenii* and *G. pasteurianus*,[36,37] many genera such as *Agrobacterium*, *Rhizobium*, *Pseudomonas* and *Alcaligenes* have the capability to synthesize BC.[34,38,39]

Any form of BC, whether it is natural, synthetic, or unmodified, has a high capacity for water absorption and is exceedingly porous. It also has a high permeability to both liquids and gases. These qualities are a result of the ultrafine network of ribbon-shaped microfibrils and nanofibrils that make up BC. These nanofibrils and microfibrils are approximately one hundred times thinner than the fibers that make up PC. BC is appropriate for use as a formulation stabilizer, for thickening, and for scouring and exfoliating without causing damage to the skin because of its soft, small-fiber texture.[34,40] Biocompatibility, low immunogenic potential, and the capacity to fold in place all make BC even more useful in the area of biomedicine. BC is a good choice for cosmetics, and topical and transdermal drug delivery because it has these properties and is strong enough to be stretched.[32] Figure 9.3 attempts to illustrate the distinctive qualities of BC.

Figure 9.4 shows how biological integration, continuous spray, and *in situ* fermentation work in *in situ* BC composite synthesis. With the help of reinforcing chemicals, the *in situ* BC fermentation makes it possible for BC composites to self-assemble in a controlled way. This makes the BC composites better in how they work. The *in situ*-fabricated BC-based composites show dramatically enhanced mechanical characteristics, crystallinity, and thermal stability.

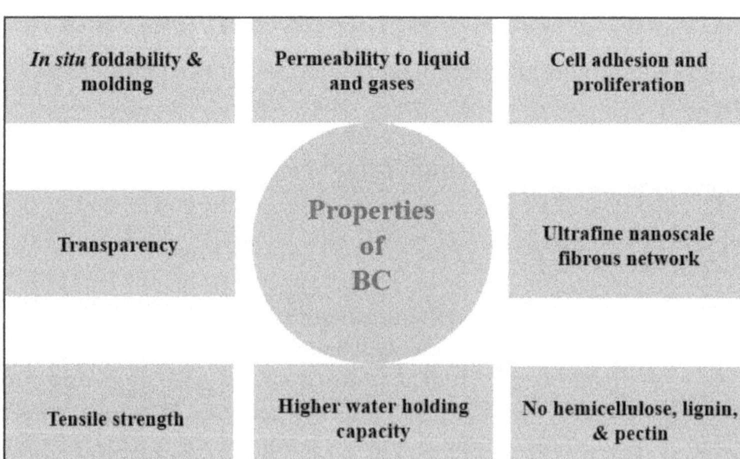


**Figure 9.3** Material properties of bacterial cellulose (BC). Reproduced from ref. 32 with permission from Springer Nature, Copyright 2016 Springer Science Business Media Dordrecht.

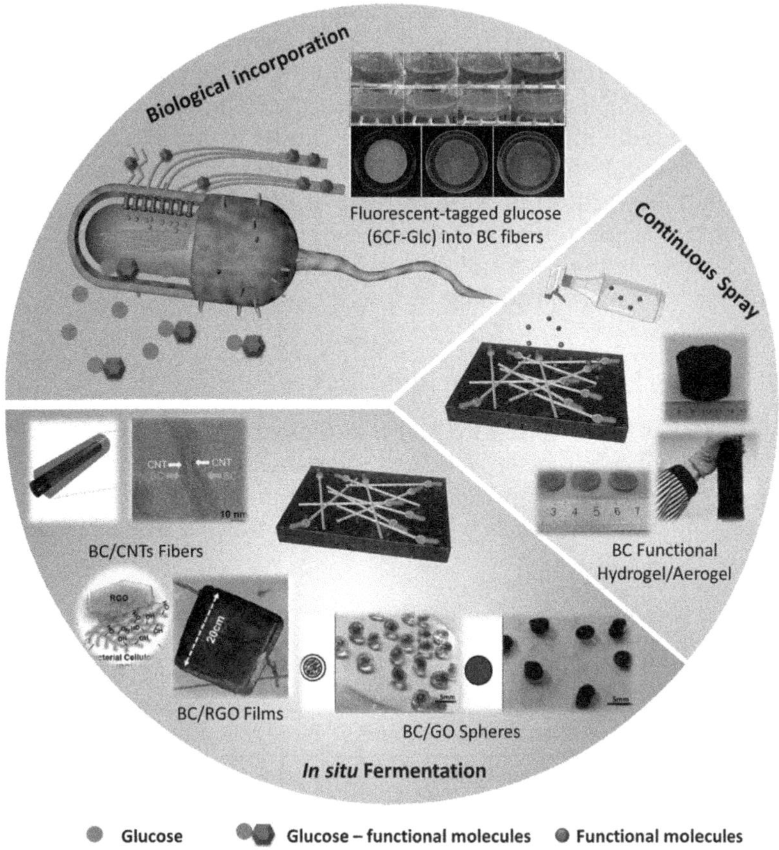


**Figure 9.4** Schematic representation of *in situ* BC composite synthesis: *in situ* fermentation, biological incorporation, and continuous spray. Reproduced from ref. 43 with permission from Elsevier, Copyright 2021.

In addition, this procedure is simple and cost-effective, which is why it has been hailed as the method of producing BC that has the highest success rate.[41] During the BC production process, reinforcing ingredients are added to the culture medium. The reinforcement materials are then incorporated into the BC matrix.[42]

## 9.3.2 Secondary Metabolites from Cyanobacteria

Historically, cyanobacteria have been known to create potentially dangerous secondary metabolites that are toxic to both vertebrates and invertebrates. But because of the wide range of secondary metabolites, new discoveries have shown that they have different bioactivities. More research has been done on filamentous marine cyanobacteria that make secondary metabolites with favorable bioactive substances. This is because they grow in colonies and are easy to collect and study directly from samples of the environment.[44] They have been found to possess a broad spectrum of potential actions, including those that are anti-bacterial, anti-cancer, anti-viral, and anti-inflammatory. Table 9.1 provides a list of some known bioactive compunds.

Secondary metabolites of cyanobacteria can be utilized in the cosmetics industry as natural components. Photoprotective mycosporine-like amino acids (MAAs) are utilized in sunscreens to shield the skin from UVR damage. Carotenoids and phycobiliproteins are two examples of pigments with potential usage as both natural colorants and anti-oxidants to protect the skin from ultraviolet (UV) damage.[45]

### 9.3.3 Bacterial Polysaccharides

Hyaluronic acid (HA, see Figure 9.5) is the most important bioactive bacterial polysaccharide presently in use in cosmetic formulations. HA is a linear, high-molecular-weight glycosaminoglycan consisting of alternating *N*-acetyl-D-glucosamine (GlcNAc) and D-glucuronic acid disaccharide units (GlcA).[46]

**Table 9.1** Secondary cyanobacterial metabolites and their potential uses in cosmeceuticals.

| Species | Bioactive compounds | Biological activities |
|---|---|---|
| *Fischerella muscicola* | Fischerellin | Anti-fungal |
| | Scytonemin | Anti-inflammatory and anti-proliferation properties |
| *Lyngbya majuscule* | Apratoxins | Anti-cancer potential |
| *Spirulina platensis* | Phycocyanin | Cosmetic colorants |
| | γ-Linolenic acid | Precursor to prostaglandins |
| | Spirulan | Anti-viral activities |
| *Nostoc commune* | Carotenoids | Anti-oxidant |
| | Nostodione | Anti-fungal |
| | MAAs | Sunscreen function |
| *Anabaena circinalis* | Anatoxin-a | Anti-inflammatory properties |
| *Leptolyngbya* sp. | Phenolic extracts | Anti-oxidant potential |
| *Nostoc commune* | Exopolysaccharides | Moisture retention |

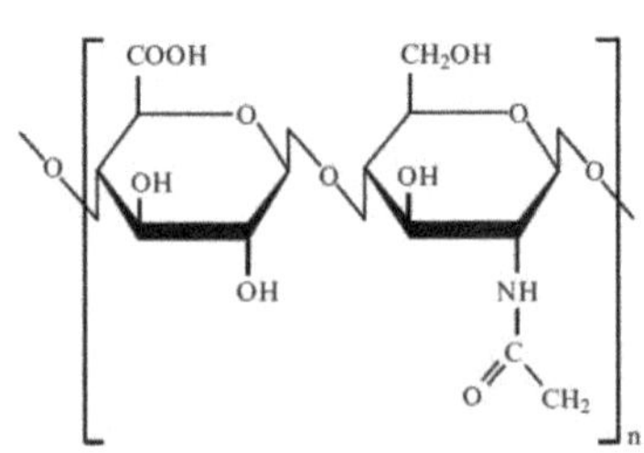


**Figure 9.5** Chemical structure of hyaluronic acid.

It is a key part of the structure of the intercellular matrix of connective tissues in mammals, in which it controls the permeability and water content of tissues, the movement of large molecules between cells, and the ability of bacteria to invade. HA may bind up to 1000 times its own volume in water, giving it a large swelling capacity.[47] Catabolism of HA may occur *via* enzymatic or chemical mechanisms. As depicted in Figure 9.6, the bioactivity and skin penetration of HA are dependent on its molecular weight, which has varying effects on the skin.

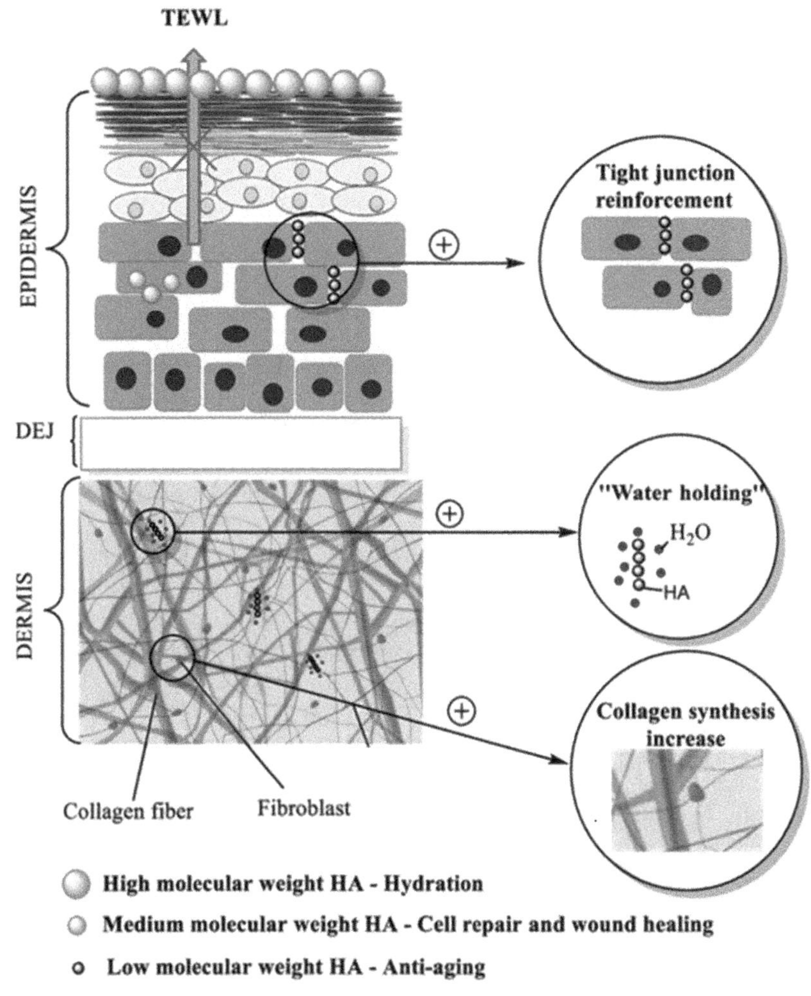


**Figure 9.6** Activity, molecular weight-dependence, and claimed effects of HA (TEWL: transepidermal water loss; DEJ: dermoepidermal junction). Reproduced from ref. 48, https://doi.org/10.3390/molecules26154429, under the terms of the CC BY 4.0 license https://creativecommons.org/licenses/by/4.0/.

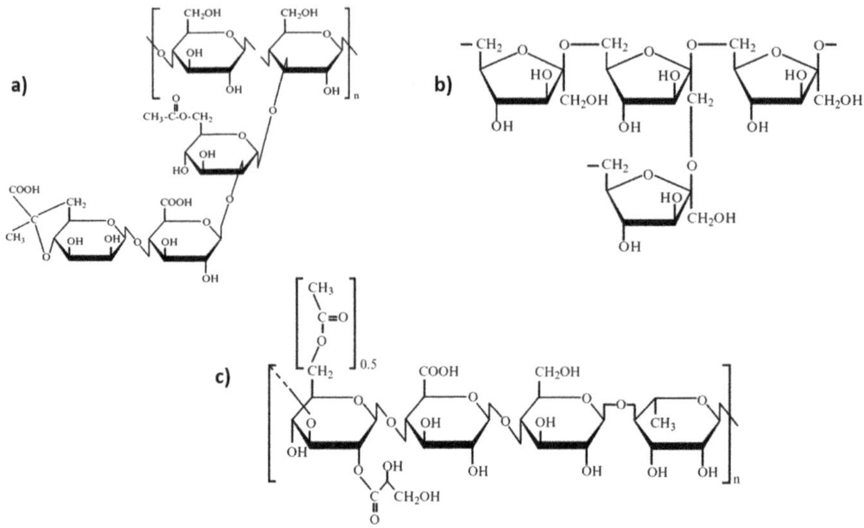


**Figure 9.7** Chemical structures of (a) xanthan gum, (b) levan, and (c) gellan gum.

Due to their solubility, polysaccharides, such as xanthan gum and gellan gum (Figure 9.7), are frequently utilized in cosmeceutical products as compounds (powdered form) for viscosity control, emulsion stabilization, or moisturization. They are a type of heteropolysaccharide that has a glucose foundation connected by β-1,4 glycosidic linkages and trisaccharide side chains attached to every other glucose residue. Highly viscous aqueous solutions can be formed from low concentrations of xanthan gum, and it can be dissolved in either hot or cold water. The aqueous solutions of this material exhibit a strong shear-thinning characteristic and a high low-shear viscosity.[49] *Xanthomonas* sp. is responsible for the production of xanthan gum. Because it is able to obtain high productivities and convert substrates into biopolymers, *X. campestris* is a very effective polysaccharide provider.[46] Because of its excellent viscosity-enhancing capabilities at low concentrations, xanthan gum is the most widely utilized bacterial polysaccharide. In addition, it is the only significant bacterial biopolymer in the global hydrocolloids industrial sector.[49]

Marine bacteria produce bioactive compounds with a high commercial value for use in the cosmeceutical industry. There are few examples of marine bacterium applications. The exopolymers produced by *Planococcus maitriensis* Anita I might be exploited as cosmetic additives.[50] Marine microorganisms isolated from a hydrothermal vent at the depths of the ocean are what are responsible for the production of exopolysaccharides (EPS). They are involved in many biological processes. *Alvinella pompejana*, a marine bacterium, makes an exopolysaccharide that is made up of neutral sugars (glucose, fructose, galactose, rhamnose, mannose, *etc.*) as well as glucuronic and galacturonic acids and a diacidic hexose that is known to be a 3-*O*-(1-carboxyethyl)-D-glucuronic acid.[50]

Levan is a naturally occurring homopolysaccharide that is made up of fructose residues connected by β-2,6 links with β-2,1 side branches.[51] Through the catalytic activity of the extracellular enzyme known as levansucrase, microbial levan can be generated from sucrose.[52] Levan sucrase is produced by various bacteria including the genera *Erwinia*, *Streptococcus*, *Pseudomonas*, and *Zymomonas*.[53]

Gellan is a heteropolysaccharide made up of rhamnose, glucose, glucuronic acid, and glucose monomers with acetyl and glyceryl substituents on the side chains.[54] The bacterial strain used to produce gellan gum commercially is *Sphingomonas paucimobilis* ATCC 31461.[55] When gellan gum is allowed to cool, a solid gel is formed. The number of acyl groups in gellan is what determines its characteristics.[56]

### 9.3.4 Oligosaccharides

The cyclic oligosaccharides known as cyclodextrins, which contain an α-(1,4)-connected glucopyranose moiety bonded together in a ring, play an important role in cosmetic formulations.[57] Enzymatic transformation is a more common method for cyclodextrin manufacturing in the commercial sector than chemical synthesis. *Bacillus subtilis* strain 313, *Brevibacterium* sp. strain 9605, *Brevibacillus brevis* strain CD 162, and *Microbacterium terrae* KNR 9 are some of the prominent strains used for the production of cyclodextrin.[58–61]

### 9.3.5 Bio-surfactants

Biosurfactants, in general, can be classified as either fatty acids, neutral lipids, glycolipids, or lipopeptides. In addition, the United States the Environmental Protection Agency (EPA) has reported that certain biosurfactants, such as rhamnolipids, are safe for use in food goods, cosmetics, and pharmaceuticals. In addition, biosurfactants possess prebiotic properties. Prebiotic compounds can be described as those elements that have the ability to rebalance the skin's microflora in the context of the cosmeceutical industry.[62]

## 9.4 Traditional Uses of Bioactive Cosmeceuticals

Cosmetics are now an important part of everyday life. Many studies have been done on how cosmeceuticals can be used and how to make new products that are better and cost less. New cosmeceuticals are supposed to have a pharmacological effect in addition to their main purpose. Cosmeceuticals have been developed for a variety of objectives, including the enhancement of skin texture, brightness, smoothness, tone, and pigmentation. By applying occlusive and humectant components to the skin's surface, the primary advantage of the majority of cosmeceutical formulations is a decrease in transepidermal water loss.[25] Cosmeceuticals can improve the way skin works if the active ingredients are put into a topical vehicle that keeps the active ingredient's integrity, delivers the active in a biologically suitable form, gets to the target site in a sufficient amount to have an impact, and releases the constituent in the right way. To prove the significance of the cosmeceutical, clinical testing using a scientific method to measure effectiveness must be done.[24] The majority of cosmeceuticals are composed of substances with an established market safety record. This may be because they are made from foods, like lycopene, which comes from tomatoes, and avocadin, which comes from avocados. Conversely, the raw material manufacturer may do rigorous animal research to verify whether the new chemical is safe for human consumption. To circumvent regulatory concerns, the majority of cosmeceuticals are created with components that are generally recognized as safe.[24] Before putting a product on the market, most companies evaluate it on a large number of people. Then, dermatologists test the item to see how well it works and to make sure it doesn't cause any problems, like allergic contact dermatitis, irritant contact dermatitis, comedogenicity, or acnegenicity. Sometimes, testing is undertaken in conjunction with an ophthalmologist to confirm that inadvertent ocular instillation does not cause any eye problems.[63] More than ever, people are looking for a miracle cream that is easy to use, works well, is cheap, and can stop the aging process. People frequently see dermatologists in search of answers to questions like "does this product really work?" or "is there a new product that gets rid of wrinkles in a week?" To give clinically and scientifically correct answers to these questions, it is essential that a clear knowledge on the construction and mode of action of human skin be known.[64] As indicated by Kligman, it is necessary to ask three questions when evaluating a new cosmeceutical product that claims to have a good physiological effect.[65]

A. Can the active substance permeate the stratum corneum and be delivered at sufficient quantities to its designated destination in the skin over a period of time commensurate with its mode of action?
B. Is there a specific biochemical way that the active ingredient works on the skin cell or tissue that it is meant to affect?
C. Are there peer-reviewed, double-blind, placebo-controlled clinical trials that can back up the claims of effectiveness?

## 9.5 Pharmacological Activities

In addition to the medical, pharmaceutical, and food sectors, biological substances also find varied use in the cosmetics industry. Several biological molecules have played an important role in the development of numerous compounds, including esters, aroma compounds, and active agents, which are widely employed in the cosmetics industry. The main benefit of employing microbial components is their biocompatibility, but they also offer other advantages including a streamlined production process, improved and consistent product quality, and less environmental impact. A study has shown that many bioactive secondary metabolites produced by bacteria have been adopted by the pharmaceutical and cosmetics industries due to their great economic value.[66] This section discusses the pharmacological activities of bacterial secondary metabolites in the formulations of cosmeceuticals.

Bacterial polysaccharides have physicochemical features that make them ideal for use in cosmetics. They are biocompatible, biodegradable, and typically nontoxic. HA, BC, and levan are some that are employed as active agents in cosmeceutical formulations because of their biological qualities, such as skin regeneration and protection. Besides, other bacterial polysaccharides like xanthan gum and gellan gum are utilized in cosmetic vehicles as emulsions, gels, and suspensions to regulate viscosity and as psychosensorial agents.[46] As a result of its high strength and low density nanofibrous porous network structure, BC is a promising material for use in the creation of membranes in a wide range of industries, including biomedicine, cosmetics, and electronics.[67]

However, BC has also been used as a structural ingredient in personal care product formulations or in natural scrub cosmetics.[32,68] BC's ability to regenerate skin has been documented, and it has been suggested that the biopolymer could be used to make products that keep skin moist, repair skin, and help wounds heal.[69–72] A study has also shown that BC is used in moisturizing creams as a result of its great hydrating capacity.[73] BC is also often used to make masks for face treatments because it is easy to shape. These masks can also be used to carry skincare ingredients[74] (Table 9.2).

Because of functional qualities, levan is also appropriate for usage in cosmetics and pharmaceuticals. Skin whitening effects[75] and cell proliferation[76] are among the documented biological activities of levan. Along with anti-inflammatory and anti-irritant effects, this polysaccharide is very effective in keeping skin hydrated.[51]

Xanthan gum is a heteropolysaccharide that has a backbone made of glucose and side chains made of trisaccharides on every other glucose residue. Both cold and hot water will dissolve xanthan gum, and the resulting aqueous solutions will be rather thick despite their low concentration.[77] *X. campestris*

**Table 9.2** The pharmacological activities of BC in cosmeceuticals.

| Products | Form | Activities |
|---|---|---|
| Facial mask | BC sheets | Moisturizing |
| Facial mask | BC–glycerine composites | Moisturizing |
| Facial mask | BC membrane | Moisturizing |
| Facial mask | BC–ginseng | Moisturizer and carrier |
| Facial scrub | BC powder | Viscosity enhancer |
| Personal cleansing product | BC fibers | Surfactant-free emulsion for sensitive skin |
| Personal cleansing product | BC particles | Cleansing and exfoliation |
| Foundation make-up | BC fragments | Stable make-up |
| Contact lenses | Regenerated BC sheet | Film-forming agent |
| Contact lenses | BC-based hydrogel | Film-forming agent |

is a very efficient polysaccharide producer because it can turn substrates into biopolymers and has a high production rate. The polymer can be used in food, beauty products, and oil recovery, among other things.[56] Xanthan gum is used in almost all kinds of cosmetics, like oral hygiene products, leave-on formulas, deodorants, products for babies, *etc.*[78] In concentrations between 4 and 6%, it functions as a binder, emulsion stabilizer, skin-conditioning agent, surfactant/emulsifier, and viscosity enhancer in dermal and mucous contact products, hair colorants, and nail treatments.[78]

Gellan gum is primarily utilized as an emulsion stabilizer and a viscosity-increasing ingredient in cosmetic compositions. It typically has concentrations of 0.3–0.5% in leave-on and dermal contact products. Other applications include hair products, powders, and eye and mucous contact products with concentrations up to 0.0004%.[78]

Cyclodextrin glucanotransferase (EC 2.4.1.19), obtained from alkalophilic *Bacillus agaradhaerensis*, is a popular enzyme in possible cosmetic formulations.[79] Most often, cyclodextrin is used in perfumes to make the volatile esters more stable. Smaller cyclodextrin powders are used to control odors in talcum, diapers, menstrual discs, pads, napkins, and other products.[80]

EPS are produced by marine microorganisms that have been isolated from a hydrothermal vent in the deep sea. These EPS are involved in a wide variety of biological processes. This EPS is used in the production of cosmetics under the commercial name DEEPSANE™.[81] It is produced from the marine bacterium *Alteromonas macleodii*, used for relaxing and decreasing skin irritation caused by chemical, mechanical, and UVB aggression.[82,83] Dextran is one of the most well-known EPS derived from glucose polymer. Dextran is derived by microorganisms belonging to the Leuconostocaceae family, such as *Leuconostoc mesenteriodes* and *Streptococcus mutans*. The unique biocompatibility and non-toxicity of microbial EPS has greatly aided in their application in the cosmetics sector. Additionally, the high water retention capacity of hydrophilic EPS aids in keeping skin compositions moist.[84] Dextran is a skin-smoothing and brightening ingredient used in cosmetics because it improves skin radiance, firmness, and the appearance of wrinkles. Moreover, dextrans increase blood flow and nitric oxide production in human epidermal keratinocytes, which show its anti-inflammatory properties.[85] In addition, bacteria belonging to the *Pseudomonas* genera, such as *Pseudomonas aeruginosa* and *Azotobacter vinelandii*, are well known to produce abundant alginate as EPS that can retain water and serve as a thickener and gelling agent in the cosmetics industry.[86] Hence, the bacterial genus *Xanthomonas* is known to produce EPS such as Xanthan that are used as emulsifiers and foaming agents in skin formulations as well as assist in skin-smoothing and moisturizing.[87–89]

Cosmetics also use proteins and peptides to improve skin health, hair growth, and nail resilience. Extensive studies have found several protein uses in the cosmetic sector. Enzymes such as superoxide dismutase (SOD) and peroxidase operate together as an exfoliate and serve as scavengers of free radicals to protect the skin from ultraviolet (UV) radiation.[90]

Lactate dehydrogenase (LDH) is another enzyme that may catalyze the reduction of NADH and pyruvate, resulting in NAD+ and lactate as the end product. UV light reduces the aforementioned process, however LDH prevents the breakdown of the reaction's subunits inside the cell, allowing it to continue operating normally.[90] Earlier reports suggest the use of SOD from extremophiles such as *Thermus thermophilus*, for developing commercial products that successfully repair sun and heat-induced skin damage.[91] Currently, with the advancement of technology, genetically engineered lactic acid bacteria is the preferred choice because of its high yield and stability.[92,93] Also, it is well-known that proteases are enzymes that break down the peptide bonds of keratin, collagen, and elastin in the skin. Bacterial alkaline aspartic proteases from different alkaliphilic bacteria have been used to treat skin diseases like xerosis, ichthyoses, and psoriasis (skin flaking and inflammation).[94] Proteases, such as keratinases, are commonly used as ointments and creams for heels, knees, and elbows, and they have the added benefit of curing stretch marks and scar tissue, and regenerating the epithelial cells to speed up the healing process. Keratinase is also a key ingredient in hair-removal treatments and medicines that slow hair growth. For the commercial production of keratinase, *Bacillus licheniformis* has been used.[95] Aside from thermophilic bacteria, Gram-positive and Gram-negative bacteria are known to manufacture keratinase enzyme.[96,97]

Peptides are another important group of biochemicals, that have a promising role in the development of a cosmeceutical. Gels, emulsions, powders, lotions, and face masks can all be made with either water- or oil-soluble peptides. Soluble peptides can also be employed.[98] Wrinkles and skin roughness may be reduced with the help of pentapeptides.[99] These commercial peptides are produced by the regulated action of proteases, which are primarily released by numerous *Bacillus* spp.[96]

HA is a popular biomaterial in the beauty, health, and pharmaceutical industries due to its broad range of useful properties, including its biocompatibility, non-immunogenicity, biodegradability, and viscoelasticity.[100] Large quantities of HA are derived from rooster combs and animal tissues. However, the *Streptococcus* group of bacteria can also manufacture HA. Recently, *Bacillus* species cultures that have undergone genetic modification have been employed to produce HA.[101] The different effects of HA are illustrated in Figure 9.8.

Because of its capability to offer water retention to the skin as well as hydration to the skin, the primary commercial use of HA is in anti-aging creams, which account for around 15 tons per year of the total production.[102] In several categories of cosmetic products, such as cosmetics, moisturizers, lotions, sprays, and lipsticks, the usage of HA and its salts has been documented at concentrations of up to 1% for HA or up to 2% for sodium hyaluronate.[103] Moreover, it has been reported that HA has an anti-wrinkle effect because it efficiently moisturizes and enhances the skin's suppleness.[104] It is extensively used in cosmetic surgery as a dermal filter and in skin lotions to boost moisture retention, ease skin wrinkles and improve skin firmness and elasticity.[105]

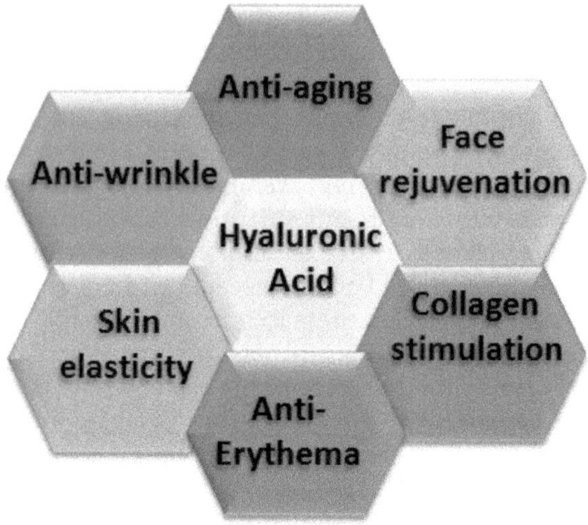


**Figure 9.8** Cosmeceutical effects of hyaluronic acid. Reproduced from ref. 48, https://doi.org/10.3390/molecules26154429, under the terms of the CC BY 4.0 license, https://creativecommons.org/licenses/by/4.0/.

A pigment called biopterin glucose, which is generated from a marine planktonic cyanobacterium, can shield the skin from the potentially detrimental effects of exposure to UV-A light. Consequently, it is used in the production of sunblock cosmetics.[106] Parabens are found in a broad range of cosmetics, as well as foods and pharmaceuticals. Parabens may be found in a wide variety of cosmetics, including foundation, moisturizer, hair care products, and shaving products. Parabens, which are alkyl esters of hydroxybenzoate, are non-carcinogenic and anti-bacterial, and have a low toxicity level for humans. 4-Hydroxybenzoate is produced by marine bacteria of the genus *Microbulbifer*.[66] Because of their distinct physiochemical properties, paraben derivatives have been included in the creation of a wide range of cosmetic goods. Parabens are useful and effective preservatives, and their low production costs make them a good economic choice.[107]

Researchers have tried to get compounds out of cyanobacteria that could be used in cosmetics and cosmeceuticals. Along with plants and algae, cyanobacteria are now being looked at as another source of compounds that can be used in skin care. Some of the compounds (*e.g.* flavonoids, pigments, phenols, saponins, steroids, tannins, terpenes *etc.*) that are found in cyanobacteria have been found to exhibit photo-protective, moisturizing, anti-oxidant, anti-inflammatory, and regenerative properties. These features make these compounds appealing for use in health and wellness therapies like thalassotherapy, as well as for the protection and equilibrium of the integumentary system and its derivatives.[108] Additionally, cyanobacteria are capable of producing chemicals that are able to defend against UV radiation, as well as molecules that are capable of alleviating severe dehydration and oxidative stress.[109]

Scytonemin is a relatively small hydrophobic alkaloid pigment molecule that functions as a defense mechanism against the short-wavelength UV light produced by the sun. It is located in the sheath that surrounds the extracellular organelles of many cyanobacteria. It is very effective in lowering reactive oxygen species generation and preventing DNA damage. Scytonemin serves as both a powerful UV sunscreen and anti-oxidant, and it has the potential to be used in the cosmetics and pharmaceuticals sectors to create innovative new cosmeceuticals.[110]

Lactic acid is an organic acid with a low molecular weight that is often synthesized or fermented by lactic acid bacteria. This refers to a large group of beneficial bacteria with probiotic effects.[111] There are several possible health advantages associated with this type of bacterium, which is present naturally in the gastrointestinal tract (GIT) of humans and animals alike.[112,113] Natural lactic acid is a useful component in cosmetics because of its properties. pH regulation, skin hydration, antimicrobial activity, and skin lightening are some of the functions of lactic acid. The ability of lactate to retain water is directly related to its moisturizing efficacy, and lactic acid's skin-lightening impact is the result of its suppression of tyrosine production, both of which contribute to its high moisture retention capacity. Studying lactic acid and its salt made perfect sense since they were already present in the human body and they had beneficial benefits like reviving and brightening the skin.[114] Moreover, lactic acid plays a crucial part in skin regeneration and skin care. Studies have also shown that the skin-brightening agent lactic acid is crucial for the re-genomic organization of the melanin pigment, which determines the degree of skin color.[115] Lactic acid peels are a type of facial exfoliation that help speed up the natural turnover rate of skin cells by removing dead skin cells from the surface of the skin. Getting rid of skin cells can help the skin make more of its own elastin and collagen. This, in turn, may help make wrinkles and fine lines look less noticeable.[116,117] Lactic acid is one of the alpha-hydroxy acids (AHAs) that helps the skin in important ways. It slows down the aging process by making fine lines and wrinkles look less noticeable. Lactic acid helps the body makes more collagen, which makes the skin firmer as well as removes age spots.[118]

Cosmetics containing saproxanthin or myxol may aid in strengthening cellular membranes by reducing oxygen permeability and increasing anti-oxidant defense. It's interesting to note that saproxanthin and myxol have even higher anti-oxidant activity than zeaxanthin and β-carotene.[119] Moreover, by blocking the production of tyrosinase, the methylene chloride generated by novel species of marine bacteria (*Pseudomonas* sp.) may diminish the pigmentation of human melanocytes and cultured skin cells.[120] Ectoines, which are produced by halophilic bacteria, may be utilized to protect human skin from harmful environmental factors. Ectoine has several uses in cosmetics, including protecting the immune system, cells and membranes from sun damage.[121] Along with the plant sources, some bacteria can also produce fatty acid esters, which play a significant role in the cosmetics industry. Ethyl oleate was produced from actinomycetes *Nocardiopsis dassonvillei*, which is extensively utilized as an emollient and perfume as well as displaying anti-inflammatory activity in many cosmetic items.[122]

## 9.6 Cosmeceutical-based Formulations and Market Products

The price of a cosmetic product can help determine how it is positioned on the market, and businesses who sell cosmetics have three options to choose from when doing so: (1) premium products; (2) mass premium products; or (3) mass-market products. When it comes to marketing, it is common knowledge that the price plays a psychological function in the collective imagination. Consumers tend to attach a higher value to a product that has a higher price tag. The term "premium cosmetics" refers to items that may be purchased at a higher cost, whereas the term "mass-market cosmetics" refers to items that can be purchased at a lower cost.[48] Table 9.3 shows details of some commercially available bio-based cosmeceuticals.

**Table 9.3** Commercially available bioactive cosmeceuticals. Reproduced from ref. 48, https://doi.org/10.3390/molecules26154429, under the terms of the CC BY 4.0 license, https://creativecommons.org/licenses/by/4.0/.

| Category | Brand | Cosmeceutical claim | Market segment | Origin |
|---|---|---|---|---|
| Skincare | FRESH deep hydration face serum | 24 h moisturization | Premium | USA |
| | The Ordinary – DECIEM multi-technology peptide serum | Anti-ageing | Premium | Canada |
| | The Ordinary – DECIEM hyaluronic acid 2% + B5 serum | Anti-ageing and moisturizing | Premium | Canada |
| | APIVITA eye serum | Anti-ageing and moisturizing | Premium | Greece |
| | FARMEC ampoules 5% serum | Anti-wrinkle and intensive moisturizing | Mass | Romania |
| | BALANCE ME tinted wonder eye cream; day cream | Soothing and anti-ageing | Mass | UK |
| | AVIVA Cosmetics infinitum cellular regenerating cream | Firming, rejuvenating, and regenerating | Premium | Romania |
| | GEROCOSSEN anti-wrinkle cream with SPF 10 | Anti-ageing and moisturizing | Mass | Romania |
| | ARBONNE RE9 anti-aging moisturizer for men SPF 15 | | Premium | USA |

## 9.7 Conclusions and Future Prospects

In the last few years, as the demand for bio-based cosmeceuticals has grown, the unique properties of bacterial metabolites and compounds, as well as the fact that they are biodegradable and can be made over and over again, have attracted more and more attention from both academia and the cosmeceutical industry. Cosmeceutical manufacturers are consistently looking for new approaches to the extraction and widespread use of microbial components like those they use in their products. Biotechnology has come a long way, and organisms can now be made better by changing their genes. There is also a huge variety of microorganisms, which has increased the use of new compounds that come from biology. Cosmeceuticals work best when they interact with multiple layers of the skin and different types of cells. This is why biosurfactants, anti-oxidants, anti-aging ingredients, *etc.* are used. The best replacement for chemical entities in the market is a compound with similar properties made by microorganisms. Some of these biologically-derived products may have side effects, so they need to be tested thoroughly and methodically through clinical research to find out what their real potential is. Since people are still interested in "natural" ingredients, we can expect to see more skin care products that say they use ingredients from biological sources. An expanding category of naturals, marine-based cosmeceuticals, are an exciting new prospect for cosmeceutical producers as they represent a new frontier in the world of naturals. Anti-oxidants, polysaccharides, bioactive peptides, and chitin obtained from marine sources are examples of potentially useful functional components for use in cosmeceuticals. In this constantly developing sector, new synthetic substances will also play an important role. The goal of sustainable production is to make sure that every component that goes into the process comes out in a good way for the environment. When it comes to cosmeceuticals, sustainability is important for both the ingredients and the way they are packaged. Good agricultural practices are important for the long-term health of agriculture. Challenges, such as discovering novel ingredients, developing better delivery systems, and manufacturing goods that are more environmentally friendly, are faced by producers all around the world. To maintain one's total health in a holistic manner, new horizons include the application of cosmeceuticals in conjunction with the use of functional foods, nutritional supplements, and technologies that may be used at home.

## References

1. A. Zesch, *Hautarzt*, 1999, **50**, 243.
2. H. S. Datta and R. Paramesh, *J. Ayurveda Integr. Med.*, 2010, **1**, 110.
3. I. A. Khan and E. A. Abourashed, *Leung's Encyclopedia of Common Natural Ingredients: Used in Food, Drugs and Cosmetics*, John Wiley & Sons, Hoboken, New Jersey, 1st edn, 2011.
4. A. Martins, H. Vieira, H. Gaspar and S. Santos, *Mar. Drugs*, 2014, **12**, 1066.

5. T. F. Alves, M. Morsink, F. Batain, M. V. Chaud, T. Almeida, D. A. Fernandes, C. Silva, E. B. Souto and P. Severino, *Cosmetics*, 2020, **7**, 75.
6. J. C. Sharad, Cosmeceuticals, in *Advances in Integrative Dermatology*, ed. K. França and T. Lotti, 2019.
7. X. H. Gao, L. Zhang, H. Wei and H. D. Chen, *Clin. Dermatol.*, 2008, **26**, 367.
8. M. J. Waites, N. L. Morgan, J. S. Rockeyand and G. Higton, *Industrial Microbiology: An Introduction*, John Wiley & Sons, 1st edn, 2009.
9. W. Naser, *Pharmacologyonline*, 2020, **2**, 243.
10. R. Gupta, R. Sharma and Q. K. Beg, *Crit. Rev. Biotechnol.*, 2013, **33**, 216.
11. J. F. Imhoff, A. Labes and J. Wiese, *Biotechnol. Adv.*, 2011, **29**, 468.
12. C. Corinaldesi, G. Barone, F. Marcellini, A. Dell'Ann and R. Danovaro, *Mar. Drugs*, 2017, **15**, 118.
13. R. T. Bianchet, A. L. V. Cubas, M. M. Machado and E. H. S. Moecke, *Biotechnol. Rep.*, 2020, **27**, e00502.
14. J. Choksi, J. Vora and N. Shrivastava, *Indian J. Microbiol.*, 2020, **60**, 379.
15. R. M. Almukhtar and S. G. Fabi, Cosmeceuticals and Delivery Mechanisms: Skin Function and Skin Barrier, in *Cosmeceuticals*, ed. J. Comstock and M. H. Gold, Karger Publishers, Basel, 5th edn, 2021.
16. Z. D. Draelos, *Clin. Dermatol.*, 2009, **27**, 431.
17. H. Dureja, D. Kaushik, M. Gupta, V. Kumar and V. Lather, *Indian J. Pharmacol.*, 2005, **37**, 155.
18. U. FDA, *Is it a Cosmetic, a Drug, or Both (Or Is it Soap)*, FDA, White Oak, Maryland, 1st edn, 2002.
19. L. E. Millikan, *Clin. Dermatol.*, 2001, **19**, 371.
20. P. Elsner and H. I. Maibach, *Cosmeceuticals: Drugs vs. Cosmetics*, Marcel Dekker, Inc., New York, 1st edn, 2000.
21. A. Pieroni, C. L. Quave, M. L. Villanelli, P. Mangino, G. Sabbatini, L. Santini, T. Boccetti, M. Profili, T. Ciccioli, L. G. Rampa, G. Antonini, C. Girolamini, M. Cecchi and M. Tomasi, *J. Ethnopharmacol.*, 2004, **91**, 331.
22. T. B. Clinic, *Cosmetic Skincare vs. Cosmeceutical Skincare*, The Banwell Clinic, East Grinstead, United Kingdom, 2018.
23. J. Ultee, *The Difference between Cosmetics and Cosmeceuticals*, Dr. Jetske Ultee Blog, 2016.
24. Z. D. Draelos, *Dermatol. Clin.*, 2014, **32**(2), 137.
25. Z. D. Draelos, *Cosmeceuticals: In Evidence-Based Procedural Dermatology*, Springer, Berlin, Germany, 1st edn, 2019.
26. X. Liu, J. E. Grice, J. Lademann, N. Otberg, S. Trauer, A. Patzelt and M. S. Roberts, *Br. J. Clin. Pharmacol.*, 2011, **72**, 768.
27. L. A. Zaleski-Larsen and S. G. Fabi, *Dermatol. Surg.*, 2016, **42**, 919.
28. A. Naik, Y. N. Kalia and R. H. Guy, *Pharm. Sci. Technol. Today*, 2000, **3**, 318.
29. J. S. Rokem, A. E. Lantz and J. Nielsen, *Nat. Prod. Rep.*, 2007, **24**, 1262.
30. B. S. Moore and J. N. Hopke, *ChemBioChem*, 2001, **2**, 35.
31. S. O. Bruce, Secondary metabolites from natural products, in *Secondary Metabolites - Trends and Reviews*, ed. R. Vijayakumar and S. S. S. Raja, IntechOpen, London, 1st edn, 2022.

32. H. Ullah, H. A. Santos and T. Khan, *Cellulose*, 2016, **23**, 2291.
33. O. Shezad, S. Khan, T. Khan and J. K. Park, *Carbohydr. Polym.*, 2010, **82**, 173.
34. P. R. Chawla, I. B. Bajaj, S. A. Survase and R. S. Singhal, *Food Technol. Biotechnol.*, 2009, **47**, 107.
35. S. Valla, H. Ertesvåg, N. Tonouchi and E. Fjaervik, *Adv. Microb. Physiol.*, 2020, **77**, 89.
36. P. C. O. Esin and B. Halil, *Afr. J. Microbiol. Res.*, 2011, **5**, 1037.
37. B. V. Mohite, B. K. Salunke and S. V. Patil, *Appl. Biochem. Biotechnol.*, 2013, **169**, 1497.
38. D. M. Barnhart, S. Su and S. K. Farrand, *J. Bacteriol.*, 2014, **196**, 1257.
39. M. C. Laus, A. A. N. Van Brussel and J. W. Kijne, *Mol. Plant-Microbe Interact.*, 2005, **18**, 533.
40. Y. C. Lin, Y. C. Wey, M. L. Lee and P. C. Lin, inventors, NYMPHEAS INTERNATIONAL CORP, assignee, Cosmetic composition containing fragments of bacterial cellulose film and method for manufacturing thereof, *US pat.*, US14/614855, 2015.
41. G. Gao, Y. Cao, Y. Zhang, M. Wu, T. Ma and T. Li, *Carbohydr. Polym.*, 2020, **248**, 116788.
42. N. Shah, M. Ul-Islam and W. A. J. K. KhattakPark, Overview of bacterial cellulose composites: a multipurpose advanced material, *Carbohydr. Polym.*, 2013, **98**, 1585.
43. B. Mbituyimana, L. Liu, W. Ye, B. O. O. Boni, K. Zhang, J. Chen, S. Thomas, R. V. Vasilievich, Z. Shi and G. Yang, *Carbohydr. Polym.*, 2021, **273**, 118565.
44. J. Martins, P. N. Leão, V. Ramos and V. Vasconcelos, *Mar. Drugs*, 2013, **11**, 4902.
45. B. Kultschar and C. Llewellyn, Secondary metabolites in cyanobacteria, in *Secondary Metabolites—Sources and Applications*, ed. R. Vijayakumar and S. S. S. Raja, IntechOpen, London, 1st edn, 2018.
46. F. Freitas, V. D. Alves and M. A. Reis, Bacterial polysaccharides: production and applications in cosmetic industry, in *Polysaccharides*, Springer, Berlin, Germany, 1st edn, 2015.
47. I. B. Allemann and L. Baumann, *Clin. Interventions Aging*, 2008, **3**, 629.
48. A. M. Juncan, D. G. Moisă, A. Santini, C. Morgovan, L. Rus, A. L. Vonica-Ţincu and F. Loghin, *Molecules*, 2021, **26**, 4429.
49. A. Imeson, *Food Stabilisers, Thickening and Gelling Agents*, Wiley-Blackwell, West Sussex, UK, 1st edn, 2010.
50. D. J. Pettitt, Xanthan gum, in *Food Hydrocolloids*, CRC Press, Boca Raton, Florida, 2020.
51. S. A. Kang, J. H. Jang, J. W. Seo, K. H. Kim, Y. H. Kim, D. Rairakhwada, M. Y. Seo, J. O. Lee, S. D. Ha, C. H. Kim and S. K. Rhee, Levan, applications and perspectives, *Microbial Production of Biopolymers and Polymer Precursors*, Caister Academic Press, Norfolk, UK, 2009, p. 145e161.
52. I. L. Shih and Y. T. Yu, *Biotechnol. Lett.*, 2005, **27**, 103.
53. V. Senthilkumar and P. Gunasekaran, *Indian J. Biotechnol.*, 2015, **4**, 491.

54. R. J. Coleman, Y. N. Patel and N. E. Harding, *J. Ind. Microbiol. Biotechnol.*, 2008, **35**, 263.
55. I. B. Bajaj, S. A. Survase, P. S. Saudagar and R. S. Singhal, *Food Technol. Biotechnol.*, 2007, **45**, 341.
56. F. Freitas, V. D. Alves, I. Coelhoso and M. A. Reis, Production and food applications of microbial biopolymers, *Engineering Aspects of Food Biotechnology. Part I: Use of Biotechnology in the Development of Food Processes and Products*, 2013, p. 61.
57. E. M. Del Valle, *Process Biochem.*, 2004, **39**, 1033.
58. T. Kato and K. Horikoshi, *J. Jpn. Soc. Starch Sci.*, 1986, **33**, 137.
59. M. H. Kim, C. B. Sohn and T. K. Oh, *FEMS Microbiol. Lett.*, 1998, **164**, 411.
60. S. Mori, S. Hirose, T. Oya and S. Kitahata, *Biosci., Biotechnol., Biochem.*, 1994, **58**, 1968.
61. K. N. Rajput, K. C. Patel and U. B. Trivedi, *Biotechnology*, 2016, **6**, 1.
62. S. A. Adu, P. J. Naughton, R. Marchant and I. M. Banat, *Pharmaceutics*, 2020, **12**, 1099.
63. E. M. Jackson, *Dermatitis*, 1993, **4**, 108.
64. E. D. O. Monteiro and L. S. Baumann, *Expert Rev. Dermatol.*, 2006, **1**, 379.
65. D. Kligman, *Dermatol. Clin.*, 2000, **18**, 609–615.
66. S. K. Kim, Y. D. Ravichandran, S. B. Khan and Y. T. Kim, *Biotechnol. Bioprocess Eng.*, 2008, **13**, 511.
67. K. Qiu and A. N. Netravali, *Polym. Rev.*, 2014, **54**, 598.
68. A. J. Silvestre, C. S. Freire and C. P. Neto, *Expert Opin. Drug Delivery*, 2014, **11**, 1113.
69. W. Czaja, A. Krystynowicz, S. Bielecki and R. M. Brown Jr., *Biomaterials*, 2006, **27**, 145.
70. D. R. Solway, M. Consalter and D. J. Levinson, *Wounds*, 2010, **22**, 17.
71. L. Fu, J. Zhang and G. Yang, Present status and applications of bacterial cellulose-based materials for skin tissue repair, *Carbohydr. Polym.*, 2013, **92**, 1432.
72. T. Amnuaikit, T. Chusuit, P. Raknam and P. Boonme, *Biomed. Devices*, 2011, **4**, 77.
73. M. Ioelovich, *BioResources*, 2008, **3**, 1403.
74. M. Gama, P. Gatenholm and D. Klemm, *Bacterial Nanocellulose: A Sophisticated Multifunctional Material*, CRC Press, Boca Raton, Florida, 1st edn, 2012.
75. M. Furukawa and T. Tsuboi, Beautifully whitening agent, *Japanese pat.*, 2006052146, Japan, 2006.
76. K. Kim, K. Kim, O. Ryo, T. Lee and T. Kim, Cosmetic composition containing levan having cell proliferation, skin-moisturizing and irritation-alleviating effects, *Japanese pat.*, 2003277225, Japan, 2003.
77. F. Garcıa-Ochoa, V. E. Santos, J. A. Casas and E. Gómez, *Biotechnol. Adv.*, 2000, **18**, 549.
78. W. F. Bergfeld, D. V. Belsito, R. A. Hill, C. D. Klaassen, D. C. Liebler, J. G. G. Marks Jr, R. C. Shank, T. J. Slaga, P. W. Synder and F. A. Andersen,

Safety assessment of microbial polysaccharide gums as used in cosmetics, *Int. J. Toxicol.*, 2016, **35**(1), 4S.
79. R. F. Martins and R. Hatti-Kaul, *Enzyme Microb. Technol.*, 2002, **30**, 116.
80. H. J. Buschmann and E. Schollmeyer, *J. Cosmet. Sci*, 2002, **53**, 185.
81. M. A. Cambon-Bonavita, G. Raguenes, J. Jean, P. Vincent and J. Guezennec, *J. Appl. Microbiol.*, 2002, **93**, 310.
82. T. Le Costaouëc, S. Cérantola, D. Ropartz, J. Ratiskol, C. Sinquin, S. Colliec-Jouault and C. Boisset, *Carbohydr. Polym.*, 2012, **90**, 49.
83. A. Courtois, C. Berthou, J. Guézennec, C. Boisset and A. Bordron, *PLoS One*, 2014, **9**, e94965.
84. H. Yildiz and N. Karatas, *Process Biochem.*, 2018, **72**, 41.
85. I. Kato, *Int. J. Cosmet. Sci.*, 2007, **29**, 68.
86. B. H. A. Rehm and S. Valla, *Appl. Microbiol. Biotechnol.*, 1997, **48**, 281.
87. Z. Wang, J. Wu, L. Zhu and X. Zhan, *Carbohydr. Polym.*, 2017, **157**, 521.
88. D. Bilanovic, J. Starosvetsky and R. H. Armon, *Carbohydr. Polym.*, 2016, **148**, 243.
89. A. Kumar, K. M. Rao and S. S. Han, *Carbohydr. Polym.*, 2018, **180**, 128.
90. L. M. Lods, C. Dres, C. Johnson, D. B. Scholz and G. J. Brooks, *Int. J. Cosmet. Sci.*, 2000, **22**, 85.
91. K. Lintner, F. Lamy, C. Mas-Chamberlin, P. Mondon, S. Scocci, P. Buche and P. Girard, *IFSCC Mag.*, 2002, **5**, 195.
92. A. Bafana, S. Dutt, A. Kumar, S. Kumar and P. S. Ahuja, The basic and applied aspects of superoxide dismutase, *J. Mol. Catal. B: Enzym.*, 2011, **68**, 129.
93. J. G. LeBlanc, S. del Carmen, A. Miyoshi, V. Azevedo, F. Sesma, P. Langella, L. G. Bermudez-Humaran, L. Watterlot, G. Perdigon and A. D. M. de LeBlanc, *J. Biotechnol.*, 2011, **151**, 287.
94. D. Bernard and B. Mehul, inventors, S. A. LOreal, assignee. *US pat.*, US7521422, 2009.
95. X. Lin, C. G. Lee, E. S. Casale and J. C. Shih, *Appl. Environ. Microbiol.*, 1992, **58**, 3271.
96. R. Gupta, Q. Beg and P. Lorenz, *Appl. Microbiol. Biotechnol.*, 2013, **59**, 15.
97. S. C. Gopinath, P. Anbu, T. Lakshmipriya, T. Tang, Y. Chen, U. Hashim, A. R. Ruslinda and M. K. Arshad, *BioMed Res. Int.*, 2015, 1.
98. G. Secchi, Role of protein in cosmetics, *Clin. Dermatol.*, 2008, **26**, 321.
99. L. R. Robinson, N. C. Fitzgerald, D. G. Doughty, N. C. Dawes, C. A. Berge and D. L. Bissett, *Int. J. Cosmet. Sci.*, 2005, **27**, 155.
100. S. Berkó, M. Maroda, M. Bodnár, G. Erős, P. Hartmann, K. Szentner, P. Szabó-Révész, L. Kemény, J. Borbély and E. Csányi, *Eur. Polym. J.*, 2013, **49**, 2511.
101. B. Widner, R. Behr, S. Von Dollen, M. Tang, T. Heu, A. Sloma, D. Sternberg, P. L. DeAngelis, P. H. Weigel and S. Brown, *Appl. Environ. Microbiol.*, 2005, **71**, 3747.
102. E. Marcellin, W. Chen and L. K. Nielsen, Microbial Hyaluronic Acid Biosynthesis, in *Microbial Production of Biopolymers and Polymer Precursors:*

*Applications and Perspectives*, ed. B. H. A. Rehm, Caister Academic Press, Norwich, UK, 2009, p. 163.
103. A. Becker and F. J. Vorhölter, Xanthan biosynthesis by Xanthomonas bacteria: an overview of the current biochemical and genomic data, in *Microbial Production of Biopolymers and Polymer Precursors*, Caister Academic Press, Norwich, UK, 2009, p. 265.
104. I. B. Allemann and L. Baumann, *Clin. Interventions Aging*, 2008, **3**(4), 629.
105. A. K. Leshchiner, N. E. Larsen and E. G. Parent, *U.S. pat.*, 9907741B2, 6 March 2018.
106. T. Matsunaga, J. G. Burgess, N. Yamada, K. Komatsu, S. Yoshida and Y. Wachi, *Appl. Microbiol. Biotechnol.*, 1993, **39**, 250.
107. L. K. Al-Halaseh, S. Al-Adaileh, A. Mbaideen, M. N. A. Hajleh, A. Al-Samydai, Z. Z. Zakaraya and W. A. Dayyih, Implication of parabens in cosmetics and cosmeceuticals: Advantages and limitations, *J. Cosmet. Dermatol.*, 2011, **21**, 3265.
108. M. L. Mourelle, C. P. Gómez and J. L. Legido, *Cosmetics*, 2017, **4**, 46.
109. Y. Tamaru, Y. Takani, T. Yoshida and T. Sakamoto, *Appl. Environ. Microbiol.*, 2005, **71**, 7327.
110. R. P. Rastogi, R. R. Sonani and D. Madamwar, *Appl. Biochem. Biotechnol.*, 2015, **176**, 1551.
111. A. L. Woiciechowski, S. Nitsche, A. Pandey and C. R. Soccol, *Braz. Arch. Biol. Technol.*, 2002, **45**, 393.
112. Z. O. Nor, H. A. El-Enshasy, A. M. Roslinda, M. R. Sarmidi and A. A. Ramlan, *Dtsch. Lebensm.-Rundsch.*, 2009, **105**, 444.
113. M. R. Sarmidi and H. A. El Enshasy, *Int. J. Biotechnol. Wellness Ind.*, 2012, **1**, 3.
114. S. K. Singh, S. U. Ahmed and A. Pandey, *Process Biochem.*, 2006, **41**, 991.
115. M. Ochi, N. Adachi, H. Nobuto, S. Yanada, Y. Ito and M. Agung, *Artif. Organs*, 2004, **28**, 28.
116. V. Natarajan, N. Krithica, B. Madhan and P. K. Sehgal, *J. Pharm. Sci.*, 2011, **100**, 195.
117. M. van Dijk, T. H. Smit, S. Sugihara, E. H. Burger and P. I. Wuisman, *Spine*, 2002, **27**, 682.
118. J. A. Bouwstra and M. Ponec, *Biochim. Biophys. Acta, Biomembr.*, 2006, **1758**, 2080.
119. K. Shindo, K. Kikuta, A. Suzuki, A. Katsuta, H. Kasai, M. Yasumoto-Hirose, Y. Matsuo, N. Misawa and T. S. Takaichi, *Appl. Microbiol. Biotechnol.*, 2007, **74**, 1350.
120. H. Y. Kang, T. J. Yoon and G. J. Lee, Whitening effects of marine pseudomonas extract, *Ann. Dermatol.*, 2011, **23**, 144.
121. V. N. Khmelenina, W. G. Sakharovskiĭ, A. S. Reshetnikov and T. IuA, Synthesis of osmoprotectors by halophilic and alkalophilic methanotrophs, *Mikrobiologiia*, 2000, **69**, 465–470.
122. T. R. A. Thomas, D. P. Kavlekar and P. A. LokaBharathi, Marine drugs from sponge-microbe association—A review, *Mar. Drugs*, 2010, **8**, 1417.

CHAPTER 10

# *Marine Natural Products as a Bioresource for Cosmeceuticals*

YOUSEF RASMI[a,b], KEVSER KÜBRA KIRBOĞA[c], TAZEEN RAO[d], MAZHAR ALI[d] AND MUHAMMAD ZEESHAN AHMED[d*] (0000-0003-1118-3693)

[a]Cellular and Molecular Research Center, Cellular and Molecular Medicine Institute, Urmia University of Medical Sciences, Urmia, Iran; [b]Department of Biochemistry, School of Medicine, Urmia University of Medical Sciences, Urmia, Iran; [c]Bilecik Seyh Edebali University, Bioengineering Department, Bilecik, 11100, Turkey; [d]Department of Biochemistry, Bahauddin Zakariya University, Multan 60800, Pakistan
*E-mail: mzeeshanahmed121@gmail.com

## 10.1 Introduction

Oceans are a diverse resource, covering over 70% of the surface of the earth. Undoubtedly, the marine environment, in which 34 out of 36 phyla of life are represented, is the largest source of chemical and biological diversity in the world.[1] For the past 30–40 years, marine invertebrates have been a fascinating subject of study for scientists around the world. A small number of marine animals, plants, and marine microbes have produced about 14 000 natural products.[2] The discovery of novel biologically active compounds has been largely attributed to marine natural products.

The research on the chemistry and biological activities of marine natural products has produced many bioactive compounds that exhibit various pharmaceutical properties. Many of them have been used clinically or in the

Bioprospecting of Natural Sources for Cosmeceuticals
Edited by Deepika Kathuria, Ajay Sharma, Meenakshi Verma and Gulzar Ahmad Nayik

Published by the Royal Society of Chemistry, www.rsc.org

treatment of diseases such as cancer, Alzheimer's disease, and infectious diseases. Also, they have been used in the development of painkillers and cosmeceuticals.[3] A "cosmeceutical" is a cosmetic product with benefits similar to those found in pharmaceuticals. Kilgman popularized the term by combining "cosmetics" and "pharmaceuticals".[4] Due to their unique biological and chemical properties, the demand for cosmeceuticals derived from marine resources has increased exponentially in recent years.[5] In addition to providing valuable bulk compounds such as carrageenan and agar that are used in the cosmetics industry as gelling and thickening agents, marine organisms also contribute valuable bulk compounds to the cosmetics industry. These substances also promote skin hydration (ectoine), prevent product alteration caused by microbial contamination (trichodin A), and act as coloring agents (mytiloxanthin). The active ingredients in cosmeceuticals can also be marine-derived, as they determine various functions of cosmeceuticals like anti-acne (sargafuran), whitening (chrysophanol), anti-tyrosinase (kojic acid), UV protection (scytonemin), anti-wrinkle (astaxanthin), and anti-oxidant activity.[5]

Cosmeceuticals contain enzymes, minerals, vitamins, and phytochemically active ingredients. These ingredients can be formulated into cream, ointment, and lotion formulations.[6] Natural bioactive substances are derived from a variety of sources, including microorganisms, terrestrial plants, and marine organisms. In addition to their functional roles in human health,[7] these substances can also promote healthy skin, nails, and hair at a cellular level.[8] Despite the popularity and wide use of plant-based cosmeceuticals, there are limitations since plants generally grow very slowly and their chemical composition varies by region and season. Essentially, modern aquaculture techniques allow the flora and fauna of the sea to produce a wide range of chemically unique biomolecules that are not found in terrestrial resources.[9]

Overall, the present chapter discusses the importance of natural products derived from the marine environment. This chapter also discusses the chemical entities with cosmeceutical potential from marine resources and the mechanisms by which these compounds exert their effects on the functions of the body, and focuses on potential applications in the future.

## 10.2 Discovery of Bioresource Marine Natural Products and Their Biodiversity

The two spongouridine and spongothymidine nucleosides as new potential therapeutical agents were first discovered from the Caribbean marine sponge *Cryptotethia crypta* in the early 1950s.[10] These discoveries stimulated new interest in discovering and/or synthesizing therapeutic agents from natural sources, and/or chemical synthesis. Traditionally, natural bioactive products are first extracted, screened, isolated, fractionated, purified, and then characterized from the sources. Despite being a slow, less efficient, and labor intensive method, this traditional process is still widely used. However, due to the advancement in drug discovery methods, the bioactive natural product discovery method demands new approaches for rapid screening, hit identification, target validation, lead compound identification, and optimization.[11–13] Subsequently, natural product discovery programs have driven some companies to develop and design synthetic analogs from natural products.[13–15] For these, high-resolution mass spectrometry (MS), nuclear magnetic resonance (NMR), high-performance liquid chromatography (HPLC), and spectroscopic methods have particularly contributed to analytical refinement and *de novo* structure determination of natural products from crude extracts.[11,16,17] Recently, metagenomics has opened a new gate for the discovery of unreachable natural products by the computational analysis of uncultivated microorganisms' genomes.[18] Genome mining is a recent computational approach that analyzes the genome sequences to identify protein-encoding genes for the discovery of novel natural bioactive products, secondary metabolites, gene clusters, and biosynthetic pathways.[19,20]

More than 20 thousand unique bioactive marine natural products have been explored in the last 50 years.[21] Initially, some compounds obtained from marine invertebrates were either found to be toxic or ineffective for disease treatment. However, such compounds were very useful as agrochemicals, biological tools, and cosmetic ingredients.[22] 1241 new bioactive compounds were identified in the marine environment in the single year 2012.[23] To date, several marine natural bioactive molecules, such as alkaloids, anthraquinones, flavonoids, saponins, and volatile oils, have been discovered and this number is increasing on a daily basis due to the cosmeceutical industry. Related compounds derived from marine organisms act as antibiotics, anticancer agents, antivirals, neurotoxins, painkillers, and many others. In parallel, marine-derived compounds are potentially useful in cosmetics (sea fan, crustacean, and algae compounds), artificial bone (corals), industrial applications (jellyfish-based fluorescent compounds, novel mussel glues, and heat-resistant enzymes from deep-sea bacteria), agrochemicals, and nutritional supplements (algae compounds).[24,25] Marine natural products are in high demand for skin care products with a variety of benefits. From marine resources, seven main types of cosmeceutical ingredients have been identified and commercialized.[26] Hence, many of the compounds isolated from marine organisms have been tested for cosmeceutical properties and are currently available on the market. Several active crude solvent extracts have been isolated from marine species of cosmetics through screening.

## 10.3 Developing Marine Natural Products: Challenges and Opportunities

The challenges are categorized into three types, namely biodiversity, supply and technical, and market challenges. The emphasis of "biodiversity" is placed on the difficulties associated with secure access to marine resources, the accurate recognition of biological materials, and the screening of compounds and samples effectively. "Supply and technical" focuses on the challenges associated with isolating and producing a pure bioactive, and understanding its mechanisms of action. As a final consideration, the process and costs involved in emergent bioactive natural products are discussed in the topic of "market challenges".[27]

### 10.3.1 Biodiversity Challenges

Our knowledge of the moon is considerably greater than our knowledge of the sea. To fully evaluate the ocean's biodiversity, application of robotics and engineering should be used to gain access to the ocean's deepest spots. Due to the lack of traditional therapeutic information related to marine creatures, a random selection of organisms has been used in the exploration of marine-derived compounds with biological activity. To improve the chances of identifying novel molecules, researchers often use three different sampling approaches: (1) exploring taxonomical or geological groups that are unexplored, (2) studying new taxa and chemically diverse regions, and (3) incorporating both.[28,29]

It is essential, however, that a solid understanding of biodiversity and chemo-diversity be acquired. A large number of potential organisms were found in untapped, but inaccessible, marine ecosystems, such as thermal vents and ocean mounts, that were not studied due to a lack of appropriate resources.[30] Scuba diving cannot reach deeper than 30 m, so trawling becomes the most common method of sampling in difficult to access spots. In addition to being unselective, this method may damage organisms and their ecosystems, obscuring important details regarding the organism's environment.[31] It is ideal to sample marine organisms in deep waters with modern or innovative instruments, such as autonomous submersibles, and remotely operated vehicles. The use of such equipment allows the observation of an organism's habitat and the prediction of its bioactivity. A sponge's metabolites likely prevent the growth of competitors if it is observed to lay by itself in its environment and if there is no litter around it.[31] Unfortunately, sampling facilities like these are extremely expensive, making them accessible to just a few laboratories, a problem that is much more challenging when it is considered that most biological diversity can be found in tropical and subtropical regions of underdeveloped countries.[32] It is, for this reason, that collaboration with other countries is so crucial to this research. Nevertheless, the Convention on Biological Diversity (CBD) now covers all aspects of biodiversity in relation to the use of natural resources. Implementation levels of the Nagoya Protocol in different regions and the difficulty of working within a still indistinct controlling framework may cause some industries to leave the market for natural products.

One of the major challenges for programs involving marine natural products lies in a scarcity of taxonomic knowledge about marine species. Pharmaceutically, microorganisms and macro-organisms, both terrestrial and marine, should be identified and classified correctly. This complication can arise from the misclassification of species, which makes it impossible to replicate the isolation of an extract or metabolite, as well as from ambiguity in the dereplication process. Various methods are used to categorize marine macro- and microorganisms (algae, fungi, invertebrates, and bacteria). Despite advances in taxonomy, unambiguous species classification is still difficult for most marine macro-organisms.[33] There are still many undescribed macroinvertebrates, and similar species require delicate physical characteristics, which makes macroinvertebrate research extremely challenging.[34]

Developing screening assays is the next step in the drug discovery process after identifying and validating a target. For natural products, high throughput screening is one of the most widely used screening paradigms.[35] Natural product applications require the construction of high-quality libraries to be successful. Research at Pfizer found that the outcome of high throughput screening is influenced by the compound library quality, target, and screening process.[36] For a balanced natural product detection program, simplified extract fractions, crude extracts, and pure compounds should be included in the library.[37] Compared with pure compound libraries, libraries of the crude extract are more convenient and inexpensive with higher diversity, and also moderate in size. A crude extract is a complex mixture of several compounds that may interact synergistically, which is why the bioactivity disappears from purified fractions and finally from pure compounds.[38] However, false negative results may also be obtained if the crude extract contains only a small percentage of an active metabolite, or if tannins[37] interfere with its activity by binding to other metabolites. Because of these reasons, drug discovery programs in the past discouraged this approach.[37] It is possible to avoid these problems by screening pre-fractionated libraries.[39] There are several methods for pre-fractionating crude extracts, and the subsequent fractions can differ extensively in complexity from a blend of several compounds to a single key compound that is >90% pure depending on how the fractionation is performed and how many compounds are in the crude extract. It is possible to identify hits more quickly by pre-fractionating several undesired compounds. Combined with robust natural product libraries, representative of wide biodiversity, modern high throughput screening methods can streamline the discovery of cosmeceutical and therapeutic drugs.

### 10.3.2 Supply and Technical Challenges

Technical and supply issues can cause several problems. Firstly, there is variability within the organism itself. As an example, sponges produce high quantities of bioactive metabolites as a chemical defense against predation, overgrown fouling organisms, and competition for space, among other environmental stresses. Sponge metabolites are more likely to occur in environments categorized by strong opposition and nourishing pressure, such as coral reefs. Due to the variable nature of these environmental conditions, the resupply of the same organism may not deliver a similar metabolite.

Additionally, when it comes to marine invertebrates, microorganisms sometimes produce the bioactive themselves.

As soon as a natural product has been isolated and identified as a lead compound, there are challenges related to its sustainable supply. Often, compounds that are of interest are only found in very small quantities and are difficult to purify.[40] It can be even more challenging to extract tissues from marine invertebrates because they have high salt and water content. In preclinical development, grams to tons are needed, whereas in clinical phases, multikilogram quantities are needed, and in cosmetic applications, tons are needed.

To solve the problem of a sustainable supply of macro-organisms, mariculture and aquaculture have been tried. Mariculture and aquatic culture are both cultivation methods that favor organism growth in their natural environment. Isolated samples, however, are difficult to cultivate or maintain due to uncommon and exclusive conditions in the sea. In general, sponges and their microbiota are difficult to cultivate, so the desired compound may have to be isolated and purified from wild specimens.[28] In addition to contributing to the loss of marine biodiversity, these constraints also represent a major bottleneck in supplying natural compounds sustainably. Several marine compounds show great promise but have not been developed further because there is not enough sustainably available substance. Synthetic or hemi-synthetic analogs, improved derivatives, or reduced complexity pharmacophores have been designed to overcome this barrier.[15] These approaches, however, are not without their challenges. Chemists still have a long way to go before they can efficiently synthesize any molecule, and complete synthesis is by no means an easy undertaking. There are often one or more stereocenters on a natural product, making their synthesis difficult due to the difficulties involved in enantioselective or diastereoselective processes. Sometimes, hemi-synthesis can be a good solution for supplying compounds. A lead is produced by converting a biosynthetic intermediate rather than the lead itself from a natural source. There are two advantages to this approach: (1) it may be easier to extract the intermediate with a higher yield than the final product, and (2) additionally, analogs of the final product may be synthesized.

Furthermore, it is essential to identify the compound isolated from the biological source when synthesizing or hemi-synthesizing a bioactive natural compound. It is impossible to eliminate errors in structural elucidation, despite modern methodologies that are well-advanced. Several structural revisions have been reported in the literature, even for recently elucidated NPs. The average number of misassigned natural products and marine natural products per 5-year period is 369 and 135, respectively.[41] There are several categories of misassignments that can occur due to structural complexity and small amounts of samples, including wrong formulation, structure, double bond configuration, absolute configuration, and stereocenters belonging to multiple locations.[42] There need to be improved strategies for dereplication used to prevent the rediscovery of the known. During bioassay-guided fractionation, more than 150 000 molecules derived naturally are often re-isolated, and that should be avoided.[13] LC-UV, LC-MS, LC-MS/MS, and LC-NMR are phenomenal technologies useful in dereplication, particularly during prefractionation.[43] Natural product research that needs to replicate crude extracts quickly requires access to suitable databases. There are several commercial databases available for dereplication, including NAPRALERT, AntiBase,[44] Beilstein, MarinLit,[45] NAPROC-13,[46] and Chapman & Hall's Natural Products Dictionary.[47] In addition to libraries of unadulterated natural compounds, there is a possibility of virtual screening. The macromolecular targets' three-dimensional model or pharmacophore model can be used as a search criterion in virtual screening. *In vitro* screening has many advantages: advanced capacity, no requirement to isolate compounds physically, less expensive, and less time consuming, with theoretical calculations and predictions being possible for every structurally defined compound with entire targets. Furthermore, it is also possible to evaluate pharmacokinetics like distribution, absorption, metabolism, toxicity and excretion processes.[48] To optimize the discovery of new drug leads, virtual screening, since it is only a predictive tool, should be integrated with traditional avenues, such as a bio-assay based fractional process, analytical action profiling, chemoinformatics and ethnopharmacological screening.[48]

### 10.3.3 Market Challenges

The final step in identifying a lead is to understand how it works against the specific target. As part of this, cell and molecule-based approaches are typically useful for secondary tests. The pharmaceutical and cosmeceutical industries are requiring this identity gradually, which is the main challenge. Last but not least, commercial or market issues are very relevant and often overlooked in natural product development. Significant questions must be asked by investigators or corporations at the start of the development process: (i) which industry applications and market needs might be met by the specific activity, (ii) how much is the market price for each kilogram of the bioactive product, (iii) how will it be formulated and administered, (iv) how sustainable will the supply chain be, and (v) how will the product be delivered to market.[31] High throughput screening technologies have produced a high number of natural product hits and leads, which highlights the need for a focused strategy in this area. Technologies and manufacturing processes can have high costs and low yields, which may drive up the price per kilogram, rendering these products uneconomical. Recombinant technologies, especially in private care commerce, are too expensive with very low margins for their packaging. It is clear that the majority of discoveries in the natural product pipeline are hardly available on the market for consumers, so academia should give more attention to market issues. As part of this effort, industry–academia partnerships have been encouraged at both national and transnational levels. These alliances benefit both partners as the industry grows with novel natural products and greater chances of market accomplishment, while knowledge, funding, and publications go to academia.[31]

Natural product designers can discover novel models *via* surpassing or minimizing the challenges just mentioned.

## 10.4 Bioresources of Cosmeceuticals from Marine Environments

In recent years, consumers have become increasingly aware of the importance of natural cosmetics for their skin.[49] The cosmetics industry is now progressively launching cosmetics containing compounds or extracts from marine sources because consumers are demanding natural products.[8] Cosmetics companies have been investing in the development and research of novel extracts of natural substances which are beneficial for health, sustainable for the environment, and biologically friendly.[50] Due to their numerous beneficial effects, customers are demanding such bioactive cosmetics derived from natural resources, rather than non-natural ingredients. Agrochemicals, fine chemicals, cosmeceuticals, nutrition supplements, molecular probes, and pharmaceuticals have great potential in the marine environment, which has extraordinary biodiversity.[51] The vast potential offered by the marine environment will need to be explored further since a large number of new species are still to be discovered. Biochemical and physiological mechanisms have evolved by which marine organisms reproduce, communicate, and protect themselves against predation and infection.[5] Due to activities against oxidants, inflammation, allergens, aging, wrinkles and tyrosinase activation, as well as in matrix metalloproteinase activation, and UV protection, many of these compounds could be used as cosmeceuticals or nutricosmetics.[52] Figure 10.1 shows the bioresources of cosmeceuticals which are derived from marine environments and are involved in the production of natural compounds that are useful in the cosmetics industry.

The current method of discovering novel products from marine environments depends on harvesting samples that are not sustainable or replicable. Problems of replicability are linked to variations in the environment at the community level, whereas sustainability problems involve large amounts of biomass being needed for drug discovery.[30] A metabolite supply may not be guaranteed between samples of identical species taken in diverse zones or time periods because of different chemical configurations.[53] For marine

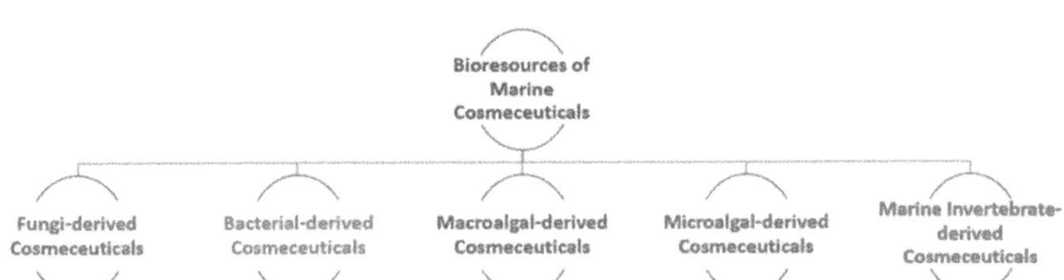


**Figure 10.1** Bioresources of marine cosmeceuticals.

invertebrates, however, alternate aquaculture may be possible to decrease such issues, since homogenous environmental conditions may allow continuous production of animal biomass.[54]

### 10.4.1 Fungi-derived Cosmeceuticals

Marine fungi produce some secondary metabolites that may have cosmeceutical applications. Some species of bacteria produce mycosporine amino acids, like *Trimmatostroma salinum*, *Aureobasidium pullulans*, *Hortaea werneckii*, and *Phaeotheca triangularis*.[55] In contrast to oxybenzone, a positive control used in sunscreen formulations, circumdatins, derived from *Exophiala* species, which are sponge-associated marine fungi, demonstrated greater UVA protective activity.[56] A *Myrothecium* species of algicolous fungus isolated from *Enteromorpha compressa*, which is a marine green alga, has higher activity against tyrosinase than kojic acid.[57] In addition, *Botrytis* sp., isolated from *Hyalosiphonia caespitosa* underwater, has robust anti-tyrosinase activity.[58] H1-7 culture of *Trichoderma viridae* inhibits mushroom tyrosinase competitively. It appears that these compounds prevent enzyme activity by binding to the active site of copper.[59] Nevertheless, chrysophanol is one of the few marine fungal metabolites to make its way into cosmetics.[60]

$CuCl_2$ was added to a culture of the marine-derived fungus *Pestalotiopsis* species Z233, isolated from *Sargassum horneri*, and two previously unreported sesquiterpenes, 1β,5α,6α,14-tetraacetoxy-9α-benzoyloxy-7βH-eudesman-2β,11-diol and 4α,5α-diacetoxy-9α-benzoyloxy-7βH-eudesman-1β,2β,11,14-tetraol, were obtained. *Alternaria* species produce two derivatives of kojic acid (showing activity against tyrosinase), kojic acid "dimethyl ether" and "monomethyl ether".[61] Squalene is found in the liver oil of sharks, and can also be found in *Protista thraustochytriales*-like microorganisms. Skin lubrication, cell structure, and protection are all enhanced by squalene, a lipid that is produced by sebaceous glands. In cosmetics, squalene keeps skin moist. Creams containing squalene are non-toxic, non-irritating, and non-sensitizing, as well as anti-static and emollient.[62] However, fatty acids are used as nutritional supplements and cosmeceuticals due to their ability to stimulate collagen production, repair soft tissues and nourish the skin, as well as healing wounds and reducing inflammation.[63]

### 10.4.2 Bacteria-derived Cosmeceuticals

Most marine bacteria are linked to organic compounds (as a substrate), and their abundance decreases with increasing depth. Secondary metabolites produced by marine bacteria can assist as the best source of bioactive complexes due to their prolific production and biodefense against other microorganisms in harsh oceanic climates.[64] The pharmaceutical and cosmetic industries have found a huge amount of bioactive secondary compounds/metabolites produced by bacteria to be highly valuable.[7] Marine bacteria produce many compounds that have antimicrobial, photoprotective, moisturizing, anti-aging, and anti-oxidant properties, including lipids, alkaloids, mycosporine amino acids, proteins, and peptides.[65]

There are many cosmeceutical products with anti-aging properties that are derived from marine sources, including polysaccharides,[66] and bacteria are known to produce higher polysaccharides.[67] Known as Abyssine,[68] deepsane (an exopolysaccharide) isolated from *Alteromonas macleodii*, which is a marine bacterium, is used to soothe and reduce irritation of sensitive skin.[31,69] Anti-aging products contain a mixture of polysaccharides derived from *Pseudoalteromonas* sp., which is isolated from Antarctic waters. Through fermentation, this mixture enhances collagen I synthesis, enhancing the structural properties of the skin.[31] Exopolysaccharide (HE 800) produced by *Vibrio diabolicus* (a marine bacterium) resembles hyaluronic acid in structure and function, stimulating collagen synthesis.[70]

*Pseudomonas* species reduce the pigmentation of human melanocytes *via* the production of methylene chloride and cultured skin cells by inhibiting the expression of tyrosinase.[71] Thalassotalic acids derived from *Thalassotalea* sp., a bivalve bacterium, exhibit antityrosinase activity. The commercially available arbutin compound is similar to thalassotalic acid A and their activity is comparable to tyrosinase. A high salt concentration was used to isolate ectoine from *Ectothiorhodospira halochloris* and some Actinobacteridae.[72] Increased spacing in inter-molecules and high lipid head group mobility in cell membranes allow this compound to improve the hydration of the cells' surfaces,[73] and human tolerance has been reported to be good.[74,75] Consequently, ectoine prevents epidermis dehydration by acting as a long-term moisturizer.[73] As well as reducing inflammation of the skin, it is now being examined for the treatment of atopic dermatitis.[74]

### 10.4.3 Macroalgal-derived Cosmeceuticals

In cosmetic products, marine algae like red and brown are used.[76] Agar, carrageenan, and alginates have traditionally been produced using macroalgae/seaweed. The brown and red macroalgae cosmetics include minerals, sugars, amino acids, vitamins, and lipids.[77,78] Common macroalgae are *Laminaria longicruris*, *Ascophyllum nodosum*, *Ulva lactuca*, *Mastocarpus stellatus*, *P. atlantis*, *Chondrus crispus* and *Porphyra*. They can be used as cosmeceuticals for skin care to protect against ultraviolet rays, aging, and oxidative stress to smooth and whiten the skin.[50] Agar and carrageenan, among others, have been used to gel, thicken, and stabilize cosmetics and nutraceuticals.[79]

The total amount of macroalgae cultivated in the world is estimated to be 59%, 40%, and <1% for brown algae, red algae, and green algae, respectively. It is possible to cultivate macroalgae on seashores very rapidly and to manipulate the culture conditions to control their production of bioactive compounds, like proteins, polyphenolic compounds, and pigments.[80] It was found that cutaneous progerin can be reduced with an *Alaria esculenta* lipophilic extract of brown alga,[81] whose overproduction is triggered by progressive telomere injury and cellular senescence.[82] However, 7-phloroeckol is an effective anti-pigmentation agent, likely due to its tyrosinase inhibitory properties, and has been recommended for skin whitening.[79,83] An isolated phlorotannin, diphlorethohydroxycarmalol, was found to be highly effective at whitening the skin,[84] as well as protecting DNA against damage caused by radiation like UVB by causing changes in fibroblast morphology. Diphlorethohydroxycarmalol is an interesting cosmeceutical candidate because of its dual biological properties.[85] A phologlucol derivative from brown algae also inhibits the activity of tyrosinase owing to its ability to chelate copper.[86] This suggests that seaweeds contain bioactive compounds that can be used in skincare.[87]

In addition, *Laminaria japonica*, also known as "kombu" in Japan, is another important seaweed that contains high concentrations of marine minerals and trace elements, which are used in the production of algae derivatives against UV radiation. The extract of this alga has binding agents that make it a potential skin moisturizer.[88] As a source of fucoxanthin, this alga also has properties for skin, like anti-tyrosinase, anti-oxidant and anti-melanogenic properties, and prevents UVB-induced skin coloration. *L. saccharina*, a species of *Laminaria* that contains proteins, vitamins, minerals, and carbohydrates, also has cosmetic potential. Algae extracts have anti-inflammatory, healing, and sebaceous gland-regulating properties.[89] *Hizikia fusiformis* (Sargassaceae family) is a brown macroalga, the flavonoid glycosides of which were found to have anti-tyrosinase activity.[86] Various cosmetic products contain extracts of green algae. In addition to supplying glucuronic acid to the skin, *Codium tomentosum* extract can also protect the skin from the harmful effects of dryness in the environment.[7] Snow alga (*Chlamydocapsa* sp.) is involved in skincare and hair care products to prevent photoaging.[90] These studies indicate that marine red and brown algae, whether as extracts or in pure form, can act as active cosmeceuticals.

### 10.4.4 Microalgal-derived Cosmeceuticals

Microalgae contain a variety of potential compounds that have bioactive applications in cosmeceuticals. Microalgae are constituents of various food products due to their high content of chlorophyll, tocopherols, carotenoids, sterols, fatty acids, proteins, vitamins, carbohydrates, antioxidants, and minerals.[91] A variety of microalgae species are involved in cosmetic products, such as *Spirulina*, *Odontella*, *Dunaliella*, and *Chlorella*.[92] Microalgae offer great potential for cosmeceuticals while a few of them absorb ultraviolet radiation, preventing skin damage, pigmentation, wrinkles, coarseness, and laxity. Figure 10.2 shows the properties of microalgal-derived cosmeceuticals.

Scytonemin, a sunscreen pigment from cyanobacteria, absorbs radiation like UVA/UVB.[93] Among the carotenoid pigments produced by microalgae is astaxanthin. Researchers have extensively studied this compound for its role in the well-being of skin and its photoprotective properties,[94] and it also has strong antioxidant and anti-inflammatory properties. Astaxanthin influences the various stages of the oxidative impairment cascade and overpowers

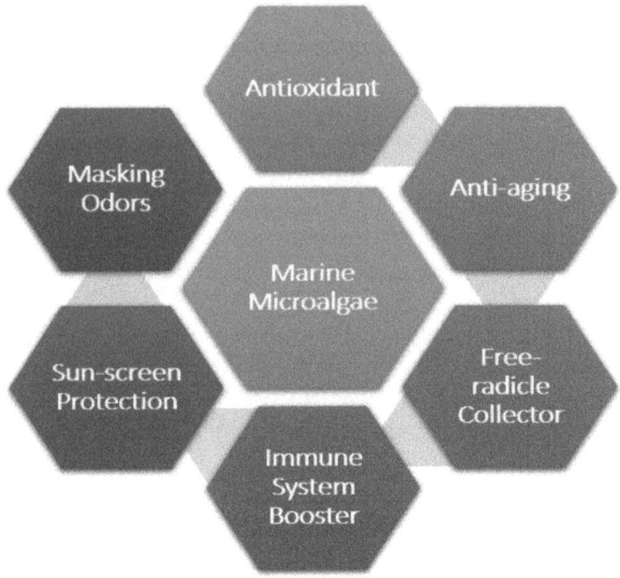


**Figure 10.2** Properties of microalgal-derived cosmeceuticals.

inflammatory mediators.[95] Furthermore, its DNA repair and immunomodulatory properties support its usage as a cosmeceutical and harm prevention agent.[94] Several other carotenoids, including lutein, canthaxanthin, lycopene, and zeaxanthin, are also significant in health and cosmetics.[96] A moisturizing property of microalgae keeps the skin and hair healthy by improving and maintaining their barrier function. Some *Spirulina* proteins have moisturizing properties in hair products, giving water retention, and are suggested for dry skin conditions such as atopic dermatitis.[50] The oils obtained from some microalgae, especially from the *Chlorella* genus, soften the skin and smoothen the hair.[97] A new marine cosmeceutical for skin care cosmetics applications is alguronic acid. Solazyme produces alguronic acid by combining undetermined polysaccharides produced by microalgae. An anti-aging skincare formula called Algenist was introduced to the market in 2011 that contains alguronic acid as an active ingredient.[98]

## 10.4.5 Marine Invertebrate-derived Cosmeceuticals

### *10.4.5.1 Marine Sponge-derived Cosmeceuticals*

Bioactive secondary metabolites are isolated from marine sponges like microorganisms/phytoplankton, especially through their enzymes.[99] These microorganisms are renewable resources of natural products that could have very important applications in pharmaceuticals, cosmeceuticals, and nutraceuticals.[100] By providing their hosts with metabolic products, bacteria offer properties like nitrification and autotrophy. Furthermore, enzymes of marine sponges are involved in delayed bacterial and fungal growth and can be used in cosmetic formulations as skin-whitening agents.[52]

Various sponge metabolites in the cosmetics industry, like halistanol trisulphate, have properties against tyrosinase.[101] A highly oxygenated diterpene called gagunin D represses melanosome expression. Gagunin D and its derivatives could be considered latent candidates for skin-whitening cosmeceuticals because of their multi-functional properties.[102] More than 40 carotenoids are also produced by marine sponges.[103] As a result, these compounds could be useful as cosmeceuticals.

### *10.4.5.2 Coral-derived Cosmeceuticals*

The physical, chemical and textural characteristics of coral powder make it a sustainable material for cosmetics.[52] Minerals other than heavy metals, such as calcium carbonate, may also be present, depending on their chemical composition. The use of coral powder on the skin provides minerals and protects the skin from ultraviolet radiation, and it has anti-acne, anti-aging, anti-oxidant, and skin softening activity, as well as being used as a cosmetic ingredient.[52] In the cosmetics industry, the diterpene glycosides pseudopterosins A to D are among the potential marine natural products.[104] In addition to their biological properties, these compounds also have anti-inflammatory and analgesic properties,[105] anti-bacterial properties,[106] anti-acne properties,[107] and wound healing properties.[108]

## 10.5 Properties of Marine Natural Products for Cosmeceutical Applications

Cosmeceuticals contain several biologically active ingredients such as carbohydrates (carrageenans, exopolysaccharides, and chitosan), enzymes, essential oils, phytochemicals (phlorotannins, astaxanthins, and pseudopterosins), mycosporine-like amino acids (MAAs), vitamins, and minerals that have significant medical applications.[5] Researchers are currently interested in exploring biologically active cosmeceutical ingredients of marine origin. Marine-derived cosmeceuticals have been identified to have anti-aging, anti-melanogenic, anti-acne, anti-oxidant, film-forming, wound healing, bioadhesive, anti-inflammatory, anti-microbial, analgesic, moisturizing, repairing, anti-cancerous, anti-viral, anti-bacterial, and many other pharmacologic and cosmetic properties.[109,110]

## 10.6 Secondary Metabolite Marine Natural Products for Cosmeceutical Applications

Natural products (NP) are small molecules having a molecular weight of less than 3000 Da and must be synthesized from living organisms like microorganisms, plants, and animals.[31] Different natural products are restricted to a specific organism, species, genus, or taxonomic family. They are frequently known as secondary metabolites as they have no direct roles in normal growth, reproduction, and development, and they are not even synthesized by common metabolic pathways.[39,111] Carbohydrates, proteins, lipids, and nucleic acids are examples of primary metabolites, along with terpenoids, alkaloids, and any other compound whose absence could cause very minor impairment, like varied morphological characteristics, for an organism.[112,113] Marine microorganisms are a rich and productive source of secondary metabolites. Actinobacteria, specifically, have appeared as hugely prolific as compared to their terrestrial analogs.[114] The compounds synthesized by living organisms present in immensely vast marine habitats are called marine natural products (MNPs). The label "marine natural products" is also used to mention cosmetics, nutritional food additives, and other substances derived independently from natural sources without adding any supplement of non-natural origin.[31]

### 10.6.1 Carbohydrates

Marine organisms are an extensive source of polysaccharides and many researchers have studied them very comprehensively in recent years. Sulfated and non-sulfated polysaccharides are the forms that marine organisms contain as a rich source of carbohydrates.[8] Polysaccharides in the form of fucoidans and carrageenan are found abundantly in brown algae and red algae, respectively.[115] Carbohydrates having a higher molecular weight that form an extracellular matrix in the majority of microbial cells living in the marine environment are known as exopolysaccharides (EPSs). A large number of exopolysaccharide (EPS) producing bacteria have been isolated from diverse marine environments recently.[116,117] Mostly, repeating units of three or four different monosaccharides mainly in the form of linear polysaccharides organized in a group of ten or fewer are produced by marine bacteria as a source of EPSs.[118] After cellulose, the most abundantly found organic compound in nature is chitin. Chitosan and other derivatives are produced solely from chitin.[119] Table 10.1 presents the sources and advantages of carbohydrate-based cosmeceutical agents.

**Table 10.1** Carbohydrate-based cosmeceutical agents.

| Type of carbohydrate | Source | Cosmeceutical advantage | Reference |
|---|---|---|---|
| Fucoidans | *Costaria costata* | Anti-aging | 120 |
| Fucoidan | *Ascophyllum nodosum* | Anti-aging | 121 |
| Carrageenan | *Porphyridium* | Skin-whitening | 122 |
| Chitosan oligosaccharides | Mushroom | Anti-photoaging | 123 |

### 10.6.1.1 *Fucoidans*

Fucoidans are present in the cell wall of several brown algae as vital sulfated polysaccharides (SPs).[124,125] It has been reported that human foreskin fibroblast HS68 cells and HaCaT cells showed anti-aging activity in response to fucoidan obtained from *Costaria costata*.[126] The mRNA and protein expression of downregulated type 1 pro-collagen and protein expression of upregulated MMP-1 that were previously initiated by UVB are suppressed by fucoidan through the inactivation of ERK and JNK. In addition, interleukin-1 (IL-1)-induced secretion of MMP-9 and -3 and deterioration of tissue inhibitor of metalloproteinase inhibitor 1 (TIMP-1) in HDFs are inhibited by fucoidan obtained from mekabu.[127] In the treatment of the skin of mice, fucoidan reduced MMP-1 plus interferon activation, the thickness of the prickle cell layer, and edema that were previously induced by UV-B, and positive correlations were shown.[128] In human skin, anti-aging potential has been exhibited by Senni *et al.* by inhibiting leukocyte elastase activity and elastic fiber degradation using fucoidan (16 kDa) obtained from *Ascophyllum nodosum* after acidic hydrolysis.[127] This shows that fucoidans provide an anti-aging advantage through different mechanisms of action.

### 10.6.1.2 *Carrageenans*

A linear sulfated polysaccharide made up of D-galactose and 6-anhydro-D-galactose (D-AHG) obtained from red algae is known as carrageenan. The kappa family (having esterification in the fourth position) and the lambda family (having no esterification at all) are the two most prominent families of carrageenans.[115,129] Due to their physical and chemical properties, carrageenans have been widely employed in cosmetic products as a moisturizer, stabilizer, and emulsifier. Due to this, they can be employed in a wide range of beauty products that include sprays, lotions, foams, creams, and sticks.[130,131] Red microalgae *Porphyridium* produces carrageenan, which has been identified as a substance toxic to macrophages. The melanin/melanosomes present in the dermis of guinea pigs were eliminated efficiently after injecting carrageenan, thus reflecting the skin-whitening capability of carrageenan. However, the use of carrageenans in different customer products has not displayed any detrimental effect on human skin.[115,132]

The complex of collagen peptide (CP) and κ-carrageenan oligosaccharides (κ-COSs) has been reported by Ren *et al.* to produce photoprotective and anti-oxidative effects in mouse embryonic fibroblasts (MEFs) and HaCaT cells.[133] Apoptosis and UV-induced cell death could be considerably reduced in HaCat and MEF cells by the complex of κ-COSs and CP (100 g $mL^{-1}$) by decreasing the level of intracellular ROS. The UV-induced elevation in MMP-1 and reduction in type-1 procollagen are frequently inhibited by a complex of κ-COSs and CP through the suppression of the mitogen-activated protein kinases (MAPKs) signaling pathway. In short, a photoprotective effect against skin aging can be achieved by a complex of κ-COSs and CP.[115,134]

### 10.6.1.3 Exopolysaccharides

Polysaccharides originating from microbes living in marine environments, specifically exopolysaccharides (EPSs), are very prominent among bioactive substances having anti-wrinkling action. In addition to bacteria, various other microorganisms such as microalgae and fungi also produce exopolysaccharides (EPSs). However, considerable amounts of exopolysaccharides are synthesized by bacteria. EPPs form a wide range of products that have thickening, gel-forming, absorbing and emulsifying properties.[135]

The synthesis of hyaluronic acid (HA) is boosted in the skin by applying Hyanify™, a marine-derived exopolysaccharide, to decrease nasolabial folds. The balance between degradation and synthesis of HA gets disturbed due to a decrease in the availability of HA with aging, thus resulting in volume loss and dehydration in the skin. Among cosmetic products, destran, pullulan, and xanthan gum have found extensive applications. Pullulan is used in creams, lotions, *etc.* due to its property of having stable viscosity towards changing pH and heating. Shampoos, creams, and toothpaste contain xanthan gum because it has high stability and it increases the suspension life of constituents. Due to its viscosifier, thickening, and emulsifying properties, dextran is widely used in moisturizers.[136]

### 10.6.1.4 Chitosan

In nature, the second most abundant polysaccharide after cellulose is chitin. It is extensively distributed in the vegetal and animal kingdoms. It also appears in the exoskeleton of invertebrates such as crustaceans, mollusks, or insects. The exoskeletons of invertebrates like insects, crustaceans, or mollusks chiefly contain chitin. It is a copolymer chemically consisting of glucosamide and *N*-acetyl glucosamine. In a few species of fungi, an *N*-deacetylated derivative of chitin known as chitosan is produced. Chitosan is also synthesized industrially by *N*-deacetylation of chitin extracted from crustaceans. Solubility is a major parameter to differentiate chitosan and chitin. Chitosan is efficiently soluble in an acidic medium, whereas chitin is unable to dissolve in an aqueous medium.[137]

Hair formulations like hair tonics, hair colorants, shampoos, hair sprays, permanent wave agents, rinses, and styling lotions contain chitosan and its derivatives.[138] The interaction of chitosan and its cationic derivatives with keratin can form elastic and transparent films over fibers of hair. Preventing hair damage and increasing its strength and softness are the main advantages provided by these films. The development of films by blending chitosan with collagen and hyaluronic acid was studied extensively on hair. The thickness and mechanical strength of hairs were increased after wrapping them with films. It also enhances the conditioning and overall appearance of the hair.[139]

Prolonged exposure to UV radiation causes early damage in the form of skin aging known as skin-photoaging. Dryness, laxity, coarse wrinkling, and random pigmentation of the skin are the most common symptoms of photo-aging, typically characterized by inflammatory disequilibrium and oxidative stress. One of the primary causes of skin cancer is directly linked to UV exposure. Chitosan films and chitosan can be used as sunscreens because their UV spectrum of absorption is below 400 nm. An *in vitro* chitosan gel has been reported with a sun protection factor (SPF) of 0.89.[140] The transmittance of a chitosan film from mushrooms was reported at 300–250 nm for UVA-UVB, which was comparatively much lower than a chitosan film obtained from shrimp. This difference suggested that the UV-resistance capacity of the extract taken from mushrooms was greater than the extract taken from shrimp.[123] In addition, skin photo-aging induced by UV exposure for 10 weeks in the dorsal skin of hairless mice has been reduced considerably by applying chitosan oligosaccharides.[141]

## 10.6.2 Marine Phytochemical Compounds

Secondary metabolites in the form of phenolic compounds represent the main group of skin-benefitting cosmetics.[142] The occurrence of phenolic compounds in algae, also called marine plants, has been extensively studied. Jellyfish mucins and snail mucins (snail filtrate) are abundant sources of glycoproteins like phlorotannins, pseudopterosins, and mycosporine, which are antioxidants, have moisture-retaining capability, and efficiently absorb ultraviolet radiation. The detrimental effects produced by acne, stretch marks, rosacea, and aging are reduced by using these compounds.[143] Different cosmeceutical advantages are achieved by using varied phenolic groups; for example, skin whitening, anti-wrinkling, anti-allergic, and anti-aging effects are provided by using phlorotannin acting as a hyaluronidase inhibitor.[144] Table 10.2 shows the action mechanisms and cosmeceutical effects of phytochemical-based cosmeceutical agents.

### *10.6.2.1 Phlorotannins*

Phlorotannins are naturally occurring polyphenolic compounds that are synthesized by the acetate–malonate pathway [polyketide pathway].[145] Their molecular sizes vary between 125 kDa and 650 KDa and they are highly

**Table 10.2** Phytochemical-based cosmeceutical agents.

| Phytochemical | Mechanism of action | Cosmeceutical effect | Reference |
|---|---|---|---|
| Phlorotannins | Tyrosinase inhibition, anti-elastase activity, preventing chronic inflammation | Used in anti-aging creams and sunscreens | 146 |
| Astaxanthin | Decreasing hyperpigmentation, inhibiting melanin synthesis | Wrinkle reduction and photo-aging inhibition | 159 |
| Pseudopterosins | Anti-inflammatory activity, initiation of the release of intracellular calcium ions | Anti-acne, wound healing, analgesic | 167 |

hydrophilic. In the cosmetics industry, anti-wrinkling, anti-aging, bleaching, and anti-allergic benefits have been exclusively provided by phlorotannins.[26]

Brown algae produce phloroglucinol-based polyphenols, known as phlorotannins, that are regarded as a secondary metabolite of high importance having diverse biological functions and extensive applications in the cosmetic, nutraceutical, and pharmaceutical industries, such as antioxidant, melanogenesis, and anti-aging.[144,146–148] The phlorotannins of plant origin are derived from flavones or gallic acid, while oligomers and polymers of phloroglucinol (1,3,5 trihydroxy benzene) are phlorotannins of brown algae origin.[149] An extract from *Ascophyllum nodosum* contains natural phlorotannin having inhibitory action against senescence, oxidative stress, and inflammation.[148] The prevention of cytokine release and ROS production and the ability to stimulate SIRT1 are effective results shown by these phlorotannins in the form of potent anti-inflammatory, anti-oxidant, and pro-sirtuin1 (SIRT1) activities. SIRT1 is an NAD-dependent deacetylase involved in maintaining the longevity of DNA-repairing proteins, nuclear histones, and transcription factors by removing the acetyl group from them.[150] Tyrosinase inhibition, anti-elastase activity, preventing chronic inflammation, and protecting skin cells from mutation induced by UVB are cosmeceutical applications of phlorotannins extracted from *Ascophyllum nodosum*, brown algae. Consequently, anti-aging creams or sunscreens contain these compounds as vital ingredients in their cosmetic preparations.[142,147]

#### *10.6.2.2 Astaxanthin*

Sea creatures and aquatic animals like lobster, trout, red sea bream shrimp, and salmon are an important source of a top-value carotenoid known as astaxanthin (3,3-dihydroxy-b-b0-4-40-carotene-dione, ASX).[151] However, the chief natural origin of astaxanthin is *Haematococcus pluvialis* (Chlorophyceae, Volvocales), a microalgae strain.[152]

Two β-ionone rings connected by a polyene chain and also containing hydroxyl and oxygenated keto moieties form the structure of astaxanthin.[153] The ability of astaxanthin to stabilize and reduce free radicals has been enhanced because its structure allows it to bind biological membranes easily.[154] Stress induced by oxidation has been considered to be the major cause of various skin diseases. However, the lipid peroxidation potential and singlet oxygen quenching capability of natural astaxanthin have been revealed to be 1000× that of vitamin E and 40 times that of β-carotene.[155]

This red pigment has been involved in the suppression of various inflammatory mediators and affects different steps in the cascade causing oxidative damage, and thus is found to play a role in improving skin health.[156] Wrinkle reduction and photo-aging inhibition have been reported as benefits of astaxanthin after various clinical studies. Diminished hyperpigmentation and a decrease in melanin synthesis have been found by assessing studies after taking astaxanthin orally.[94]

Several changes in the extracellular matrix (ECM) structure, like elastin, collagen, and glycosaminoglycans (GAG), are the cause of skin aging.[157] Skin damage and skin aging take place due to oxidative stress as reported by different studies. Positive effects have been generated on the skin based on the antioxidant activity of astaxanthin as reported by several studies. The transformation of skin from an aged one to a younger one was credited to astaxanthin because a sufficient decrease in the level of malondialdehyde (MDA) proved the antioxidant activity of astaxanthin. The activation of nuclear factor erythroid 2-related factor (Nrf2) has been reported by the action of astaxanthin and the antioxidant response generated as a result of binding of these factors to enzymatic elements.[158] So, age spots, fine lines, and wrinkles are decreased by astaxanthin as published in various studies.[158,159] The prevention of saggy skin and increased elasticity of the skin have proven to be added advantages of this natural molecule. Thus, astaxanthin can be safely used and is said to be an anti-aging agent. A reduction in transepidermal water loss (TEWL) and an increase in the moisture content of the skin have been identified as benefits of applying astaxanthin. Therefore, problems arising due to seasonal stress and dry skin can be solved by utilizing astaxanthin.[94]

#### *10.6.2.3 Pseudopterosins*

A soft coral *Antillogorgia elisabethae*, previously named *Pseudopterogorgia elisabethae*, produces a class of diterpene glucosides having 31 reported derivatives called pseudopterosins.[160] Pseudopterosins are highly involved in the nourishment of the skin, protecting the skin from sun damage, and have recently become a regular part of several skin care formulations.[7]

*Pseudopterogorgia elisabethae* has become a rich source of providing a unique extracellular extract to an Estée Lauder company having a series of skin care products known as Resilience®. This extracellular extract chiefly consists of tricyclic diterpene glycosides, also called pseudopterosins.[106,161] The biological functions performed by these compounds vary from wound healing, analgesic, anti-inflammatory and anti-acne to anti-bacterial. It is a highly fortunate natural compound because it has achieved utilization in skin care very quickly as compared to other natural compounds having the aim to achieve higher pharmaceutical targets.[162,163] Moreover, pseudopterosin A is a highly reported member, due to its anti-inflammatory activity in this class of compounds. It participates in binding to G protein-coupled receptors for initiating the release of intracellular calcium and suppressing the formation of phagosomes.[164–166]

### 10.6.3 Mycosporines

Molecules having low molecular weight, enriched by nitrogen, soluble in water, and having absorption maxima of 310–365 nm in the UV region are known as mycosporine-like amino acids (MAAs).[168] Owing to the dissipation of energy in the form of heat, high thermal- and photo-stability, an excited state of short existence, thus preventing the possibility of them participating in any unnecessary photochemical reactions, and strong absorption in the UV range, they are ideal sunscreens. Eukaryotic phytoplankton, macro- and microalgae, cyanobacteria, and fungi are the organisms in which MAAs are synthesized.[168,169] MAAs are employed as biological photo protectors because their various features and biological functions make them highly compatible candidates for cosmeceutical formulations. These molecules are excellent UV absorbers having strong anti-oxidant activity, high stability, and low toxicity.[170]

Compounds having the potential to absorb UV radiation are known as photoprotectants. It is interesting to note that radiation in the range of 315–400 nm (UVA) and 280–315 nm (UVB) can reach ground surface level after crossing the atmosphere in the UV spectrum.[171] MAAs have developed potency to absorb highly energetic and even lethal doses of this UV radiation. In cyanobacteria, it has been revealed that cytoplasmic targets are prevented from being hit by every 3 out of 10 photons by MAAs.[172] In addition, to absorb UV radiation, MAAs also contribute to vanishing needless energy in the form of mild heat radiation without forming reactive oxygen species (ROS). When comparing the data of various reports, it has been shown that the amount of MAAs becomes elevated in response to UVB, UVA, and PAR radiation exposure in cyanobacteria. A relation between the content of MAAs and the level of irradiance has encouraged the use of MAAs as sunscreens and in photoprotection.[169]

Matrix metalloproteinase (MMP) activation takes place because of elastase secretion and dystrophy of elastic fibers as a result of increased production of reactive oxygen species. Elastin fibers and collagen are degraded by elastase and MMPs, respectively. The aging is stimulated and wrinkles are formed due to a decrease in elastin and collagen.[173] Thus, MMPs are highly effective in preventing photo-aging because they suppress both (i) activation of MMPs and (ii) production of ROS. The anti-photoaging character of MAA porphyra-334 has been studied by Ryu *et al.*[174] The number of matrix metalloproteinases (MMPs) and production of reactive oxygen species have been greatly reduced by this MAA in fibroblasts of humans under UV-A exposure.

### 10.6.4 Neo Fatty (Carboxylic) Acid Derivatives

Fruit acids can be either polyhydroxy acids like gluconolactone, α-hydroxy acids like glycolic acid and lactic acid, or β-hydroxy acids like salicylic acid. They play an important role in retaining moisture in the epidermis by promoting skin peeling and chelating calcium ions. They improve texture and act as antioxidants, and hence are known as humectants. The levels of collagen and glycosaminoglycans are also increased by them. Glycolic and lactic acids have the smallest size and penetrate the most. Lactic and glycolic acids have high penetrating power because they are very small in size. The low potency of α-hydroxy acids has been rectified by polyhydroxy acids.

They are mostly distributed with a concentration of 5–12 percent. However, up to a 20 percent concentration is recommended as harmless for self-use.[175,176]

To retain water in the skin for keeping transepidermal water loss (TEWL) to its optimum level, gamma-linolenic acid (GLA) and linoleic acid are frequently used as water or oil emulsions.[177] Recently, various biosurfactants derived from marine microorganisms have been reported for their potential activities in the cosmetics industry, including dispersing, wetting, emulsifying, foaming, and solubilizing. These are sophorolipids, mannosylerythritol, and rhamnolipids. They are not only engaged in aiding the delivery of hydrophobic ingredients through the skin barrier, but also increase the solubility of these ingredients in formulations.[178] As these marine-based biosurfactants are comparatively very low irritants to skin over their synthetic equivalents, they have an added advantage as an ultimate part of anti-wrinkling products.[179,180]

## 10.7 Applications of Marine Natural Products for Cosmeceuticals

### 10.7.1 Antiaging Activities

Aging, which affects physical appearance, is becoming one of the biggest daily problems in the cosmetics field.[181] In the aging process, the activity of mitochondria in the body decreases as an energy producer in cells and tissues that are renewed. While most aging effects are affected by external factors such as drugs, diet, and environmental situations, some are controlled by genes. Therefore, the interaction of various internal and external factors should be studied to fully understand the process. Many theories have been put forward regarding this process, such as immune theory, radical damage and genetic control theory, mitochondrial damage, chromosome mutation theory, and telomere theory.[182]

Marine natural products promise better performance and more substances with anti-aging characteristics, but were relatively unexplored in detail until recently. Based on the results of various studies, it is seen that seafood has industrial potential as a cosmetic ingredient and is a valuable source of unique chemical compounds.[68,183,184]

Plankton, sponges, fish, crustaceans, sea anemones, sea stars, tunicates, and soft corals, which are sources of active components of wide biodiversity in the ocean, have been reported to be effective in skin care through many anti-oxidant, moisturizing and anti-aging effects.[68,185] However, with the impact of their benefits, the high demand for these bioactive compounds from the cosmetics industry has led to the development of advanced techniques for their production.

Given the enormous social burden, investments in health care, and the growing trend of serious aging in today's society, strengthening the development and research of antiaging medicine is of great social importance to society and science. The study of the anti-aging activities of marine

**Table 10.3** Effects of marine natural products on aging.

| Species | Natural product | Mechanism of action | Reference |
|---|---|---|---|
| *Sargassum fusiforme* | Carbohydrate | SOD(31%↑) MDA(20%↓) T-AOC(26%↑) | 186 |
| *Porphyra haitanensis* | Carbohydrate | p53(35%↓) p21(24%↓) | 187 |
| *Ulva lactuca* and *U. prolifera* | Carbohydrate | p53 and FOXO1 ↓IFN-γ, TNF-α, and IL-6↓ | 188 |
| *Chlorella pyrenoidosa* | Carbohydrate | SOD(31%↑) GSH-Px(28%↑) CAT(43%↑) | 189 |
| *Chlorella vulga* | Carbohydrate | ALP(24%↑) T-SOD(91%↑) CAT(92%↑) | 190 |
| Marine mollusks and arthropods | Carbohydrate | UA(20%↓)AST(27%↓)ALT(29%↓) ALP(32%↓) CREA(15%↓) | 191 |
| *Sipunculus nudus* | Carbohydrate | SOD(35%↑)T-AOC(22%↑) NO(17%↑)MDA(16%↓) GSH-Px(52%↑) | 192 |
| *Spirulina* | Protein | SPF(58%↓) | 193 |
| *Ulva rigida* | Protein | SOD(65%↑) CAT(72%↑) | 194 |
| *Argopecten irradias* | Protein | SOD(27%↑) GSH-Px(24%↑) CAT(36%↑)MDA(53%↓) | 195 |
| *Cucumaria frondosa* | Protein | SOD(48%↑) MDA(51%↓) GSH-Px(32%↑) | 196 |

organisms from a wide range of marine resources has become a hopeful and exciting research focus of scientists. Table 10.3 shows how proteins and carbohydrates obtained from sea creatures act on aging.

The most extensive research, in short, focused on the antiaging properties of polysaccharides from marine organisms. Numerous studies have been done on cell models, including *Drosophila*, mice, and nematodes, and the process has been extensively studied. Many polysaccharides have anti-aging use on the skin, which is worth investigating in-depth. The most advanced and extensively researched polysaccharides, chitin, and chitosan derivatives from marine animals, are also the most commonly used in the cosmetics market.

## 10.7.2 Antimelanogenic Activities

The demand for skin care products is motivated by removing localized hyperpigmentation and brightening the skin tone.[197] For example, skin lightening uses natural or synthetic ingredients that reduce the concentration of melanin in the skin to reduce pigmentation. This application may be prompted by various dermatological reasons such as UV exposure, skin hyperpigmentation, genetic factors, and hormonal changes.[135] One or more steps in the melanogenesis pathway may be involved in the process of melanosome translocation, post-translocation pigmentation, degradation, and depigmentation. Therefore, melanocyte metabolism and proliferation or removal of melanin and inhibition of tyrosinase enzyme can inhibit melanin production.[198] Various mechanisms can cause skin lightening, including inhibition of microphthalmia-associated transcription factors, decreasing the expression of melanocortin-1 receptor activity, impaired maturation and transmission of melanosome, loss of melanocytes, and inhibition of tyrosinase enzymes.[199] In the past, plant-derived substances like azelaic acid and hydroquinone glycoside arbutin and fungal-derived substances such as kojic acid were often used as skin-lightening agents in cosmetics.[135]

Depigmentation agents can have essential effects by affecting one or more steps of the melanogenesis pathway, transport of melanosome or pigmentation, and post-transport degradation. Many depigmentation agents use similar transmission pathways. Tyrosinase-related melanogenic enzymes reduce skin pigmentation by acting on the transcriptional activity of tyrosinase-related protein-1 (TYRP-1), tyrosinase-related protein-2 (TYRP-2), or peroxidase. In addition, the expression of tyrosinase and related melanogenic enzymes is initially regulated at the transcriptional level. This enzyme catalyzes the rate-limiting step of pigmentation and is therefore one of the most common methods used to inhibit tyrosinase and induce skin hypopigmentation.[200]

Consumers' increasing dependence on natural cosmetics has given rise to a drive to study natural resources to produce therapeutic solid natural compounds for cosmetic usage. Utilizing marine microorganisms as depigmentation agents is a novel way to develop new industrial uses while exploring the cosmeceutical potential of marine-based compounds. However, only a very tiny portion of the various tyrosinase inhibitors found in sea animals have thus far been employed for aesthetic purposes. Thus, there are still so many undiscovered marine species that will be required to use the immense potential that marine life gives us correctly.

### 10.7.3 Anti-acne Activities

Acne or acne vulgaris usually occurs in young adolescents with hormonal imbalances. It also remains an increasingly common skin condition in the adult population. Acne can be an important risk factor for psychological problems in men and women of all ages.[201]

To prevent acne and cure it medically, *Staphylococcus epidermidis* and *P. acnes* are significant targets.[202,203] These anaerobic bacteria promote the synthesis of proinflammatory cytokines, and their overproduction triggers the release of ROS, a devastating phenomenon that leads to scarring.[204] They produce lipases to digest excess sebum and skin oil, triggering a severe local inflammatory response that destroys hair follicles. Consequently, the cosmetics industry identified the suppression of *P. acne* growth as a strategic approach to acne treatment. Sargafuran is among the antibacterial compounds that are being researched to prevent acne and develop new natural cosmetic products.[205] *Sargassum macrocarpum* isolated from marine brown algae extract was found to exhibit a potent effect. It showed anti-acne activity against *P. acnes* with a minimum inhibitory concentration (MIC) of 15 μg $mL^{-1}$ While evaluating the antibacterial activity of several macroalgae species often found on the coasts of South Korea, *Ecklonia kurome*, *E. cava*, and *Ishige sinicola* showed potent growth inhibitory and anti-inflammatory activity against *P. acnes*.[206]

### 10.7.4 Antioxidant Activities

All aerobic organisms rely on oxygen as an essential part of their metabolic processes. Free radicals, chemicals with unpaired valence electrons, are more easily produced in environments with high redox potential.[207] Any molecular entity that can exist independently and has an unpaired electron in an atomic orbital is called a free radical. Because of this condition, many radicals are unstable and highly reactive. They can act as oxidants or reductants by giving or taking electrons from other molecules.

The ocean covers about half of the earth's surface and is home to many macro-organisms and microorganisms. While providing a renewable source of potential therapeutic agents, diverse marine environments trigger the production of many bioactive compounds. Organisms in the sea can be resistant to extreme environmental conditions such as cold, heat, high pressure, high salinity, and low nutrients.[208] Seaweeds, which exhibit biological activities, constitute a rich source of various natural products structurally.[209]

Algae reproduce and thrive in sunlight and high oxygen levels. In addition, the absence of oxidative damage to structural fatty acid membranes suggests that these organisms have compounds that provide antioxidant activity.[210] There are many antioxidant compounds derived from green (Chlorophyta), brown (Ochrophyta), and red algae (Rhodophyta). According to the results of a study, algal metabolites with antioxidant activity can be listed as follows, fluorotannins, bromophenols, flavonoids, steroids and carotenoids, peptides, meroterpenoids, alkaloids, chlorophyll-related pigments, polysaccharides and carbohydrates.[211]

Additionally, it has been noted that eating fish has nutritional advantages due to the high biological quality of its protein and the availability of its rich mineral and vitamin elements.[212] In addition, seafood byproducts and products are a comprehensive resource for the mining of bioactive peptides. Furthermore, marine proteins feature unique biofunctional peptides in their basic structure that can be exploited in culinary and medicinal applications because of their significant structural diversity.[213] Figure 10.3 depicts the antioxidant activity of fucoin (*Ascophyllum nodosum*).

### 10.7.5 Sunscreens

Sunscreen constitutes a significant reservoir of inorganic and metallic nutrients crucial to the ecological dynamics of coastal water ecosystems. This poses potential risks to the marine environment, as sunscreens are made with organic and inorganic chemical UV filters that disrupt the delicate balance of our waterways. Homosalate, octocrylene, octisalate, octinoxate, and avobenzone are some of the most well-liked petrochemical sunscreen UV

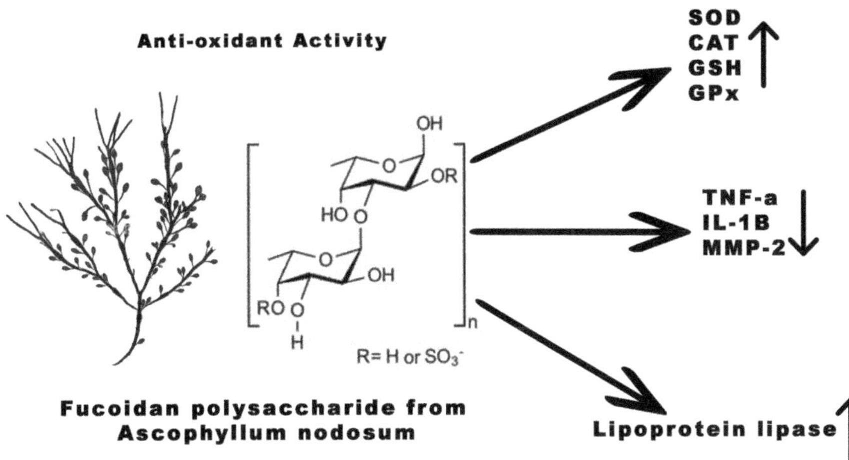


**Figure 10.3** Antioxidant activity of fucoin (*Ascophyllum nodosum*).

filters. In environmentally relevant amounts, these chemicals can be dangerous to wildlife. However, these UV filters are not only dangerous in their own right – they may also interact negatively or positively with other global pollutants, such as marine plastic waste, and have an impact on climate change.[214–217] Other UV absorbers have also been studied to create UV filters and sunscreen products, reduce photodamage, and other uses.[218–220]

Oligosaccharides and peptides produced from marine organisms are another exciting class of natural antioxidants. Carbohydrates produced from algae have been hypothesized to have antioxidant, anti-melanogenic, and antiaging effects in addition to their thickening and moisturizing effects, making them valuable cosmetic products.[115,221]

### 10.7.6 Wound Healing, Anti-inflammatory, and Analgesic Activities

Increased fibroblast and keratinocyte migration in the healing of skin wounds, vascularization, and epidermal development are among the wound-healing processes.[222–224] Marine collagen, which is at the forefront of wound healing, has been shown to have anti-aging properties and accelerate wound healing by reducing the aging of mice.[225–228] Human studies have also revealed that marine collagen can increase skin elasticity, minimize wrinkles, and improve the overall structure and appearance of the skin.

Marine collagen peptides produced from collagen are more easily absorbed. Both chemical and enzymatic hydrolysis take place during their production. They also have lower molecular weight and greater water solubility.[229,230]

Wang *et al.* found that marine collagen peptides (MCPs) isolated from salmon skin significantly improved the tensile strength in skin wounds in rat experiments.[222] In addition, hydroxyproline levels were found to increase significantly and proportionally with time and dose when the collagen-treated and control groups were compared.[222]

These results are similar to the results of Zhang *et al.*'s experiment,[223] and it has been reported that marine collagen peptides increase wound closure rates, tensile strength at the incision site, and collagen deposition in rats. In the results of histological analysis, fibroblast infiltration, vascularization, and epithelialization were increased in the groups treated with collagen.

To repair the tissue's integrity, therapies may involve steps including cell division, inflammation, differentiation, and vascularization. Endothelial permeability enables cell adhesion, which is then followed by cell maturation and differentiation.[231,232]

Marine collagen has been proven to work well as a biomaterial to heal wounds. Collagen can be used in various forms, including collagen fibers, peptides and hydroxylates, and scaffold-like structures.[224,229]

By nuclear factor (NF)-B signaling pathway activation, LPS can potentially increase the production of several proinflammatory mediators, including enzymes such as iNOS and COX2, cytokines, and chemokines (Figure 10.3).[233,234]

Numerous studies have shown that agents with anti-inflammatory properties can reduce inflammatory reactions, oedema, discomfort, and leukocyte infiltration in animals injected with carrageenan by inhibiting COX-2 or iNOS production in LPS-stimulated macrophages. In addition, in a study on a carrageenan-induced paw edema model, it was discovered that excavatolid B reduced iNOS and COX-2 mRNA expression and had anti-inflammatory and analgesic effects.[235–238]

### 10.7.7 Moisturizing Activities

A disruption in the regular exfoliation process, as well as an imbalance or reduction in the stratum corneum's (SC) natural moisturizing factor (NMF), can contribute to dry skin, a hallmark of age.[239,240]

Amino acids are responsible for the majority of NMF function to preserve epidermal moisture, providing proper exfoliation and healthy skin.[241,242] Topical moisturizers containing chemicals similar to those found in NMF or designed to stimulate congestion or attraction and hydrate the epidermis are common treatments for dry skin.[243]

Ceramide lipid molecules are an essential component of the SC permeability barrier and make up about half of the extracellular lipid matrix and are produced from the omega-6 fatty acid linoleic acid (SCPB). Dry skin may result from the loss of lipid components, as SCPB reduces pathogenic invasion and transepidermal water loss (TEWL).[185,244,245] Marine-derived lipids may help maintain skin moisture by preserving the lipid matrix of SC.[246,247]

Docosahexaenoic acid (DHA) and eicosapentaenoic acid (EPA), two naturally abundant omega-3 fats in both EPA and DHA, may be readily extracted from a wide range of freshwater and marine fish.[248–250] After 30 and 60 days of treatment, Barcelos *et al.* investigated changes in dermal sensitivity to skin irritation in rats receiving daily doses of omega-3 fish oil (FO). Both discomfort and TEWL were significantly reduced in response to acetone exposure after 60 days of supplementation.[247]

Similar moisturizing and emollient properties have been observed in the oil of the deep sea bass *Hoplostethus atlanticus*, which is equivalent to petroleum-based products such as Vaseline®[251] alcohols and fatty acids,[252] and can reduce skin dryness by up to 70% as effectively as petrolatum products.[251]

However, despite the moisturizing ability of *H. atlanticus*, it could not be considered a sustainable source for skin moisturizing products in the field of cosmetics due to its low fertility, late maturation, and slow growth.[253]

The sustainability of targeted organisms should always be at the forefront, as the marine environment is a new and limited resource for new cosmetic discoveries. Similarly, squid oil (*Loligo loligo*) has been touted as a new source of omega-6 fatty acids, and contains high levels of EPA and linoleic acid accounting for 13% of the adult squid's wet weight in oil.[254] Squid culture has not yet been explored, but *Octopus vulgaris* aquaculture is under investigation and has had some early success.[255]

Cephalopods are considered suitable aquaculture candidates due to their short lifespan, rapid maturation, and easy adaptation.[255,256] As a result, additional research into this method of producing omega-3 and omega-6 fats is warranted. Although collagen and hydrolyzate are popular active ingredients in moisturizing products, there is little scientific data to support their moisturizing effects.[257] Dry skin, one of the signs of aging, is caused by the loss of glycosaminoglycan (GAG) and hyaluronic acid (HA), a vital component of the dermal skin matrix found in all tissues and body fluids.[258]

Alternative water-absorbing compounds from marine sources have been investigated recently. For example, water retention, anti-inflammatory properties, non-toxicity, and broad antibacterial activity are some of the desired aesthetic properties shown to be exhibited by polysaccharides from Crustacea, Phaeophyta, and Rhodophyta.[259]

In contrast, exopolysaccharides (EPS) of marine bacteria may have aesthetic potential comparable to their plant and animal counterparts, as they are readily available and have a wide range of biochemical properties due to living in harsh conditions. In addition, an EPS produced by *Vibrio diabolicus* has been shown to increase the development of HA and keratinocytes, improve the skin's moisture content, and prevent neuronal exocytosis. This process promotes the appearance and deepening of wrinkles.[260] Several examples of marine bacterial EPS go into products like Hyadisine® and Hyanify™ and stimulate HA production, leading to antiaging and moisturizing effects.[261,262] However, the mechanisms underlying this are not well documented.

Due to their largely unexplored biochemical diversity and the scarcity of commercially available bacterial EPS samples, the chances of discovering new EPS from extremophiles are high. This could be a viable avenue to develop new bioactive compounds with moisturizing or anti-aging properties.

### 10.7.8 Repairing Activities

The skin's primary task, as the first barrier of the immune system, is to separate the inside of the body from the outside. The skin barrier prevents the entry of hazardous chemicals and reduces water loss.[263] The skin gets dry when the skin barrier is compromised. Therefore, it is more susceptible to invasion by external bacteria and irritants that cause inflammatory responses and additional symptoms.[263] The cuticle, which has keratinocytes and cells rich in protein and lipids, forms the majority of the skin barrier. Free fatty acids, ceramides, and cholesterol are among these lipids.[264]

A structural skin protein called collagen is often used as an active ingredient in moisturizers.[183] In 2002, the effects of collagen produced from sponges and collagen already present in human skin were studied by obtaining collagen from kidney-shaped sponges made of cartilage.[265]

Chemicals with antioxidant properties can also be used to increase the skin cells' vitality and connection, improve immune function, and decrease skin cell aging caused by free radicals. Squalene is an antioxidant, and cosmetics employ hydrogenated squalene.[266] Also, biologically active indole derivatives from sea sponge *Rhopaloeides odorabile* and the fungus *Hyrtios* sp. were isolated, and their antioxidant strength was assessed using DPPH (a stable free radical).[267] The aforementioned hypotheses offer a method for skin restoration, but more research is required to demonstrate how these functional molecules might be successfully combined to produce novel skin healing solutions.

### 10.7.9 Skin Lipid Activities

Normal human skin produces approximately 1 mg per 10 $cm^2$ of sebum every 3 hours.[268] According to Yvergnaux,[269] lipase can accelerate the breakdown of insoluble triglycerides into glycerol diesters and monomers, creating cleansing facial oil. Most of these current commercial products contain enzymes, as enzymes are gentler and less irritating than chemical reagents.[270] There are several different sources of lipases, including plants, animals, and microbes. Significantly, the benefits of microorganisms are numerous and include ease of cultivation, rapid production, and resistance to external environmental factors.[271] Because of their unique habitats, marine microorganisms can produce a wide variety of lipases that can withstand various situations, and better meet commercial production demands and application requirements, so humans choose them.

### 10.7.10 Antitumor Activities

Algae are one of the most valuable ocean resources, both commercially and environmentally, among marine species.[272] Polyunsaturated fatty acids, fiber, protein, vitamins, and macro and trace elements are all abundant in algae. Phycobilins, polysaccharides, polyunsaturated fatty acids, proteins, vitamins, minerals, carotenoids, sterols, tocopherols, terpenes, and phycocyanins have all recently been discovered to be fascinating sources of beneficial bioactive compounds.

While used in potential applications as therapeutic agents in the biomedical field, these substances have also been proven to be effective in nutrition and have various types of functional importance.[273–276] However, algae's multifunctional qualities, distinctive structures, and metabolic properties should be used entirely (Figure 10.4).

In a few cellular studies on tumors, the lethal effect of chemicals extracted from algae has been observed, but studies investigating intracellular signal transmission pathways affecting this process are insufficient. When compounds from algae with antitumor activity are analyzed, they may likely mediate some of the tumor treatments. However, some chemicals obtained from algae that have anticancer properties can also be used in targeted diseases.

One of the treatment goals in solid tumors is angiogenesis. This is the system that focuses on the formation of blood vessels and the continuous growth and development of tumors. Inhibiting this system improves the outcome of tumors and metastatic diseases. The commercial use of these inhibitory drugs is one of the leading aspects of anticancer research.[277]

Studies of algae-derived compounds with antitumor activity on *in vitro* human tumor models are quite extensive. However, extensive preclinical and clinical trial work is necessary to demonstrate the potential value of

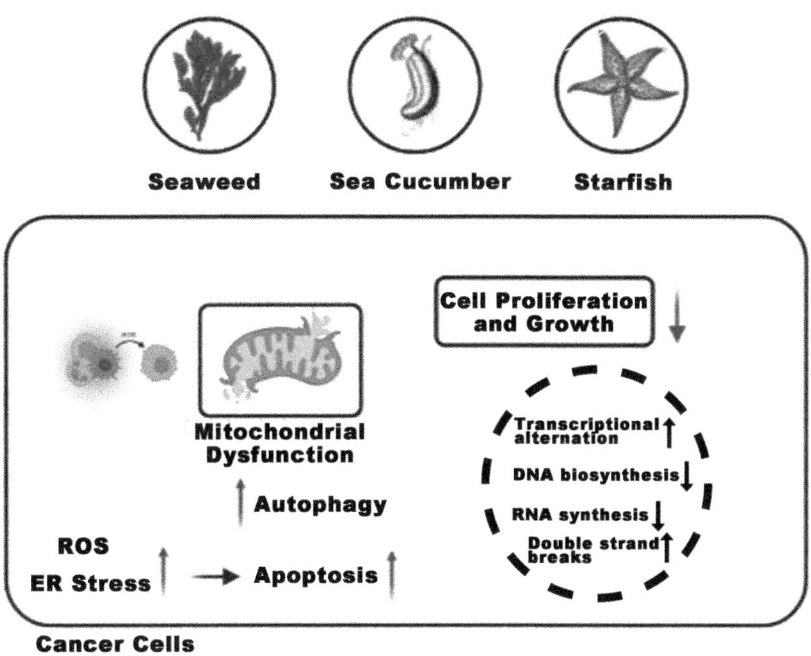


**Figure 10.4** Molecular effects of marine creatures in cancer cells.

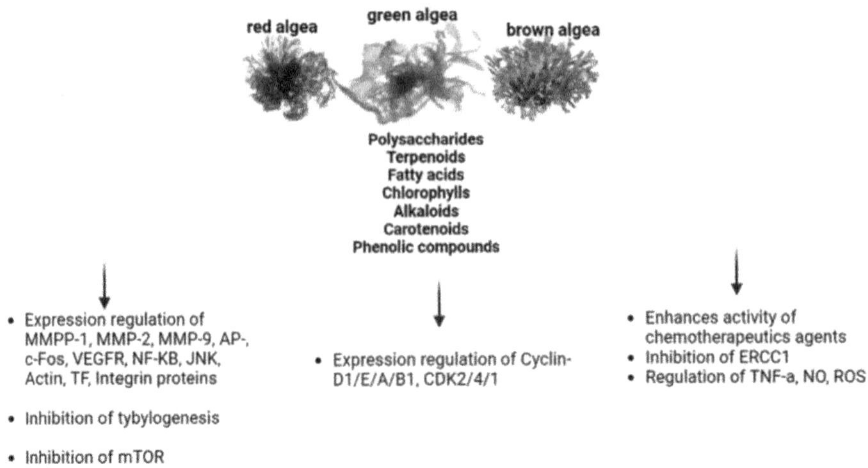


**Figure 10.5** Mechanisms of the antitumor activities of algal species.

these compounds as anti-tumor agents and adjuvants. The most prominent among the chemical compounds derived from algae and supported by *in vivo* studies are sulfated polysaccharides. The potential of these structures to be used in antitumor treatments has been revealed. Figure 10.5 describes the mechanisms of the anti-tumor activities of different algal species.

In addition, it has been revealed that when standard hormonal drugs are administered in conjugation with letrozole and tamoxifen, they are well tolerated in patients with breast cancer. However, these drugs do not affect steady-state plasma concentrations.[278] The use of fucoidans as an anticancer drug is being investigated in clinical trials for adjunctive therapy. This is because fucoidan preparations from different sources induce different anticancer activities *in vivo*. These different activities and responses appear to be related to structural features. To permit fucoidan's prospective usage as a therapeutic agent, it will be crucial to pinpoint the structural aspects of the compound that are accountable for its observed *in vivo* anticancer effects.[279]

The production of nanoformulated drugs derived from algae-derived compounds that can be used in drug delivery systems may be a promising approach to improve their anticancer properties. Nanoformulations may also help overcome some of the restrictions brought on by the unique characteristics of each molecule, such as systemic toxicity, non-specific biodistribution, low water solubility, subpar therapeutic impact, lack of targeting ability, and restricted bioavailability.[280,281]

The promising activity of compounds derived from algae is predicted to be seen in the next few years. Some compounds also assist in the development of new compounds that allow them to be converted into clinically useful drugs as they reach clinical trial stages. Moreover, using some of these compounds as an adjuvant in pre-existing therapeutic regimens seems an excellent approach to improve the therapeutic effects of antitumor drugs and reduce their side effects.[282]

The development of anticancer medications generated from algae might be sped up, limits lifted, and some algae-derived chemical qualities improved by using cutting-edge techniques, including nanotechnology, to build novel nanoformulations. However, additional technological advancements are required to prove that their anticancer actions are boosted when included in nanoformulations. This search demonstrates that this region has not yet been well investigated. By forming multidisciplinary teams, it is possible to expedite the bio-screening procedure from the separation and identification of chemicals originating from algae to the thorough evaluation of their anticancer properties.

### 10.7.11 Antiviral Activities

Marine antiviral agents (MAVAs) are involved in the biological control of human enteropathogenic virus contamination and disease transmission in sewage-contaminated waters. This role is very important for the food industry and coastal waters.[283] One of the earliest sources of naturally occurring chemicals with *in vitro* anti-HIV activity came from micro and macro alga.[284] Table 10.4 presents the organisms and use of marine-derived antiviral agents.

The first large antiproliferative and antiviral screening of ascidians and extracts from a gorgonian sea sponge from Mexico's Yucatan Peninsula was conducted. Based on these screening results, extracts from sponges tunicate *E. amanitum* and *H. (Rhizoniera) curacaoensis*, *C. Kuekenthali*, and *L. floridana* proved to be the most interesting extracts for future studies due to their high potency against other tested ones.[290]

**Table 10.4** Antiviral compounds derived from marine sources.

| Compound | Drug | Use | Reference |
|---|---|---|---|
| *Tethya cripta* synthetically produced arabinosyl nucleosides derived from sponge | Acyclovir | Antiviral – HSV | 285 |
| *Tethya cripta*, synthetically produced and sponge-derived arabinosyl nucleosides | Ara-A (vidarabine) | Antiviral – HSV | 286 |
| Sponge, *Tethya cripta* | Ara-C (cytarabine) | Antiviral – HSV | 286 |
| Sponge, *Disidea avara* | Avarol | Antiviral – HIV | 287 |
| *Tethya cripta* synthetically produced arabinosyl nucleosides derived from sponge | Azidothymidine (zidovudine) | Antiviral – HSV | 288 |
| Cyanobacteria, *Nostoc ellipsosporum* | Cyanovirin-N | Antiviral – HIV-1, HIV-2, SIV | 289 |

## 10.8 Side Effects of Cosmeceutical Marine Natural Products

Cosmeceuticals are products that can cause adverse effects on human health such as acute toxicity, subchronic toxicity, eye irritation, photoirradiation, photosensitization, phototoxicity, genotoxicity, skin sensitization, percutaneous absorption, skin irritation, and many other effects.[291] Studies have confirmed the incidence of side effects of different cosmetics to be in the range from 8 to 38%.[292–296] Table 10.5 lists the side effects of different natural products of marine-based cosmeceuticals.

## 10.9 Conclusion and Future Perspectives

With the growing industry of cosmeceuticals, the demand for new green, natural, and cruelty-free products has urged researchers to discover new active ingredients from living systems. For this, scientists have extensively been studying marine life to find such biologically active ingredients to be used as cosmeceuticals. Recent development in the field of cosmeceuticals has exploited the design of new skincare products with *in vitro* studies to understand the physiology and molecular pathology of healthy skin. In the foreseeable future, the utilization of animal studies may redefine the associations between drug targets and cosmeceuticals, contributing to the verification of safety assessments and authenticity claims. In this respect, regulation

**Table 10.5** Side effects of different natural products of marine-based cosmeceuticals.

| Cosmeceuticals | Side effects | Reference |
|---|---|---|
| Carrageenan | Chronic low-grade inflammation in obesity, diabetes, and metabolic syndrome | 297 and 298 |
| Exopolysaccharides | Neurotoxicity, carcinogenicity, and Alzheimer's disease | 67 |
| Chitosan | Gastrointestinal problems, such as nausea, diarrhea, and constipation | 299 and 300 |
| Phlorotannins | Extremely mild toxicity, weight loss, mild fatigue, dizziness, nausea, and abdominal distension | 146 |
| Astaxanthin | Increased bowel movements, red stool color, and stomach pain | 301 |
| Mycosporine-like amino acids | Photoallergies, and hormone disorders | 302 |
| Eicosapentaenoic acid | Nausea, heartburn, diarrhea, itching, skin rash, nosebleed, back, joint, and muscle pain | 303 |
| Docosahexaenoic acid | Loose stools, stomach upset, and belching, abdominal pain, bloating, nosebleeds, mild diarrhea, fatigue, skin irritation or burning, drowsiness, decrease in platelet count, and vitamin A and D toxicity | 303 |

laws must be followed for the *in vivo* testing of different cosmeceuticals. The European Union has restricted the testing of cosmetics on animals, instead exploiting a way of testing *via in vitro* methods.

## References


1. M. Donia and M. T. Hamann, Marine natural products and their potential applications as anti-infective agents, *Lancet Infect. Dis.*, 2003, **3**, 338–348.
2. P. Proksch, R. Edrada-Ebel and R. Ebel, Drugs from the sea-opportunities and obstacles, *Mar. Drugs*, 2003, **1**, 5–17.
3. W. Y. Lu, H. J. Li, Q. Y. Li and Y. C. Wu, Application of marine natural products in drug research, *Bioorg. Med. Chem.*, 2021, **35**, 116058.
4. A. M. Kilgman, Cosmeceuticals, A Broad-Spectrum Category between Cosmetics and Drugs, in *Cosmeceuticals and Active Cosmetics Drug Versus Cosmetics*, 2020, pp. 21–30.
5. A. Alves, E. Sousa, A. Kijjoa and M. Pinto, Marine-derived compounds with potential use as cosmeceuticals and nutricosmetics, *Molecules*, 2020, **25**, 2536.
6. L. Lopez-Hortas, N. Florez-Fernandez, M. D. Torres, T. Ferreira-Anta, M. P. Casas and E. M. Balboa, *et al.*, Applying seaweed compounds in cosmetics, cosmeceuticals and nutricosmetics, *Mar. Drugs*, 2021, **19**, 552.
7. S.-K. K. Kim, Y. D. Ravichandran, S. B. Khan and Y. T. Kim, Prospective of the cosmeceuticals derived from marine organisms, *Biotechnol. Bioprocess Eng.*, 2008, **13**, 511–523.
8. E. A. Siahaan, R. Pangestuti, H. Munandar and S.-K. K. Kim, Cosmeceuticals properties of sea cucumbers, Prospects and trends, *Cosmetics*, 2017, **4**, 26.
9. S. Agrawal, A. Adholeya, C. J. Barrow and S. K. Deshmukh, Marine fungi, An untapped bioresource for future cosmeceuticals, *Phytochem. Lett.*, 2018, **23**, 15–20.
10. A. M. S. Mayer, A. D. Rodríguez, R. G. S. Berlinck and M. T. Hamann, Marine pharmacology in 2005-6, Marine compounds with anthelmintic, antibacterial, anticoagulant, antifungal, anti-inflammatory, antimalarial, antiprotozoal, antituberculosis, and antiviral activities, affecting the cardiovascular, immune and nervous syste, *Biochim. Biophys. Acta, Gen. Subj.*, 2009, **1790**, 283–308.
11. A. G. Atanasov, S. B. Zotchev, V. M. Dirsch, I. E. Orhan, M. Banach and J. M. Rollinger, *et al.*, Natural products in drug discovery, advances and opportunities, *Nat. Rev. Drug Discovery*, 2021, **20**, 200–216.
12. J. Y. Ortholand and A. Ganesan, Natural products and combinatorial chemistry, Back to the future, *Curr. Opin. Chem. Biol.*, 2004, **8**, 271–280.
13. J. A. Beutler, Natural products as a foundation for drug discovery, *Curr. Protoc. Pharmacol.*, 2009, **49**, 9–11.

14. M. E. Maier, Design and synthesis of analogues of natural products, *Org. Biomol. Chem.*, 2015, **13**, 5302–5343.
15. O. K. Radjasa, Y. M. Vaske, G. Navarro, H. C. Vervoort, K. Tenney and R. G. Linington, *et al.*, Highlights of marine invertebrate-derived biosynthetic products, Their biomedical potential and possible production by microbial associants, *Bioorg. Med. Chem.*, 2011, 6658–6674.
16. T. Gebretsadik, W. Linert, M. Thomas, T. Berhanu and R. Frew, LC–NMR for Natural Product Analysis, A Journey from an Academic Curiosity to a Robust Analytical Tool, *Sci*, 2021, **3**, 6.
17. N. G. M. Gomes, D. M. Pereira, P. Valentão and P. B. Andrade, Hybrid MS/NMR methods on the prioritization of natural products, Applications in drug discovery, *J. Pharm. Biomed. Anal.*, 2018, **147**, 234–249.
18. L. Zhang, F. X. Chen, Z. Zeng, M. Xu, F. Sun and L. Yang, *et al.*, Advances in Metagenomics and Its Application in Environmental Microorganisms, *Front. Microbiol.*, 2021, **12**, 3847.
19. K. Scherlach and C. Hertweck, Mining and unearthing hidden biosynthetic potential, *Nat. Commun.*, 2021, **12**, 1–12.
20. L. Albarano, R. Esposito, N. Ruocco and M. Costantini, Genome mining as new challenge in natural products discovery, *Mar. Drugs*, 2020, **18**, 199.
21. A. Choudhary, L. M. Naughton, I. Montánchez, A. D. W. Dobson and D. K. Rai, Current status and future prospects of Marine Natural Products (MNPs) as antimicrobials, *Mar. Drugs*, 2017, **15**, 272.
22. K. Senthilkumar and S. K. Kim, Marine invertebrate natural products for anti-inflammatory and chronic diseases, *J. Evidence-Based Complementary Altern. Med.*, 2013, **2013**, 572859.
23. J. W. Blunt, B. R. Copp, R. A. Keyzers, M. H. G. Munro and M. R. Prinsep, Marine natural products, *Nat. Prod. Rep.*, 2014, **31**, 160–258.
24. A. Karthikeyan, A. Joseph and B. G. Nair, Promising bioactive compounds from the marine environment and their potential effects on various diseases, *J. Genet. Eng. Biotechnol.*, 2022, **20**, 1–38.
25. R. Srinivasan, A. Kannappan, C. Shi and X. Lin, Marine bacterial secondary metabolites, A treasure house for structurally unique and effective antimicrobial compounds, *Mar. Drugs*, 2021, **19**, 530.
26. D. I. S. P. Resende, M. Ferreira, C. Magalhães, J. M. S. Lobo, E. Sousa and I. F. Almeida, Trends in the use of marine ingredients in anti-aging cosmetics, *Algal Res.*, 2021, **55**, 102273.
27. A. M. Rouhi, Supply issues complicate trek of chemicals from sea to market, *Chem. Eng. News*, 1995, **73**, 42–44.
28. M. C. Leal, C. Madeira, C. A. Brandão, J. Puga and R. Calado, Bioprospecting of marine invertebrates for new natural products - A chemical and zoogeographical perspective, *Molecules*, 2012, **17**, 9842–9854.
29. R. K. Pettit, Culturability and Secondary Metabolite Diversity of Extreme Microbes, Expanding Contribution of Deep Sea and Deep-Sea Vent Microbes to Natural Product Discovery, *Mar. Biotechnol.*, 2011, **13**, 1–11.

30. R. Montaser and H. Luesch, Marine natural products, A new wave of drugs?, *Future Med. Chem.*, 2011, **3**, 1475–1489.
31. A. Martins, H. Vieira, H. Gaspar and S. Santos, Marketed marine natural products in the pharmaceutical and cosmeceutical industries, Tips for success, *Mar. Drugs*, 2014, **12**, 1066–1101.
32. G. M. Cragg, F. Katz, D. J. Newman and J. Rosenthal, The impact of the United Nations Convention on Biological Diversity on natural products research, *Nat. Prod. Rep.*, 2012, **29**, 1407–1423.
33. M. J. Costello, P. Bouchet, C. S. Emblow and A. Legakis, European marine biodiversity inventory and taxonomic resources, State of the art and gaps in knowledge, *Mar. Ecol.: Prog. Ser.*, 2006, **316**, 257–268.
34. X. Jiang, J. Xiong, Z. Song, J. C. Morse, F. C. Jones and Z. Xie, Is coarse taxonomy sufficient for detecting macroinvertebrate patterns in floodplain lakes?, *Ecol. Indic.*, 2013, **27**, 48–55.
35. J. P. Hughes, S. S. Rees, S. B. Kalindjian and K. L. Philpott, Principles of early drug discovery, *Br. J. Pharmacol.*, 2011, **162**, 1239–1249.
36. P. Gribbon and S. Andreas, High-throughput drug discovery, What can we expect from HTS?, *Drug Discovery Today*, 2005, **10**, 17–22.
37. D. G. I. Kingston, Modern natural products drug discovery and its relevance to biodiversity conservation, *J. Nat. Prod.*, 2011, **74**, 496–511.
38. C. N. Parker, J. Ottl, D. Gabriel and J.-H. Zhang, *Advances in Biological Screening for Lead Discovery*, 2009, ch. 8, pp. 243–271.
39. G. T. Carter, Natural products and Pharma 2011, Strategic changes spur new opportunities, *Nat. Prod. Rep.*, 2011, **28**, 1783–1789.
40. T. F. Molinski, D. S. Dalisay, S. L. Lievens and J. P. Saludes, Drug development from marine natural products, *Nat. Rev. Drug Discovery*, 2009, **8**, 69–85.
41. T. L. Suyama, W. H. Gerwick and K. L. McPhail, Survey of marine natural product structure revisions, A synergy of spectroscopy and chemical synthesis, in *Bioorganic and Medicinal Chemistry*, Pergamon, 2011, pp. 6675–6701.
42. M. E. Maier, Structural revisions of natural products by total synthesis, *Nat. Prod. Rep.*, 2009, **26**, 1105–1124.
43. T. Michel, M. Halabalaki and A. L. Skaltsounis, New concepts, experimental approaches, and dereplication strategies for the discovery of novel phytoestrogens from natural sources, *Planta Med.*, 2013, **79**, 514–532.
44. *AntiBase 2021 – The Natural Compound Identifier Now Available*, Wiley Science Solutions, https://sciencesolutions.wiley.com/antibase-2021-the-natural-compound-identifier-now-available/.
45. J. W. Blunt and M. H. G. Munro, MarinLit - A database of the marine natural products literature, https://marinlit.rsc.org/.
46. J. L. López-Pérez, R. Therón, E. del Olmo and D. Díaz, NAPROC-13, A database for the dereplication of natural product mixtures in bioassay-guided protocols, *Bioinformatics*, 2007, **23**, 3256–3257.

47. J. B. Harborne, Dictionary of Natural Products, *Phytochemistry*, 1995, **38**, 279.
48. *Natural Compounds as Drugs*, ed. F. Petersen, and R. Amstutz, Birkhäuser, Basel, 2008.
49. *Marine Cosmeceuticals: Trends and Prospects*, ed. S.-K. Kim, CRC Press, 2011.
50. M. B. Ariede, T. M. Candido, A. L. M. Jacome, M. V. R. Velasco, J. C. M. de Carvalho and A. R. Baby, Cosmetic attributes of algae - A review, *Algal Res.*, 2017, **25**, 483–487.
51. A. Kijjoa and P. Sawangwong, Drugs and cosmetics from the sea, *Mar. Drugs*, 2004, **2**, 73–82.
52. L. Uppala, A Review on Active Ingredients from Marine Sources used in Cosmetics, *SOJ Pharm. Pharm. Sci.*, 2015, **2**, 01–03.
53. M. E. Hay, Marine chemical ecology, What's known and what's next?, *J. Exp. Mar. Biol. Ecol.*, 1996, **200**, 103–134.
54. M. C. Leal, R. Calado, C. Sheridan, A. Alimonti and R. Osinga, Coral aquaculture to support drug discovery, *Trends Biotechnol.*, 2013, **31**, 555–561.
55. T. Kogej, C. Gostinčar, M. Volkmann, A. A. Gorbushina and N. Gunde-Cimerman, Mycosporines in extremophilic fungi - Novel complementary osmolytes?, *Environ. Chem.*, 2006, **3**, 105–110.
56. D. Zhang, X. Yang, J. S. Kang, H. D. Choi and B. W. Son, Circumdatin I, a new ultraviolet-A protecting benzodiazepine alkaloid from a marine isolate of the fungus Exophiala, *J. Antibiot.*, 2008, **61**, 40–42.
57. X. Li, M. K. Kim, U. Lee, S. K. Kim, J. S. Kang and H. D. Choi, *et al.*, Myrothenones A and B, cyclopentenone derivatives with tyrosinase inhibitory activity from the marine-derived fungus Myrothecium sp, *Chem. Pharm. Bull.*, 2005, **53**, 453–455.
58. D. Zhang, X. Li, J. S. Kang, H. D. Choi and B. W. Son, A New α-Pyrone Derivative, 6-[(E)-Hept-1-enyl]-α-pyrone, with Tyrosinase Inhibitor Activity from a Marine Isolate of the Fungus Botrytis, *Bull. Korean Chem. Soc.*, 2007, **28**, 5.
59. T. Tsuchiya, K. Yamada, K. Minoura, K. Miyamoto, Y. Usami and T. Kobayashi, *et al.*, Purification and determination of the chemical structure of the tyrosinase inhibitor produced by Trichoderma viride strain H1-7 from a marine environment, *Biol. Pharm. Bull.*, 2008, **31**, 1618–1620.
60. C. Corinaldesi, G. Barone, F. Marcellini, A. Dell'Anno and R. Danovaro, Marine microbial-derived molecules and their paotential use in cosmeceutical and cosmetic products, *Mar. Drugs*, 2017, **15**, 118.
61. X. Li, H. J. Jee, T. L. Kang, R. R. Jung, D. C. Hong and S. K. Jung, *et al.*, γ-Pyrone derivatives, kojic acid methyl ethers from a marine-derived fungus Altenaria sp, *Arch. Pharmacal Res.*, 2003, **26**, 532–534.
62. K. Kaya, A. Nakazawa, H. Matsuura, D. Honda, I. Inouye and M. M. Watanabe, Thraustochytrid aurantiochytrium sp. 18w-13a accummulates high amounts of squalene, *Biosci., Biotechnol., Biochem.*, 2011, **75**, 2246–2248.

63. V. A. Ziboh, C. C. Miller and Y. Cho, Metabolism of polyunsaturated fatty acids by skin epidermal enzymes, Generation of antiinflammatory and antiproliferative metabolites, *Am. J. Clin. Nutr.*, 2000, 361s–366s.
64. I. Bhatnagar and S. K. Kim, Immense essence of excellence, Marine microbial bioactive compounds, *Mar. Drugs*, 2010, **8**, 2673–2701.
65. J. F. Imhoff, A. Labes and J. Wiese, Bio-mining the microbial treasures of the ocean, New natural products, *Biotechnol. Adv.*, 2011, **29**, 468–482.
66. A. Balkrishna, V. Agarwal, G. Kumar and A. K. Gupta, Applications of bacterial polysaccharides with special reference to the cosmetic industry, in *Microbial Bioprospecting for Sustainable Development*, Springer, Singapore, 2018, pp. 189–202.
67. U. U. Nwodo, E. Green and A. I. Okoh, Bacterial exopolysaccharides, Functionality and prospects, *Int. J. Mol. Sci.*, 2012, **13**, 14002–14015.
68. J.-B. B. Guillerme, C. Couteau and L. Coiffard, Applications for marine resources in cosmetics, *Cosmetics*, 2017, **4**, 35.
69. T. Le Costaouëc, S. Cérantola, D. Ropartz, J. Ratiskol, C. Sinquin and S. Colliec-Jouault, *et al.*, Structural data on a bacterial exopolysaccharide produced by a deep-sea Alteromonas macleodii strain, *Carbohydr. Polym.*, 2012, **90**, 49–59.
70. A. Courtois, C. Berthou, J. Guézennec, C. Boisset and A. Bordron, Exopolysaccharides isolated from hydrothermal vent bacteria can modulate the complement system, *PLoS One*, 2014, **9**, e94965.
71. H. Y. Kang, T. J. Yoon and G. J. Lee, Whitening effects of marine pseudomonas extract, *Ann. Dermatol.*, 2011, **23**, 144–149.
72. H. Kunte, G. Lentzen and E. Galinski, Industrial Production of the Cell Protectant Ectoine, Protection Mechanisms, Processes, and Products, *Curr. Biotechnol.*, 2014, **3**, 10–25.
73. A. Bownik and Z. Stępniewska, Ectoine as a promising protective agent in humans and animals, *Arh. Hig. Rada Toksikol.*, 2016, **67**, 260–264.
74. A. Marini, K. Reinelt, J. Krutmann and A. Bilstein, Ectoine-containing cream in the treatment of mild to moderate atopic dermatitis, A randomised, comparator-controlled, intra-individual double-blind, multi-center trial, *Skin Pharmacol. Physiol.*, 2014, **27**, 57–65.
75. U. Heinrich, B. Garbe and H. Tronnier, In vivo assessment of ectoin, A randomized, vehicle-controlled clinical trial, *Skin Pharmacol. Physiol.*, 2007, **20**, 211–218.
76. V. Jesumani, H. Du, M. Aslam, P. Pei and N. Huang, Potential use of seaweed bioactive compounds in skincare—A review, *Mar. Drugs*, 2019, **17**, 688.
77. M. A. Borowitzka, High-value products from microalgae-their development and commercialisation, *J. Appl. Phycol.*, 2013, **25**, 743–756.
78. A. M. Sanghvi and Y. M. Lo, Present and Potential Industrial Applications of Macro- and Microalgae, *Recent Pat. Food, Nutr. Agric.*, 2012, **2**, 187–194.
79. S. H. Cha, S. C. Ko, D. Kim and Y. J. Jeon, Screening of marine algae for potential tyrosinase inhibitor, Those inhibitors reduced tyrosinase activity and melanin synthesis in zebrafish, *J. Dermatol.*, 2011, **38**, 354–363.

80. E. Christaki, E. Bonos, I. Giannenasa and P. Florou-Paneria, Functional properties of carotenoids originating from algae, *J. Sci. Food Agric.*, 2013, **93**, 5–11.
81. C. Verdy, J. E. Branka and N. Mekideche, Quantitative assessment of lactate and progerin production in normal human cutaneous cells during normal ageing, Effect of an Alaria esculenta extract, *Int. J. Cosmet. Sci.*, 2011, **33**, 462–466.
82. K. Cao, C. D. Blair, D. A. Faddah, J. E. Kieckhaefer, M. Olive and M. R. Erdos, *et al.*, Progerin and telomere dysfunction collaborate to trigger cellular senescence in normal human fibroblasts, *J. Clin. Invest.*, 2011, **121**, 2833–2844.
83. C. Zhang, Y. Li, X. Shi and S. K. Kim, Inhibition of the expression on MMP-2, 9 and morphological changes via human fibrosarcoma cell line by 6,6'-bieckol from marine alga Ecklonia cava, *BMB Rep.*, 2010, **43**, 62–68.
84. S. J. Heo, S. C. Ko, S. M. Kang, S. H. Cha, S. H. Lee and D. H. Kang, *et al.*, Inhibitory effect of diphlorethohydroxycarmalol on melanogenesis and its protective effect against UV-B radiation-induced cell damage, *Food Chem. Toxicol.*, 2010, **48**, 1355–1361.
85. N. Y. Yoon, T. K. Eom, M. M. Kim and S. K. Kim, Inhibitory effect of phlorotannins isolated from Ecklonia cava on mushroom tyrosinase activity and melanin formation in mouse B16F10 melanoma cells, *J. Agric. Food Chem.*, 2009, **57**, 4124–4129.
86. H. S. Kang, H. R. Kim, D. S. Byun, B. W. Son, T. J. Nam and J. S. Choi, Tyrosinase inhibitors isolated from the edible brown alga Ecklonia stolonifera, *Arch. Pharmacal Res.*, 2004, **27**, 1226–1232.
87. N. V. Thomas and S. K. Kim, Beneficial effects of marine algal compounds in cosmeceuticals, *Mar. Drugs*, 2013, **11**, 146–164.
88. J. S. Choi, W. S. Moon, J. N. Choi, K. H. Do, S. H. Moon and K. K. Cho, *et al.*, Effects of seaweed Laminaria japonica extracts on skin moisturizing activity in vivo, *J. Cosmet. Sci.*, 2013, **64**, 193–205.
89. J. H. Hitton, M. Irhimeh and N. Falk, Macroalgal Fucoidan Extracts, A New Opprotunity for Marine Cosmetics, *Cosmetics Toiletries*, 2007, vol. 122, pp. 55–64.
90. C. S. Stutz, D. Schmid and F. Zilli, *Use of an extract from snow algae in cosmetic or dermatological formulations*, *US pat.*, 8206721, 2012.
91. L. Gouveia, A. P. Batista, I. Sousa, A. Raymundo and N. M. Bandarra, Microalgae in novel food products, in *Algae, Nutrition, Pollution Control and Energy Sources*, Nova Science Publishers, 2009, pp. 265–300.
92. S. Einarsson, A. Brynjolfsdottir and J. Krutmann, Pharmaceutical and cosmetic use of extracts from algae obtainable from saline hot water sources, *US pat.*, 8795679, 2014.
93. P. Derikvand, C. A. Llewellyn and S. Purton, Cyanobacterial metabolites as a source of sunscreens and moisturizers, a comparison with current synthetic compounds, *Eur. J. Phycol.*, 2017, **52**, 43–56.

94. K. N. Singh, S. Patil and H. Barkate, Protective effects of astaxanthin on skin, Recent scientific evidence, possible mechanisms, and potential indications, *J. Cosmet. Dermatol.*, 2020, **19**, 22–27.
95. T. Komatsu, S. Sasaki, Y. Manabe, T. Hirata and T. Sugawara, Preventive effect of dietary astaxanthin on UVA-induced skin photoaging in hairless mice, *PLoS One*, 2017, **12**, e0171178.
96. R. Sathasivam and J. S. Ki, A review of the biological activities of microalgal carotenoids and their potential use in healthcare and cosmetic industries, *Mar. Drugs*, 2018, **16**, 26.
97. M. L. Mourelle, C. P. Gómez and J. L. Legido, The potential use of marine microalgae and cyanobacteria in cosmetics and thalassotherapy, *Cosmetics*, 2017, **4**, 46.
98. L. Alparslan, N. Şekeroğlu and A. Kijjoa, The Potential of Marine Resources in Cosmetics, *Curr. Perspect. Med. Aromat. Plants*, 2018, **1**, 53–66.
99. J. Kennedy, P. Baker, C. Piper, P. D. Cotter, M. Walsh and M. J. Mooij, *et al.*, Isolation and analysis of bacteria with antimicrobial activities from the marine sponge haliclona simulans collected from irish waters, *Mar. Biotechnol.*, 2009, **11**, 384–396.
100. H. Rahman, B. Austin, W. J. Mitchell, P. C. Morris, D. J. Jamieson and D. R. Adams, *et al.*, Novel anti-infective compounds from marine bacteria, *Mar. Drugs*, 2010, **8**, 498–518.
101. E. Townsend, R. Moni, R. Quinn and P. G. Parsons, Reversible depigmentation of human melanoma cells by halistanol trisulphate, a novel marine sterol, *Melanoma Res.*, 1992, **1**, 349–357.
102. H. Y. Lee, e. J. Jang, S. Y. Bae, J. e. Jeon, H. J. Park and J. Shin, *et al.*, Anti-melanogenic activity of gagunin D, a highly oxygenated diterpenoid from the marine sponge phorbas sp., via modulating tyrosinase expression and degradation, *Mar. Drugs*, 2016, **14**, 212.
103. C. Galasso, C. Corinaldesi and C. Sansone, Carotenoids from marine organisms, Biological functions and industrial applications, *Antioxidants*, 2017, **6**, 96.
104. S. A. Look, W. Fenical, R. S. Jacobs and J. Clardy, The pseudopterosins, Anti-inflammatory and analgesic natural products from the sea whip Pseudopterogorgia elisabethae, *Proc. Natl. Acad. Sci. U. S. A.*, 1986, **83**, 6238–6240.
105. H. Correa, A. L. Valenzuela, L. F. Ospina and C. Duque, Anti-inflammatory effects of the gorgonian pseudopterogorgia elisabethae collected at the islands of providencia and san andrés (SW Caribbean), *J. Inflammation*, 2009, **6**, 1–10.
106. A. Ata, H. Y. Win, D. Holt, P. Holloway, E. P. Segstro and G. S. Jayatilake, New antibacterial diterpenes from Pseudopterogorgia elisabethae, *Helv. Chim. Acta*, 2004, **87**, 1090–1098.
107. N. Onumah, A novel anti-inflammatory in treatment of acne vulgaris, The pseudopterosins, *J Drugs Dermatology*, 2013, **12**, 1177–1179.

108. A. M. S. Mayer, K. B. Glaser, C. Cuevas, R. S. Jacobs, W. Kem and R. D. Little, *et al.*, The odyssey of marine pharmaceuticals, a current pipeline perspective, *Trends Pharmacol. Sci.*, 2010, **31**, 255–265.
109. E. B. Souto, A. R. Fernandes, C. Martins-Gomes, T. E. Coutinho, A. Durazzo and M. Lucarini, *et al.*, Nanomaterials for skin delivery of cosmeceuticals and pharmaceuticals, *Appl. Sci.*, 2020, **10**, 1594.
110. A. Nowak, M. Zagórska-Dziok, M. Perużyńska, K. Cybulska, E. Kucharska and P. Ossowicz-Rupniewska, *et al.*, Assessment of the Anti-Inflammatory, Antibacterial and Anti-Aging Properties and Possible Use on the Skin of Hydrogels Containing Epilobium angustifolium L. Extracts, *Front. Pharmacol.*, 2022, **13**, 896706.
111. D. A. Dias, S. Urban and U. Roessner, A Historical overview of natural products in drug discovery, *Metabolites*, 2012, **2**, 303–336.
112. R. A. Hussein and A. A. El-Anssary, Plants Secondary Metabolites, The Key Drivers of the Pharmacological Actions of Medicinal Plants, in *Herbal Medicine*, IntechOpen, 2019, DOI: 10.5772/intechopen.76139.
113. K. Nawrot-Chorabik, M. Sułkowska and N. Gumulak, Secondary Metabolites Produced by Trees and Fungi, Achievements So Far and Challenges Remaining, *Forests*, 2022, **13**, 1338.
114. K. Jakubiec-Krzesniak, A. Rajnisz-Mateusiak, A. Guspiel, J. Ziemska and J. Solecka, Secondary metabolites of actinomycetes and their antibacterial, antifungal and antiviral properties, *Pol. J. Microbiol.*, 2018, **67**, 259–272.
115. J. H. Kim, J.-E. E. Lee, K. H. Kim and N. J. Kang, Beneficial effects of marine algae-derived carbohydrates for skin health, *Mar. Drugs*, 2018, **16**, 459.
116. A. Poli, G. Anzelmo and B. Nicolaus, Bacterial exopolysaccharides from extreme marine habitats, Production, characterization and biological activities, *Mar. Drugs*, 2010, **8**, 1779–1802.
117. J. Angelin and M. Kavitha, Exopolysaccharides from probiotic bacteria and their health potential, *Int. J. Biol. Macromol.*, 2020, **162**, 853–865.
118. A. Poli, P. Di Donato, G. R. Abbamondi and B. Nicolaus, Synthesis, production, and biotechnological applications of exopolysaccharides and polyhydroxyalkanoates by Archaea, *Archaea*, 2011, **2011**, 693253.
119. T. Pusztahelyi, Chitin and chitin-related compounds in plant–fungal interactions, *Mycology*, 2018, **9**, 189–201.
120. A. Zayed and R. Ulber, Fucoidans, Downstream processes and recent applications, *Mar. Drugs*, 2020, **18**, 170.
121. B. Venardou, J. V. O'Doherty, S. Maher, M. T. Ryan, V. Gath and R. Ravindran, *et al.*, Potential of a fucoidan-rich Ascophyllum nodosum extract to reduce Salmonella shedding and improve gastrointestinal health in weaned pigs naturally infected with Salmonella, *J. Anim. Sci. Biotechnol.*, 2022, **13**, 1–16.
122. L. Pereira, Seaweeds as source of bioactive substances and skin care therapy-Cosmeceuticals, algotheraphy, and thalassotherapy, *Cosmetics*, 2018, **5**, 68.

123. I. Aranaz, N. Acosta, C. Civera, B. Elorza, J. Mingo and C. Castro, *et al.*, Cosmetics and cosmeceutical applications of chitin, chitosan and their derivatives, *Polymers*, 2018, **10**, 213.
124. F. Hentati, L. Tounsi, D. Djomdi, G. Pierre, C. Delattre and A. V. Ursu, *et al.*, Bioactive polysaccharides from seaweeds, *Molecules*, 2020, **25**, 3152.
125. A. Zayed, M. El-Aasr, A. R. S. Ibrahim and R. Ulber, Fucoidan characterization, Determination of purity and physicochemical and chemical properties, *Mar. Drugs*, 2020, **18**, 571.
126. J. M. Hee, S. P. Kyong, J. K. Mi, S. L. Myeong, H. J. Seok and T. I. Imbs, *et al.*, Effect of Costaria costata fucoidan on expression of matrix metalloproteinase-1 promoter, mRNA, and protein, *J. Nat. Prod.*, 2009, **72**, 1731–1734.
127. K. Senni, F. Gueniche, A. Foucault-Bertaud, S. Igondjo-Tchen, F. Fioretti and S. Colliec-Jouault, *et al.*, Fucoidan a sulfated polysaccharide from brown algae is a potent modulator of connective tissue proteolysis, *Arch. Biochem. Biophys.*, 2006, **445**, 56–64.
128. H. Maruyama, H. Tamauchi, F. Kawakami, K. Yoshinaga and T. Nakano, Suppressive Effect of Dietary Fucoidan on Proinflammatory Immune Response and MMP-1 Expression in UVB-Irradiated Mouse Skin, *Planta Med.*, 2015, **81**, 1370–1374.
129. I. P. S. Fernando, K. N. Kim, D. Kim and Y. J. Jeon, Algal polysaccharides, potential bioactive substances for cosmeceutical applications, *Crit. Rev. Biotechnol.*, 2019, **39**, 99–113.
130. T. F. R. Alves, M. Morsink, F. Batain, M. V. Chaud, T. Almeida and D. A. Fernandes, *et al.*, Applications of natural, semi-synthetic, and synthetic polymers in cosmetic formulations, *Cosmetics*, 2020, **7**, 75.
131. R. D. Singh, S. Kapila, N. G. Ganesan and V. Rangarajan, A review on green nanoemulsions for cosmetic applications with special emphasis on microbial surfactants as impending emulsifying agents, *J. Surfactants Deterg.*, 2022, **25**, 303–319.
132. S. Rodrigues, L. Cunha, J. Rico, A. M. R. da Costa, A. J. Almeida and M. L. Faleiro, *et al.*, Carrageenan from red algae, an application in the development of inhalable tuberculosis therapy targeting the macrophages, *Drug Delivery Transl. Res.*, 2020, **10**, 1675–1687.
133. S. W. Ren, J. Li, W. Wang and H. S. Guan, Protective effects of κ-ca3000+CP against ultraviolet-induced damage in HaCaT and MEF cells, *J. Photochem. Photobiol., B*, 2010, **101**, 22–30.
134. E. J. Yun, I. G. Choi and K. H. Kim, Red macroalgae as a sustainable resource for bio-based products, *Trends Biotechnol.*, 2015, **33**, 247–249.
135. C. Corinaldesi, G. Barone, F. Marcellini, A. Dell'Anno, R. Danovaro and A. Dell'Anno, *et al.*, Marine microbial-derived molecules and their potential use in cosmeceutical and cosmetic products, *Mar. Drugs*, 2017, **15**, 118.
136. S. Rana and L. S. B. Upadhyay, Microbial exopolysaccharides, Synthesis pathways, types and their commercial applications, *Int. J. Biol. Macromol.*, 2020, **157**, 577–583.

137. I. Aranaz, A. R. Alcántara, M. C. Civera, C. Arias, B. Elorza and A. H. Caballero, *et al.*, Chitosan, An overview of its properties and applications, *Polymers*, 2021, **13**, 3256.
138. E. Guzmán, F. Ortega and R. G. Rubio, Chitosan, A Promising Multifunctional Cosmetic Ingredient for Skin and Hair Care, *Cosmetics*, 2022, **9**, 99.
139. A. Sionkowska, B. Kaczmarek, M. Michalska, K. Lewandowska and S. Grabska, Preparation and characterization of collagen/chitosan/hyaluronic acid thin films for application in hair care cosmetics, in *Pure and Applied Chemistry*, Walter de Gruyter GmbH, 2017, pp. 1829–1839.
140. Y. A. Gomaa, L. K. El-Khordagui, N. A. Boraei and I. A. Darwish, Chitosan microparticles incorporating a hydrophilic sunscreen agent, *Carbohydr. Polym.*, 2010, **81**, 234–242.
141. S. Z. Kong, D. D. Li, H. Luo, W. J. Li, Y. M. Huang and J. C. Li, *et al.*, Anti-photoaging effects of chitosan oligosaccharide in ultraviolet-irradiated hairless mouse skin, *Exp. Gerontol.*, 2018, **103**, 27–34.
142. H. S. Kalasariya, L. Pereira and N. B. Patel, Pioneering Role of Marine Macroalgae in Cosmeceuticals, *Phycology*, 2022, **2**, 172–203.
143. T. Sae-leaw, Y. C. O'Callaghan, S. Benjakul and N. M. O'Brien, Antioxidant activities and selected characteristics of gelatin hydrolysates from seabass (Lates calcarifer) skin as affected by production processes, *J. Food Sci. Technol.*, 2016, **53**, 197–208.
144. J. Cotas, A. Leandro, P. Monteiro, D. Pacheco, A. Figueirinha and A. M. M. Gonçalves, *et al.*, Seaweed phenolics, From extraction to applications, *Mar. Drugs*, 2020, **18**, 384.
145. J. Venkatesan, K. K. Keekan, S. Anil, I. Bhatnagar and S. K. Kim, Phlorotannins, in *Encyclopedia of Food Chemistry*, Elsevier, 2018, pp. 515–527.
146. E. S. Okeke, e. J. Nweze, C. C. Chibuogwu, e. G. Anaduaka, K. I. Chukwudozie and T. P. C. Ezeorba, Aquatic Phlorotannins and Human Health, Bioavailability, Toxicity, and Future Prospects, *Nat. Prod. Commun.*, 2021, **16**, 1934578X211056144.
147. K. Thiyagarasaiyar, B. H. Goh, Y. J. Jeon and Y. Y. Yow, Algae metabolites in cosmeceutical, An overview of current applications and challenges, *Mar. Drugs*, 2020, **18**, 323.
148. M. D. Catarino, S. J. Amarante, N. Mateus, A. M. S. Silva and S. M. Cardoso, Brown algae phlorotannins, A marine alternative to break the oxidative stress, inflammation and cancer network, *Foods*, 2021, **10**, 1478.
149. M. D. Catarino, A. M. S. Silva and S. M. Cardoso, Fucaceae, A source of bioactive phlorotannins, *Int. J. Mol. Sci.*, 2017, **18**, 1327.
150. C. K. Singh, G. Chhabra, M. A. Ndiaye, L. M. Garcia-Peterson, N. J. MacK and N. Ahmad, The Role of Sirtuins in Antioxidant and Redox Signaling, *Antioxid. Redox Signaling*, 2018, **28**, 643–661.
151. R. Sun, N. Xia and Q. Xia, Non-aqueous nanoemulsions as a new strategy for topical application of astaxanthin, *J. Dispersion Sci. Technol.*, 2020, **41**, 1777–1788.

152. M. M. R. Shah, Y. Liang, J. J. Cheng and M. Daroch, Astaxanthin-producing green microalga Haematococcus pluvialis, From single cell to high value commercial products, *Front. Plant Sci.*, 2016, **7**, 531.
153. T. H. P. Brotosudarmo, L. Limantara, E. Setiyono, and Heriyanto, Structures of Astaxanthin and Their Consequences for Therapeutic Application, *Int. J. Food Sci.*, 2020, **2020**, 2156582.
154. M. Mularczyk, I. Michalak and K. Marycz, Astaxanthin and other nutrients from haematococcus pluvialis—Multifunctional applications, *Mar. Drugs*, 2020, **18**, 459.
155. K. Miazek, K. Beton, A. Śliwińska and B. Brożek-Płuska, The Effect of β-Carotene, Tocopherols and Ascorbic Acid as Anti-Oxidant Molecules on Human and Animal In Vitro/In Vivo Studies, A Review of Research Design and Analytical Techniques Used, *Biomolecules*, 2022, **12**, 1087.
156. G. Pizzino, N. Irrera, M. Cucinotta, G. Pallio, F. Mannino and V. Arcoraci, *et al.*, Oxidative Stress, Harms and Benefits for Human Health, *Oxid. Med. Cell. Longevity*, 2017, **2017**, 8416763.
157. J. W. Shin, S. H. Kwon, J. Y. Choi, J. I. Na, C. H. Huh and H. R. Choi, *et al.*, Molecular mechanisms of dermal aging and antiaging approaches, *Int. J. Mol. Sci.*, 2019, **20**, 2126.
158. P. Alugoju, V. K. D. K. Swamy, N. V. A. Anthikapalli and T. Tencomnao, Health benefits of astaxanthin against age-related diseases of multiple organs, A comprehensive review, *Crit. Rev. Food Sci. Nutr.*, 2022, 1–66.
159. S. Davinelli, M. E. Nielsen and G. Scapagnini, Astaxanthin in skin health, repair, and disease, A comprehensive review, *Nutrients*, 2018, **10**, 522.
160. C. Duque, M. Puyana, G. Narváez, O. Osorno, N. Hara and Y. Fujimoto, P-V Pseudopterosins, new compounds from the gorgonian octocoral Pseudopterogorgia elisabethae from Providencia island, Colombian Caribbean, *Tetrahedron*, 2004, **60**, 10627–10635.
161. S. Leslie and M. Baumann, Marine ingredients and the skin, *Dermatology News*, 2016, pp. 1–2.
162. I. N. 'Izzah, S. K. Wong, I. N. Mohamed, N. Mohamed, K. Y. Chin and S. Ima-Nirwana, *et al.*, Wound healing properties of selected natural products, *Int. J. Environ. Res. Public Health*, 2018, **15**, 2360.
163. S. B. Fialová, K. Rendeková, P. Mučaji, M. Nagy and L. Slobodníková, Antibacterial activity of medicinal plants and their constituents in the context of skin and wound infections, considering European legislation and folk medicine—A review, *Int. J. Mol. Sci.*, 2021, **22**, 10746.
164. S. B. García, X. Garrell-Salat, F. Trejo-Velasco, D. Aragón-Roca, M. Á Zapata and J. García-Arumí, Prevalence and implications of the presence of intraocular silicone oil droplets in patients treated with intravitreal injections of anti-VEGF, *BMC Ophthalmol.*, 2022, **22**, 1–8.
165. Y. González, D. Torres-Mendoza, G. E. Jones and P. L. Fernandez, Marine Diterpenoids as Potential Anti-Inflammatory Agents, *Mediators Inflammation*, 2015, **2015**, 263543.

166. F. Flachsmann, K. Schellhaas, C. E. Moya, R. S. Jacobs and W. Fenical, Synthetic pseudopterosin analogues, A novel class of antiinflammatory drug candidates, *Bioorg. Med. Chem.*, 2010, **18**, 8324–8333.
167. J. Sperlich, R. Kerr and N. Teusch, The marine natural product pseudopterosin blocks cytokine release of triple-negative breast cancer and monocytic leukemia cells by inhibiting NF-κB signaling, *Mar. Drugs*, 2017, **15**, 262.
168. F. L. Figueroa, Mycosporine-Like Amino Acids from Marine Resource, *Mar. Drugs*, 2021, **19**, 18.
169. H. Kageyama and R. Waditee-Sirisattha, Mycosporine-Like Amino Acids as Multifunctional Secondary Metabolites in Cyanobacteria, From Biochemical to Application Aspects, in *Studies in Natural Products Chemistry*, Elsevier B.V., 2018, pp. 153–194.
170. J. Vega, J. Bonomi-Barufi, J. L. Gómez-Pinchetti and F. L. Figueroa, Cyanobacteria and Red Macroalgae as Potential Sources of Antioxidants and UV Radiation-Absorbing Compounds for Cosmeceutical Applications, *Mar. Drugs*, 2020, **18**, 659.
171. IARC, *Solar and Ultraviolet Radiation*, Radiation - NCBI Bookshelf, IARC Work. Gr. Eval. Carcinog. Risks to Humans, 2012, https,//www.ncbi.nlm.nih.gov/books/NBK304366/, accessed 5 October 2022.
172. R. P. Rastogi, D. Madamwar, H. Nakamoto and A. Incharoensakdi, Resilience and self-regulation processes of microalgae under UV radiation stress, *J. Photochem. Photobiol., C*, 2020, **43**, 100322.
173. P. Pittayapruek, J. Meephansan, O. Prapapan, M. Komine and M. Ohtsuki, Role of matrix metalloproteinases in Photoaging and photocarcinogenesis, *Int. J. Mol. Sci.*, 2016, **17**, 868.
174. J. Ryu, S. J. Park, I. H. Kim, Y. H. Choi and T. J. Nam, Protective effect of porphyra-334 on UVA-induced photoaging in human skin fibroblasts, *Int. J. Mol. Med.*, 2014, **34**, 796–803.
175. T. Soleymani, J. Lanoue and Z. Rahman, A practical approach to chemical peels, A review of fundamentals and step-by-step algorithmic protocol for treatment, *J. Clin. Aesthet. Dermatol.*, 2018, **11**, 21–28.
176. G. Arora, Insights into Cosmeceuticals, *Cosmoderma*, 2021, **1**, 32.
177. A. Kawamura, K. Ooyama, K. Kojima, H. Kachi, T. Abe and K. Amano, *et al.*, Dietary supplementation of gamma-linolenic acid improves skin parameters in subjects with dry skin and mild atopic dermatitis, *J. Oleo Sci.*, 2011, **60**, 597–607.
178. B. Bhattacharya, T. K. Ghosh and N. Das, Application of Bio-Surfactants in Cosmetics and Pharmaceutical Industry, *Scholars Acad. J. Pharm.*, 2017, **6**, 320–329.
179. E. Gayathiri, P. Prakash, N. Karmegam, S. Varjani, M. K. Awasthi and B. Ravindran, Biosurfactants, Potential and Eco-Friendly Material for Sustainable Agriculture and Environmental Safety—A Review, *Agronomy*, 2022, **12**, 662.
180. D. S. Pardhi, R. R. Panchal, V. H. Raval, R. G. Joshi, P. Poczai and W. H. Almalki, *et al.*, Microbial surfactants, A journey from fundamentals to recent advances, *Front. Microbiol.*, 2022, **13**, 2877.

181. J. M. Chonody and B. Teater, Why do I dread looking old?, A test of social identity theory, terror management theory, and the double standard of aging, *J Women Aging*, 2016, **28**, 112–126.
182. X. Wang, Z. Zhang, S. Zhang, F. Yang, M. Yang and J. Zhou, *et al.*, Antiaging compounds from marine organisms, *Food Res. Int.*, 2021, **143**, 110313.
183. E. G. Brunt and J. G. Burgess, The promise of marine molecules as cosmetic active ingredients, *Int. J. Cosmet. Sci.*, 2018, **40**, 1–15.
184. S.-K. Kim, I. Wijesakara and I. Wijesekara, Cosmeceuticals from marine resources, *Mar. Cosmeceuticals: Trends Prospects*, 2011, 19.
185. *Cosmeceuticals and Active Cosmetics*, ed. R. Sivamani, J. R. Jagdeo, P. Elsner, and H. I. Maibach, CRC Press, 2015, DOI: 10.1201/b18895.
186. L. Xia, B. Zhang, F. Meng and J. P. Sun, A Criss-cross Heart with Double Outlet Right Ventricle, in *Comparative Cardiac Imaging*, 2018, pp. 39–43.
187. Z. Zhang, X. Wang, H. Su, Y. Pan, J. Han and T. Zhang, *et al.*, Effect of sulfated galactan from Porphyra haitanensis on $H_2O_2$-induced premature senescence in WI-38 cells, *Int. J. Biol. Macromol.*, 2018, **106**, 1235–1239.
188. X. Y. Liu, D. Liu, G. P. Lin, Y. J. Wu, L. Y. Gao and C. Ai, *et al.*, Anti-ageing and antioxidant effects of sulfate oligosaccharides from green algae Ulva lactuca and Enteromorpha prolifera in SAMP8 mice, *Int. J. Biol. Macromol.*, 2019, **139**, 342–351.
189. Y. Chen, X. Liu, L. Wu, A. Tong, L. Zhao and B. Liu, *et al.*, Physicochemical characterization of polysaccharides from Chlorella pyrenoidosa and its anti-ageing effects in Drosophila melanogaster, *Carbohydr. Polym.*, 2018, **185**, 120–126.
190. J. Gui, W. Tong, S. Huang, X. Liang, Z. Fang and W. Wang, *et al.*, Effects of Chlorella vulgaris polysaccharides accumulation on growth characteristics of Trachemys scripta elegans, *Int. J. Biol. Macromol.*, 2019, **141**, 1304–1313.
191. S. Z. Kong, J. C. Li, S. D. Li, M. N. Liao, C. P. Li and P. J. Zheng, *et al.*, Anti-aging effect of chitosan oligosaccharide on d-galactose-induced subacute aging in mice, *Mar. Drugs*, 2018, **16**, 181.
192. G. Qian and Z. Xu, Effect of polysaccharide extracted from Glaciecola polaris on the protection of mouse macrophages from oxidative injury, *Bioresour. Technol.*, 2007, **98**, 202–206.
193. C. Souza and P. M. B. G. M. Campos, Development and photoprotective effect of a sunscreen containing the antioxidants Spirulina and dimethylmethoxy chromanol on sun-induced skin damage, *Eur. J. Pharm. Sci.*, 2017, **104**, 52–64.
194. A. Zehlila, A. Schaumann, A. B. Mlouka, I. Bourguiba, J. Hardouin and O. Masmoudi, *et al.*, Glioprotective effect of Ulva rigida extract against UVB cellular damages, *Algal Res.*, 2017, **23**, 203–215.
195. L. Ma, A. Hu, Z. Liu, R. Duan, J. Zhang and Y. Qu, *et al.*, Antioxidant function of peptides from scallop by-product in aged mice in vivo, *J. Food Saf. Food Qual.*, 2018, **9**, 5159–5163.
196. L. Lin, K. Yang, L. Zheng, M. Zhao, W. Sun and Q. Zhu, *et al.*, Anti-aging effect of sea cucumber (Cucumaria frondosa) hydrolysate on fruit flies and d-galactose-induced aging mice, *J. Funct. Foods*, 2018, **47**, 11–18.

197. S. K. Kim, Marine cosmeceuticals, *J. Cosmet. Dermatol.*, 2014, **13**, 56–67.
198. P. Burger, A. Landreau, S. Azoulay, T. Michel and X. Fernandez, Skin whitening cosmetics, Feedback and challenges in the development of natural skin lighteners, *Cosmetics*, 2016, **3**, 36.
199. J. P. Ebanks, R. R. Wickett and R. E. Boissy, Mechanisms regulating skin pigmentation, the rise and fall of complexion coloration, *Int. J. Mol. Sci.*, 2009, **10**, 4066–4087.
200. S. Babitha and E.-K. Kim, Effect of Marine Cosmeceuticals on the Pigmentation of Skin, *Mar. Cosmeceuticals*, 2011, 63–66.
201. K. A. James, C. N. Burkhart and D. S. Morrell, Emerging drugs for acne, *Expert Opin. Emerging Drugs*, 2009, **14**, 649–659.
202. S. Agrawal, A. Adholeya, C. J. Barrow and S. K. Deshmukh, In-vitro evaluation of marine derived fungi against Cutibacterium acnes, *Anaerobe*, 2018, **49**, 5–13.
203. M. T. Chomnawang, S. Surassmo, V. S. Nukoolkarn and W. Gritsanapan, Antimicrobial effects of Thai medicinal plants against acne-inducing bacteria, *J. Ethnopharmacol.*, 2005, **101**, 330–333.
204. P. Ralifo and P. Crews, A new structural theme in the imidazole-containing alkaloids from a calcareous Leucetta sponge, *J. Org. Chem.*, 2004, **69**, 9025–9029.
205. Y. Kamei, M. Sueyoshi, K. I. Hayashi, R. Terada and H. Nozaki, The novel anti-Propionibacterium acnes compound, Sargafuran, found in the marine brown alga Sargassum macrocarpum, *J. Antibiot.*, 2009, **62**, 259–263.
206. J. S. Choi, H. J. Bae, S. J. Kim and I. S. Choi, In vitro antibacterial and anti-inflammatory properties of seaweed extracts against acne inducing bacteria, Propionibacterium acnes, *J. Environ. Biol.*, 2011, **32**, 313–318.
207. J. Nordberg and E. S. J. Arnér, Reactive oxygen species, antioxidants, and the mammalian thioredoxin system, *Free Radical Biol. Med.*, 2001, **31**, 1287–1312.
208. S. K. Kim and E. Mendis, Bioactive compounds from marine processing byproducts - A review, *Food Res. Int.*, 2006, **39**, 383–393.
209. N. Barzkar, S. T. Jahromi, H. B. Poorsaheli and F. Vianello, Metabolites from marine microorganisms, micro, and macroalgae, Immense scope for pharmacology, *Mar. Drugs*, 2019, **17**, 464.
210. R. Matsukawa, Z. Dubinsky, E. Kishimoto, K. Masaki, Y. Masuda and T. Takeuchi, *et al.*, A comparison of screening methods for antioxidant activity in seaweeds, *J. Appl. Phycol.*, 1997, **9**, 29–35.
211. L.-A. A. Tziveleka, M. A. Tammam, O. Tzakou, V. Roussis and E. Ioannou, Metabolites with Antioxidant Activity from Marine Macroalgae, *Antioxidants*, 2021, **10**, 1431.
212. K. S. Sidhu, Health benefits and potential risks related to consumption of fish or fish oil, *Regul. Toxicol. Pharmacol.*, 2003, **38**, 336–344.
213. P. A. Harnedy and R. J. FitzGerald, Bioactive peptides from marine processing waste and shellfish, A review, *J. Funct. Foods*, 2012, **4**, 6–24.

214. B. Whitehouse, The effects of dissolved organic matter on the aqueous partitioning of polynuclear aromatic hydrocarbons, *Estuarine, Coastal Shelf Sci.*, 1985, **20**, 393–402.
215. R. K. Sivamani, M. Ghiya and H. I. Maibach, Shedding light on sunscreens and their labels, Testing policies need to match actual use, *Am. J. Prev. Med.*, 2010, **38**, 679–681.
216. Y. Liu, Y. Wang, N. Li and S. Jiang, Avobenzone and nanoplastics affect the development of zebrafish nervous system and retinal system and inhibit their locomotor behavior, *Sci. Total Environ.*, 2022, **806**, 150681.
217. C. A. Downs, O. T. Cruz, T. E. Remengesau Jr and T. E. Remengesau, Sunscreen pollution and tourism governance, Science and innovation are necessary for biodiversity conservation and sustainable tourism, *Aquat. Conserv. Mar. Freshwater Ecosyst.*, 2022, **32**, 896–906.
218. R. P. Rastogi, R. R. Sonani and D. Madamwar, Cyanobacterial sunscreen scytonemin, role in photoprotection and biomedical research, *Appl. Biochem. Biotechnol.*, 2015, **176**, 1551–1563.
219. A. Oren and N. Gunde-Cimerman, Mycosporines and mycosporine-like amino acids, UV protectants or multipurpose secondary metabolites?, *FEMS Microbiol. Lett.*, 2007, **269**, 1–10.
220. N. N. Rosic, Mycosporine-like amino acids, making the foundation for organic personalised sunscreens, *Mar. Drugs*, 2019, **17**, 638.
221. M. A. Gammone, G. Riccioni, N. D'Orazio and N. D'Orazio, Marine carotenoids against oxidative stress, effects on human health, *Mar. Drugs*, 2015, **13**, 6226–6246.
222. J. Wang, M. Xu, R. Liang, M. Zhao, Z. Zhang and Y. Li, Oral administration of marine collagen peptides prepared from chum salmon (Oncorhynchus keta) improves wound healing following cesarean section in rats, *Food Nutr. Res.*, 2015, **59**, 26411.
223. Z. Zhang, J. Wang, Y. Ding, X. Dai and Y. Li, Oral administration of marine collagen peptides from Chum Salmon skin enhances cutaneous wound healing and angiogenesis in rats, *J. Sci. Food Agric.*, 2011, **91**, 2173–2179.
224. J. Chen, K. Gao, S. Liu, S. Wang, J. Elango and B. Bao, *et al.*, Fish collagen surgical compress repairing characteristics on wound healing process in vivo, *Mar. Drugs*, 2019, **17**, 33.
225. N. Ito, S. Seki and F. Ueda, Effects of composite supplement containing collagen peptide and ornithine on skin conditions and plasma IGF-1 levels—A randomized, double-blind, placebo-controlled trial, *Mar. Drugs*, 2018, **16**, 482.
226. C. De Luca, E. V. Mikhal'Chik, M. V. Suprun, M. Papacharalambous, A. I. Truhanov and L. G. Korkina, Skin antiageing and systemic Redox effects of supplementation with marine collagen peptides and plant-derived antioxidants, A single-blind case-control clinical study, *Oxid. Med. Cell. Longevity*, 2016, **2016**, 4389410.
227. M. Allouche, I. Hamdi, A. Nasri, A. H. Harrath, L. Mansour and H. Beyrem, *et al.*, Laboratory bioassay exploring the effects of anti-aging

skincare products on free-living marine nematodes, a case study of collagen, *Environ. Sci. Pollut. Res.*, 2020, **27**, 11403–11412.
228. C. R. Kim, Y. M. Kim, M. K. Lee, I. H. Kim, Y. H. Choi and T. J. Nam, Pyropia yezoensis peptide promotes collagen synthesis by activating the TGF-β/Smad signaling pathway in the human dermal fibroblast cell line Hs27, *Int. J. Mol. Med.*, 2017, **39**, 31–38.
229. Z. Hu, P. Yang, C. Zhou, S. Li and P. Hong, Marine collagen peptides from the skin of Nile Tilapia (Oreochromis niloticus), Characterization and wound healing evaluation, *Mar. Drugs*, 2017, **15**, 102.
230. A. León-López, A. Morales-Peñaloza, V. M. Martínez-Juárez, A. Vargas-Torres, D. I. Zeugolis and G. Aguirre-Álvarez, Hydrolyzed collagen-sources and applications, *Molecules*, 2019, **24**, 4031.
231. C. Cerutti and A. J. Ridley, Endothelial cell-cell adhesion and signaling, *Exp. Cell Res.*, 2017, **358**, 31–38.
232. J. Han, V. V. Shuvaev and V. R. Muzykantov, Catalase and superoxide dismutase conjugated with platelet-endothelial cell adhesion molecule antibody distinctly alleviate abnormal endothelial permeability caused by exogenous reactive oxygen species and vascular endothelial growth factor, *J. Pharmacol. Exp. Ther.*, 2011, **338**, 82–91.
233. Q. Li and I. M. Verma, NF-κB regulation in the immune system, *Nat. Rev. Immunol.*, 2002, **2**, 725–734.
234. Y. Yamamoto and R. B. Gaynor, Therapeutic potential of inhibition of the NF-κB pathway in the treatment of inflammation and cancer, *J. Clin. Invest.*, 2001, **107**, 135–142.
235. Y. Y. Lin, Y. H. Jean, H. P. Lee, W. F. Chen, Y. M. Sun and J. H. Su, *et al.*, A Soft Coral-Derived Compound, 11-epi-Sinulariolide Acetate Suppresses Inflammatory Response and Bone Destruction in Adjuvant-Induced Arthritis, *PLoS One*, 2013, **8**, e62926.
236. K. Omote, K. Hazama, T. Kawamata, M. Kawamata, Y. Nakayaka and M. Toriyabe, *et al.*, Peripheral nitric oxide in carrageenan-induced inflammation, *Brain Res.*, 2001, **912**, 171–175.
237. J. G. Leme, L. Hamamura, M. P. Leite and M. R. e Silva, Pharmacological analysis of the acute inflammatory process induced in the rat's paw by local injection of carrageenin and by heating, *Br. J. Pharmacol.*, 1973, **48**, 88.
238. L. C. Chen, Y. Y. Lin, Y. H. Jean, Y. Lu, W. F. Chen and S. N. Yang, *et al.*, Anti-inflammatory and analgesic effects of the marine-derived compound comaparvin isolated from the crinoid, Comanthus bennetti, *Molecules*, 2014, **19**, 14667–14686.
239. A. Rawlings, R. Sabin, C. Harding, A. Watkinson, J. Banks and C. Ackerman, *et al.*, The effect of glycerol and humidity on desmosome degradation in stratum corneum, *Arch. Dermatol. Res.*, 1995, **287**, 457–464.
240. K. Tominaga, N. Hongo, M. Karato and E. Yamashita, Cosmetic benefits of astaxanthin on humans subjects, *Acta Biochim. Pol.*, 2012, **59**, 43–47.
241. W. Harding and S. Rawlings, Dry skin, moisturization and corneodesmolysis, *Int. J. Cosmet. Sci.*, 2000, **22**, 21–52.

242. N. Lu, P. Chandar, D. Tempesta, C. Vincent, J. Bajor and H. McGuiness, Characteristic differences in barrier and hygroscopic properties between normal and cosmetic dry skin. I. Enhanced barrier analysis with sequential tape-stripping, *Int. J. Cosmet. Sci.*, 2014, **36**, 167–174.
243. A. V. Rawlings, D. A. Canestrari and B. Dobkowski, Moisturizer technology versus clinical performance, *Dermatol. Ther.*, 2004, **17**(Suppl 1), 49–56.
244. P. M. Elias, B. E. Brown and V. A. Ziboh, The permeability barrier in essential fatty acid deficiency, evidence for a direct role for linoleic acid in barrier function, *J. Invest. Dermatol.*, 1980, **74**, 230–233.
245. P. Garidel, B. Fölting, I. Schaller and A. Kerth, The microstructure of the stratum corneum lipid barrier, mid-infrared spectroscopic studies of hydrated ceramide, palmitic acid, cholesterol model systems, *Biophys. Chem.*, 2010, **150**, 144–156.
246. S. Meguro, Y. Arai, Y. Masukawa, K. Uie and I. Tokimitsu, Relationship between covalently bound ceramides and transepidermal water loss (TEWL), *Arch. Dermatol. Res.*, 2000, **292**, 463–468.
247. R. C. S. S. Barcelos, C. de Mello-Sampayo, C. T. D. D. Antoniazzi, H. J. Segat, H. Silva and J. C. Veit, *et al.*, Oral supplementation with fish oil reduces dryness and pruritus in the acetone-induced dry skin rat model, *J. Dermatol. Sci.*, 2015, **79**, 298–304.
248. G. A. Gbogouri, M. Linder, J. Fanni and M. Parmentier, Analysis of lipids extracted from salmon (Salmo salar) heads by commercial proteolytic enzymes, *Eur. J. Lipid Sci. Technol.*, 2006, **108**, 766–775.
249. S. C. Larsson, M. Kumlin, M. Ingelman-Sundberg and A. Wolk, Dietary long-chain n− 3 fatty acids for the prevention of cancer, a review of potential mechanisms, *Am. J. Clin. Nutr.*, 2004, **79**, 935–945.
250. D. H. Wang, J. R. Jackson, C. Twining, L. G. Rudstam, E. Zollweg-Horan and C. Kraft, *et al.*, Saturated branched chain, normal odd-carbon-numbered, and n-3 (omega-3) polyunsaturated fatty acids in freshwater fish in the northeastern United States, *J. Agric. Food Chem.*, 2016, **64**, 7512–7519.
251. N. Domoto, T. Koriyama, B. S. Chu and T. Tsuji, Evaluation of the efficacy of orange roughy (Hoplostetbus atlanticus) oil in subjects with dry skin, *Int. J. Cosmet. Sci.*, 2012, **34**, 322–327.
252. D. H. Buisson, D. R. Body, G. J. Dougherty, L. Eyres and P. Vlieg, Oil from deep water fish species as a substitute for sperm whale and jojoba oils, *J. Am. Oil Chem. Soc.*, 1982, **59**, 390–395.
253. C. Minto and C. P. Nolan, Fecundity and maturity of orange roughy (Hoplostethus atlanticus Collett 1889) on the Porcupine Bank, Northeast Atlantic, *Environ. Biol. Fishes*, 2006, **77**, 39–50.
254. Y. A. Asadpour, Squid (Loligo loligo), The new source to extract omega-3 and omega-6 rich marine oils, *Iran. J. Fish. Sci.*, 2016, **15**, 100–107.
255. J. Iglesias, R. Villanueva, L. Fuentes and R. Villanueva, *Cephalopod Culture*, Springer Science & Business Media, 2014, DOI: 10.1007/978-94-017-8648-5.

256. J. C. Navarro and R. Villanueva, The fatty acid composition of Octopus vulgaris paralarvae reared with live and inert food, deviation from their natural fatty acid profile, *Aquaculture*, 2003, **219**, 613–631.
257. T. Nagai and N. Suzuki, Preparation and partial characterization of collagen from paper nautilus (Argonauta argo, Linnaeus) outer skin, *Food Chem.*, 2002, **76**, 149–153.
258. J. R. E. R. E. Fraser, T. C. Laurent and U. B. G. G. Laurent, Hyaluronan, its nature, distribution, functions and turnover, *J. Intern. Med.*, 1997, **242**, 27–33.
259. Y. Yanase, T. Hiragun, K. Uchida, K. Ishii, S. Oomizu and H. Suzuki, *et al.*, Peritoneal injection of fucoidan suppresses the increase of plasma IgE induced by OVA-sensitization, *Biochem. Biophys. Res. Commun.*, 2009, **387**, 435–439.
260. R. Delgado-González, A. S. Astlas, A. Courtois and B. Thollas, Exopolysaccharide for the treatment and/or care of the skin, mucous membranes, hair and/or nails, *EP pat.*, EP2646115B1, Brittany, 2011, https://www.google.com/patents/EP2646115B1?hl=ca&cl=en, accessed 4 October 2022.
261. Lipotec, HyanifyTM marine ingredient, 2017, http//www.lipotec.com/en/products/hyanify-trade-marine-ingredient/.
262. Lipotec, Hyadisine® marine ingredient, 2022.
263. E. Proksch, R. Fölster-Holst, M. Bräutigam, M. Sepehrmanesh, S. Pfeiffer and J. M. Jensen, Role of the epidermal barrier in atopic dermatitis, Academy CME, *J. Ger. Soc. Dermatol.*, 2009, **7**, 899–910.
264. P. M. Elias and D. S. Friend, The permeability barrier in mammalian epidermis, *J. Cell Biol.*, 1975, **65**, 180–191.
265. D. Swatschek, W. Schatton, J. Kellermann, W. E. G. Müller and J. Kreuter, Marine sponge collagen, isolation, characterization and effects on the skin parameters surface-pH, moisture and sebum, *Eur. J. Pharm. Biopharm.*, 2002, **53**, 107–113.
266. M. Stoyneva-Gärtner, B. Uzunov and G. Gärtner, Enigmatic Microalgae from Aeroterrestrial and Extreme Habitats in Cosmetics, The Potential of the Untapped Natural Sources, *Cosmetics*, 2020, **7**, 27.
267. A. Longeon, B. R. Copp, E. Quévrain, M. Roué, B. Kientz and T. Cresteil, *et al.*, Bioactive indole derivatives from the South Pacific marine sponges Rhopaloeides odorabile and Hyrtios sp, *Mar. Drugs*, 2011, **9**, 879–888.
268. J. Y. Hong, S. J. Park, S. J. Seo and K. Y. Park, Oily sensitive skin, A review of management options, *J. Cosmet. Dermatol.*, 2020, **19**, 1016–1020.
269. F. Yvergnaux, Lipases, particularly effective biocatalysts for cosmetic active ingredients, *OCL: Oilseeds Fats, Crops Lipids*, 2017, **24**, 1029.
270. J. Ding, B. Wu and L. Chen, Application of Marine Microbial Natural Products in Cosmetics, *Front. Microbiol.*, 2022, **13**, 892505.
271. F. Hasan, A. A. Shah and A. Hameed, Industrial applications of microbial lipases, *Enzyme Microb. Technol.*, 2006, **39**, 235–251.

272. *Seafood Processing By-Products: Trends and Applications*, ed. S.-K. Kim, Springer, 2013, DOI: 10.1007/978-1-4614-9590-1.
273. B. Narayan, C. S. Kumar, T. Sashima, H. Maeda, M. Hosokawa and K. Miyashita, Composition, Functionality and Potential Applications of Seaweed Lipids, *Biocatal. Bioenergy*, 2008, 463–490.
274. S. Lordan, R. P. Ross and C. Stanton, Marine bioactives as functional food ingredients, potential to reduce the incidence of chronic diseases, *Mar. Drugs*, 2011, **9**, 1056–1100.
275. K. A. Mohammed, C. F. Hossain, L. Zhang, R. K. Bruick, Y.-D. D. Zhou and D. G. Nagle, Laurenditerpenol, a new diterpene from the tropical marine alga laurencia i ntricata that potently inhibits HIF-1 mediated hypoxic signaling in breast tumor cells, *J. Nat. Prod.*, 2004, **67**, 2002–2007.
276. C. Alves, S. Pinteus, T. Simões, A. Horta, J. Silva and C. Tecelão, *et al.*, Bifurcaria bifurcata, a key macro-alga as a source of bioactive compounds and functional ingredients, *Int. J. Food Sci. Technol.*, 2016, **51**, 1638–1646.
277. N. I. Marín-Ramos, D. Alonso, S. Ortega-Gutierrez, F. J. Ortega-Nogales, M. M. Balabasquer and H. Vázquez-Villa, *et al.*, New inhibitors of angiogenesis with antitumor activity in vivo, *J. Med. Chem.*, 2015, **58**, 3757–3766.
278. S. Tocaciu, L. J. Oliver, R. M. Lowenthal, G. M. Peterson, R. Patel and M. Shastri, *et al.*, The effect of Undaria pinnatifida fucoidan on the pharmacokinetics of letrozole and tamoxifen in patients with breast cancer, *Integr. Cancer Ther.*, 2018, **17**, 99–105.
279. J. Y. Kwak, Fucoidan as a marine anticancer agent in preclinical development, *Mar. Drugs*, 2014, **12**, 851–870.
280. T. Sun, Y. S. Zhang, B. Pang, D. C. Hyun, M. Yang and Y. Xia, Engineered nanoparticles for drug delivery in cancer therapy, *Angew. Chem., Int. Ed.*, 2014, **53**, 12320–12364.
281. E. Blanco, H. Shen and M. Ferrari, Principles of nanoparticle design for overcoming biological barriers to drug delivery, *Nat. Biotechnol.*, 2015, **33**, 941–951.
282. C. Alves, J. Silva, S. Pinteus, H. Gaspar, M. C. Alpoim and L. M. Botana, *et al.*, From Marine Origin to Therapeutics, The Antitumor Potential of Marine Algae-Derived Compounds, *Front. Pharmacol*, 2018, **9**, 777.
283. J. Yasuhara-Bell and Y. Lu, Marine compounds and their antiviral activities, *Antiviral Res.*, 2010, **86**, 231–240.
284. D. J. Schaeffer and V. S. Krylov, Anti-HIV activity of extracts and compounds from algae and cyanobacteria, *Ecotoxicol. Environ. Saf.*, 2000, **45**, 208–227.
285. G. B. Elion, P. A. Furman, J. A. Fyfe, P. de Miranda, L. Beauchamp and H. J. Schaeffer, Selectivity of action of an antiherpetic agent, 9-(2-hydroxyethoxymethyl) guanine, *Proc. Natl. Acad. Sci. U. S. A.*, 1977, **74**, 5716–5720.
286. M. Privatdegarilhe and J. de Rudder, Effet de deux nucl'eosides de l'arabinose sur la multiplication des virus de l'herp'es et de la vaccine en culture cellulaire, *C. R. Hebd. Seances Acad. Sci.*, 1964, **259**, 2725–2728.

287. W. E. G. MüLler, A. Maidhof, R. K. Zahn, H. C. Schröder, M. J. Gasić and D. Heidemann, *et al.*, Potent antileukemic activity of the novel cytostatic agent avarone and its analogues in vitro and in vivo, *Cancer Res.*, 1985, **45**, 4822–4826.
288. W. E. G. Müller, C. Sobel, B. Diehl-Seifert, A. Maidhof and H. C. Schröder, Influence of the antileukemic and anti-human immunodeficiency virus agent avarol on selected immune responses in vitro and in vivo, *Biochem. Pharmacol.*, 1987, **36**, 1489–1494.
289. M. R. Boyd, K. R. Gustafson, J. B. McMahon, R. H. Shoemaker, B. R. O'Keefe and T. Mori, *et al.*, Discovery of cyanovirin-N, a novel human immunodeficiency virus-inactivating protein that binds viral surface envelope glycoprotein gp120, potential applications to microbicide development, *Antimicrob. Agents Chemother.*, 1997, **41**, 1521–1530.
290. D. Pech-Puch, J. Berastegui-Cabrera, M. Pérez-Povedano, H. Villegas-Hernández, S. Guillén-Hernández and B. Cautain, *et al.*, Antiviral and Antiproliferative Potential of Marine Organisms From the Yucatan Peninsula, Mexico, *Front. Mar. Sci.*, 2020, **7**, 607.
291. P. Nigam, Adverse reactions to cosmetics and methods of testing, *Indian J. Dermatol. Venereol. Leprol.*, 2009, **75**, 10–19.
292. J. M. Lucca, R. Joseph, Z. H. Al Kubaish, S. M. Al-Maskeen and Z. A. Alokaili, An observational study on adverse reactions of cosmetics, The need of practice the Cosmetovigilance system, *Saudi Pharm. J.*, 2020, **28**, 746–753.
293. B. L. Wubishet, Z. S. Sabe, H. Dibaba, D. Yadesa, B. Legesse and Z. Shewamene, *et al.*, Cosmetics utilization pattern and related adverse reactions among female university students, *Int. J. Pharm. Sci. Res.*, 2013, **4**, 997–1004.
294. A. I. Bilal, Z. Tilahun, E. D. Osman, A. Mulugeta, M. Shekabdulahi and D. F. Berhe, Cosmetics Use-Related Adverse Events and Determinants Among Jigjiga Town Residents, Eastern Ethiopia, *Dermatol. Ther.*, 2017, **7**, 143–153.
295. M. Norudin, M. A. Desnika and M. Rafi, Cosmetic usage in Malaysia, understanding the major determinants affecting the users, *Int. J. Bus. Soc. Sci.*, 2010, **1**, 273.
296. C. Di Giovanni, V. Arcoraci, L. Gambardella and L. Sautebin, Cosmetovigilance survey, Are cosmetics considered safe by consumers?, *Pharmacol. Res.*, 2006, **53**, 16–21.
297. S. David, C. S. Levi, L. Fahoum, Y. Ungar, E. G. Meyron-Holtz and A. Shpigelman, *et al.*, Revisiting the carrageenan controversy, Do we really understand the digestive fate and safety of carrageenan in our foods?, *Food Funct.*, 2018, **9**, 1344–1352.
298. M. L. Weiner and J. M. McKim, Comment on 'revisiting the carrageenan controversy, Do we really understand the digestive fate and safety of carrageenan in our foods?' by S. David, C. S. Levi, L. Fahoum, Y. Ungar, E. G. Meyron-Holtz, A. Shpigelman and U. Lesmes, *Food Funct.*, 2018, 9, 1, *Food Funct.*, 2019, **10**, 1760–1762.

299. Natural products Editors, *Chitosan Uses, Benefits & Side Effects*, Drugs.com Herbal Database, Nat. Prod., 2010, https//www.drugs.com/npc/chitosan.html, accessed 2 October 2022.
300. N. Moolphuerk, T. Lawson and W. Pattanagul, Chitosan mitigates the adverse effects and improves photosynthetic activity in rice (Oryza sativa L.) seedlings under drought condition, *J. Crop Improv.*, 2022, **36**, 638–655.
301. T. Brendler and E. M. Williamson, Astaxanthin, How much is too much? A safety review, *Phytother. Res.*, 2019, **33**, 3090–3111.
302. J. Vega, G. Schneider, B. R. Moreira, C. Herrera, J. Bonomi-Barufi and F. L. Figueroa, Mycosporine-like amino acids from red macroalgae, Uv-photoprotectors with potential cosmeceutical applications, *Appl. Sci.*, 2021, **11**, 5112.
303. R. Zelig and D. R. Radler, Understanding the properties of common dietary supplements, Clinical implications for healthcare practitioners, *Nutr. Clin. Pract.*, 2012, **27**, 767–776.

CHAPTER 11

# Cosmeceutical Applications of Clay Minerals

ARUNIMA KALITA[a], ARUP GIRI[b], NITISH KUMAR[c] AND PUSHPENDER BHARDWAJ[c*]

[a]CSIR – Institute of Microbial Technology, Sector-39A, Chandigarh 160036, India; [b]Department of Zoology, Baba Mastnath University, Rohtak, Haryana 124021, India; [c]Medicinal Plant Division, Defence Institute of High Altitude Research, Chandigarh 160002, India
*E-mail: pbshanw06@gmail.com

## 11.1 Introduction

Clay minerals are a group of hydrous aluminum silicates with a layered crystal structure. They are one of the most abundant minerals in the Earth's crust and are formed through weathering and alteration of rocks.[1,2] Clay minerals have a significant impact on many important geological processes, including soil formation, geotechnical properties of soil and rocks, and the transport of water and nutrients in the soil.[3,4] They also have numerous industrial applications,[5] including use as a raw material in the production of ceramics, catalysts and drilling muds,[6] and as a component of some cosmetic products.[7–9] Clay minerals are characterized by their small particle size, high surface area, and ability to absorb water, which gives them a plastic consistency when wet.[1,10] The different types of clay minerals include kaolinite, illite, smectite, and chlorite, which can be distinguished based on their chemical composition, crystal structure, and physical properties.[11,12]

Kaolinite is a white or light-colored mineral that is composed of aluminum silicate and water, making it a versatile material with potential applications

Bioprospecting of Natural Sources for Cosmeceuticals
Edited by Deepika Kathuria, Ajay Sharma, Meenakshi Verma and Gulzar Ahmad Nayik

Published by the Royal Society of Chemistry, www.rsc.org

in drug delivery and biomedicine.[13] It has a well-ordered crystal structure and is the most common type of clay mineral found in soils. Kaolinite particles can be functionalized with drug molecules or other active agents, and can be designed to release the drug at a controlled rate over time.[14] This controlled release of drugs can improve the efficacy of treatments and reduce side effects by avoiding peak concentrations of the drugs in the body. In addition, kaolinite has also been investigated for its potential use in biomedicine. Its small particle size, high surface area, and ability to absorb water make it a useful material for wound healing, as it can absorb wound exudates and promote wound healing. Kaolinite has also been studied for its potential use as a tissue engineering scaffold, as it has a porous structure that can support cell growth and tissue regeneration. Furthermore, kaolinite has demonstrated antimicrobial activity against a range of bacteria,[15,16] making it a potential candidate for use in antimicrobial coatings and wound dressings. It has also shown anti-inflammatory[17] and antioxidant properties,[18] suggesting that it may have potential as a therapeutic agent in various diseases.

Illite is another common clay mineral, typically found in shale and mudrocks. It has a flaky crystal structure and is composed of aluminum, silicon, magnesium, and iron.[19] Illite is used in the production of drilling muds and as a component in some cosmetic products. In pharmaceuticals, illite has been explored as a drug delivery system. Its ability to absorb large amounts of water makes it a suitable candidate for hydrophilic drug delivery, while its flaky structure can provide a high surface area for drug adsorption. Illite particles can be functionalized with drugs or other active agents, and their release can be controlled by adjusting their particle size and surface properties. In biomedicine, illite has demonstrated a range of biological activities, including antibacterial, anti-inflammatory, and antioxidant effects.[20,21] It has been shown to have potential as a therapeutic agent in various diseases, including cancer and skin diseases, due to its ability to selectively target specific cell types. Illite has also been explored as a material for wound healing due to its ability to absorb wound exudates and promote wound healing. Its antimicrobial activity against a range of bacteria also makes it a potential candidate for use in antimicrobial coatings and wound dressings.

However, smectite is a type of clay mineral with a highly expandable crystal structure that allows it to absorb large amounts of water. It is commonly found in volcanic ash and bentonite deposits. Smectite is used in a variety of industrial applications, including drilling muds, catalysts, and as a component in some pharmaceuticals and cosmetics.[22,23]

Clay minerals have been used for centuries due to their various therapeutic properties and are still used today as natural remedies for various health conditions. Clay minerals have also been used for centuries for their ability to absorb toxins and heavy metals from the body.[1,24] When ingested or applied topically, clay minerals can bind to toxic substances and help remove them from the body. However, clay minerals have been shown to have a soothing effect on the digestive system and can help alleviate digestive disorders such as acid reflux and irritable bowel syndrome. Meanwhile, clay minerals are commonly used in cosmetics and skincare products due to their ability to absorb oils and impurities from the skin, making them an effective ingredient in face masks and other skin treatments. They have antimicrobial properties and can help promote wound healing by reducing inflammation and promoting tissue regeneration. Clay minerals have been shown to have anti-inflammatory properties, which makes them useful for treating various inflammatory conditions, such as rheumatoid arthritis and osteoarthritis. It's important to note that the therapeutic properties of clay minerals can vary depending on the type of clay mineral, as well as its source and purity. Additionally, not all clays have the same properties or benefits, and some may contain impurities that can be harmful to health. It's important to consult with a healthcare provider before using clay minerals for therapeutic purposes, especially if you have any underlying health conditions or are taking any medications.

However, fibrous clays are a type of clay mineral that have received attention in the fields of dermopharmaceuticals and cosmetics due to their unique properties. These clays are characterized by their fibrous structure and high aspect ratio, which gives them a large surface area and high adsorption capacity. In traditional dermopharmaceutical and cosmetic applications, fibrous clays have been used as a base material for various topical formulations, including creams, lotions, and ointments.[25,26] Their ability to absorb oils and impurities from the skin makes them an effective ingredient in face masks, cleansers, and other skin treatments.

In recent years, fibrous clays have been explored for their potential as a drug delivery system in dermopharmaceuticals. Their high surface area and adsorption capacity make them a suitable candidate for the delivery of active agents, such as vitamins, minerals, and antioxidants. The release of these agents can be controlled by adjusting the particle size and surface properties of the fibrous clay particles. Fibrous clays have also been explored for their potential as a wound healing agent in cosmetic applications. Their ability to absorb wound exudates and promote tissue regeneration, as well as their antimicrobial properties, make them a promising candidate for use in wound dressings and other wound care products.[27]

Clay minerals have been used in spas for centuries due to their unique properties and benefits for the skin. They can be used in a variety of treatments, including body wraps, facials, baths, and massages, to help improve the appearance and health of the skin and to provide a relaxing and rejuvenating spa experience. Clay minerals have also been used in aesthetic medicine for a range of purposes, including skin rejuvenation, the treatment of skin conditions, exfoliation, and detoxification. However, the use of clay minerals in aesthetic medicine should be done under the guidance of a healthcare professional to ensure the safety and efficacy of the treatment.

## 11.2 History of Clay Mineral Use as Cosmetics

Clay minerals have been used as cosmetics for thousands of years, dating back to ancient civilizations such as the Egyptian, Greek, and Roman. The use of clay minerals in cosmetics was based on their unique properties and benefits for the skin, and they have been used for a range of purposes, including cleansing, exfoliation, healing, and detoxification. In ancient Egypt, clay minerals were used as part of the beauty regimen of Cleopatra, who was known for her radiant skin. Clay minerals were also used in ancient Greece and Rome as part of daily bathing rituals, as they were believed to have therapeutic benefits for the skin. In more recent history, clay minerals have been used in a variety of cosmetic products, including face masks, soap bars, and body treatments. They have also been used as ingredients in spa treatments and other forms of aesthetic medicine, due to their benefits for the skin and their ability to promote relaxation and rejuvenation. The use of clay minerals as cosmetics dates back thousands of years, and they have been used for a range of purposes, including cleansing, exfoliation, healing, and detoxification. Today, they continue to be used in a variety of cosmetic products and treatments, and they remain popular due to their unique properties and benefits for the skin.[28–30]

Clay minerals have been used in cosmetics for centuries, due to their unique properties and benefits for the skin. Clay minerals can be used to cleanse the skin, removing impurities and excess oil, and leaving the skin feeling refreshed and revitalized. They can be used in face masks, soap bars, and other cosmetic products. They can be used to exfoliate the skin, removing dead skin cells and promoting the growth of new skin cells. This can help to improve the overall appearance and health of the skin. Clay minerals have also been used for centuries to soothe and heal skin conditions, including eczema, psoriasis, and acne. They can be used topically in the form of creams, lotions, and gels, or they can be incorporated into facial and body treatments. Clay minerals are known for their ability to absorb impurities and toxins from the skin, and they have been used in cosmetics to help detoxify the skin and promote the elimination of toxins from the body. Therefore, clay minerals have a long history of use in cosmetics, due to their unique properties and benefits for the skin. They can be used for cleansing, exfoliation, healing and soothing, and absorption and detoxification, and can be incorporated into a range of cosmetic products, including face masks, soap bars, and body treatments. However, the use of clay minerals in cosmetics can vary depending on the type of clay mineral and its source, as well as the individual's skin type and the desired outcome.

However, clay minerals have been used in spas for centuries due to their unique properties and benefits for the skin. They can be used in a variety of treatments, including body wraps, facials, baths, and massages, to help improve the appearance and health of the skin and to provide a relaxing and rejuvenating spa experience. It has also been found that clay minerals have been used in aesthetic medicine for a range of purposes, including skin rejuvenation, the treatment of skin conditions, exfoliation, and detoxification. However, the use of clay minerals in aesthetic medicine should be done under the guidance of a healthcare professional to ensure the safety and efficacy of the treatment.

## 11.3 Uses of Clay Minerals in Cosmeceuticals

The use of clays in cosmetics has recently increased due their characteristic benefits in the generation of desired and useful technological properties.[31] The abundant and economic viability of clay minerals has increased the interest in the recent cosmetic industry.[32] Apart from that, clay in cosmetics imparts cosmetological properties when applied. The important technological characteristics of clay that improve the quality of the product are the emulsion stability and the viscosity of suspensions.[33] Clay possesses unique surface properties such as surface area, layer charge, cation exchange capacity *etc.*, and mechanical and physical properties including color, opacity, particle size, shape, reflectance, softness, iridescence, *etc.*[22,34,35] Clays also impart specific functions that include solar protection, cleaning of the hair and skin, and water-loss reduction.[7] The distinctive rheological properties of clay, such as plasticity, thixotropy, viscosity and rheopexy, enhance the product quality.[22] Thus, these attributes of clay make it a potent ingredient in the manufacturing of cosmetics. The ubiquitous nature of clay, availability, cheap cost, and vital intrinsic characteristics derived from the structure render it widely used in products and formulations for different purposes.

Clay minerals are naturally occurring crystalline materials composed chemically of hydrated aluminum silicates, with magnesium, calcium, iron, sodium or potassium present as essential constituents. Clay also contains organic compounds, quartz particles, soluble salts, calcite, pyrite, other non-clay minerals and amorphous components.[32]

The classification of clays and clay minerals has been proposed and further updated by the nomenclature committee of AIPEA (Association Internationale pour l'Étude des Argiles – the International Association for the Study of Clay) in 2006.[36] One or more clays can be present as the key ingredients in the cosmetic formulations. Each clay mineral or clay possesses unique therapeutic or cosmetic functions, such as wound healing, oil-absorbing skin lightening, moisturization (Figure 11.1), antisepsis, embellishment, vascularization and toxin elimination.[37] Among them, the most widely used clays include talc, mica, kaolin, bentonite, and smectites, in addition to other clays in formulations with enhanced functions.

The major clays used in haircare products, such as shampoos, hair masks *etc.*, include talc and kaolinite due to their oil-absorbing capacity. The enhanced specific surface area of talc and kaolinite promotes them as important ingredients in these formulations. Their properties, such as high viscosity and thixotropy, are desired in several cosmetic formulations. Talc is one of the most used cosmetic ingredients with >3000 related patents. The distinctive structure and properties of talc with a large surface area make it a potent constituent in makeup preparations, acting as an emulgent. Talc improves the fluidity and the texture of powdery mixtures of solids by enhancing the softness. Studies have revealed that talc acts as a color booster and improves the coverage and cohesion in skin powders.[38,39] The other formulations that include talc are aerosol-moisturizing formulations mixed

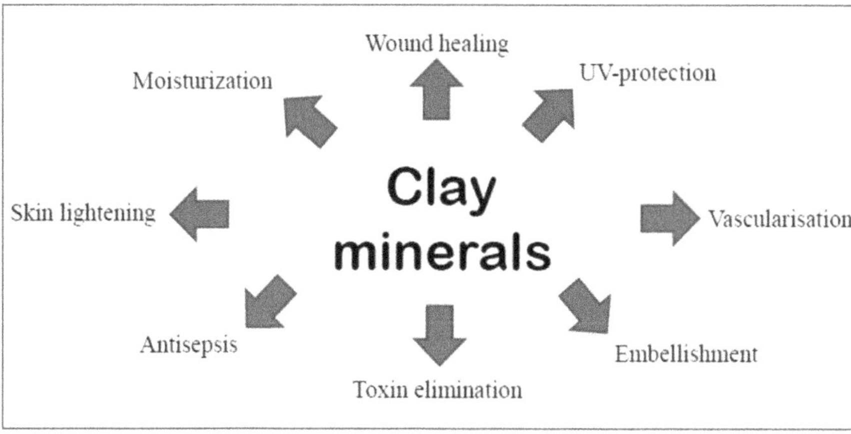


**Figure 11.1** Uses of clay minerals.

with butane.[40] Apart from this, the skin protection and opacity effects of talc are utilized in cosmetic formulations.[41] The abrasive and adsorptive features of talc make it one of the key components in shaving formulations.[42,43] Synthetic nanoparticulate talc combined with nanoparticulate silica has been reported to be a filler for polymers that is able to adsorb UV-protective active compounds found in formulations for sun protection.[44,45] The combination of talc and other minerals, such as mica, shows synergistic effects upon modification.[46,47]

Kaolin has been used in 1700 cosmetic patents. In recent years, kaolin has been used in powder masks and different makeup formulations, as well as in makeup removal products.[48–50] However, synthetic kaolin has also been reported to be used in color masks by the Sekisui Chemical Co., Ltd.[51] The effectiveness of administration of certain active ingredients with kaolinite and montmorillonite as a carrier has been reported to improve acne symptoms.[52] Mixing kaolinite or montmorillonite with aloe vera has been developed as a peel-off mask gel formulation.[53,54] Studies have revealed that leaf ethanolic extracts of *Litchi chinensis* mixed with kaolin to form an emulsion could prevent UVB-induced damage.[55]

Other clays such as dioctahedral and trioctahedral smectites have also been used in various cosmetic ingredients. Both dioctahedral and trioctahedral smectites impart optimal light dispersion and high refractive index properties. These unique properties makes it an ideal ingredient in sunscreen formulations acting as a shield from harmful solar radiation and protecting against UV-induced DNA damage.[56,57] Hectorite in both raw and organically modified forms has been used in different cosmetic formulations as a gel-forming agent, bulking additive, and rheology modifier to enhance the quality of the product. Lipid-based formulations such as lipsticks contain stearalkonium hectorite.[58]

Dioctahedral smectites including montmorillonite, nontronite and beidellite, are important ingredients in the formulations of skincare products because of their enhanced absorption/adsorption capacity. They are extensively used in facial mask formulations as they help to remove dirt, toxins, sebum and dead cells. Clay face masks are desired because of their diverse properties, including cleansing, exfoliating, astringent, and lifting effects. Utilizing the rheological behavior of smectites in skin mask formulations helps to achieve pseudoplastic properties desirable for easy application. The optimization of the smectite and aloe vera composition has been studied for facial and peel-off mask formulations.[53] Smectites are also reported to be used as a natural eye shadow and blush powder, as well as in facial makeup.[59,60]

The characteristic properties of stevensite and saponite as a cleanser have been explored for centuries as a natural shampoo and soap. The inherent properties of stevensite and saponite have been utilized as the ingredient in the formulations of various commercially available body/skin and haircare products, such as Moroccan black soap© (Savon Stories®), the Himalayan charcoal purifying glow mask© (The Body Shop®) *etc.*

The emulsification ability of bentonite has been extensively used in oil-in-water makeup (O/W), nail enamel remover, cleansing lotions and low oil-content creams.[61] A recent study reported the formulation of face masks by combining bentonite, polyvinyl alcohol and starch.[62] Quaternium-18 bentonite has also been reported to be used in the formulation eyeliner, eye lotion, eyeshadow, eyebrow pencils, and other eye makeup preparations and aerosolized deodorant formulations, and also in lipstick. A recent study showed the induction of telomerase activity of bentonite and proposed it to be a potent ingredient in anti-aging cosmetic formulations.[63]

Non-planar clay minerals, such as sepiolite (magnesium trisilicate), palygorskite (attapulgite) and halloysite, have also been used in various cosmetics formulations as carriers that absorb UV radiation.[64,65] In recent years, both natural and organically modified palygorskite have been used in suncare products and makeup formulations.[66,67]

The use of halloysite in haircare cosmetic formulations is due to the effective loading of the dye that reduces the binding of the dye with the scalp, thus helping to minimize the toxicity and possible allergic reactions.[68–70] The main user of halloysite and sepiolite as the ingredients in cosmetics is L'Oréal® and the major user of palygorskite is Procter & Gamble®.

Laponite® is a clay-like structure made synthetically that is used in cosmetics due to its ability to exchange. Laponite® has been used in two patented formulations due to its ability of delivering incompatible ingredients in a single formulation by Avon®.[71,72] A list of a few clay minerals commonly used in cosmetic formulations is given in Table 11.1 along with their functions.

**Table 11.1** Clays and clay minerals used as cosmetic ingredients.

| Minerals | INCI[a] name | Functions in formulation |
|---|---|---|
| Kaolinite | Kaolin | Abrasive, colorant, opacifying, anti-caking, bulking, absorbent |
| Kaolinite and smectite | Fuller's earth | Abrasive, opacifying, bulking, absorbent |
| Pyrophyllite | Pyrophyllite | Abrasive, opacifying, absorbent |
| Talc | Talc | Skin protective, bulking, anti-caking, absorbent, opacifying |
| Modified talc | Magnesium/potassium/silicon/fluoride/hydroxide/oxide/ceria/silica talc | Bulking, UV absorber |
| Mica | Mica | Anti-caking, opacifying, colorant, bulking |
| Sepiolite | Sepiolite | Viscosity-controlling |
| Palygorskite | Attapulgite | Absorbent, abrasive, opacifying, viscosity, bulking |
| Laponite | — | Emulsion stabilizer, thickening agent, viscosity-controlling and filler, gelling agent |
| Mineral composition not specified | Clay | Absorbent, binding, skin conditioner, viscosity-controller |
| | Clay extract | Skin conditioner |
| Dioctahedral smectites | Hectorite | Absorbent, antistatic, gel forming, bulking, viscosity-controller |
| Stevensite | Moroccan lava clay | Absorbent and hair cleanser |
| Trioctahedral smectites Montmorillonite | Bentonite | Absorbent, emulsion stabilizer, bulking, viscosity-controller |

[a]INCI = International Nomenclature Cosmetic Ingredient.

## 11.4 Cosmeceutical-based Formulations and Marketed Products that Use Clay Minerals

Clay minerals are commonly used in cosmeceutical products as they are known to have a wide range of benefits for the skin. Some of the most commonly used clay minerals in cosmeceuticals include bentonite clay, kaolin clay, activated charcoal, white kaolin clay, Amazonian white clay, Irish moor mud, Moroccan clay, and Spanish clay (see Table 11.2).

Aztec Secret's "Indian Healing Clay" is a popular face mask that uses bentonite clay to unclog pores, detoxify the skin, and reduce acne breakouts.[73] L'Oréal Paris' "Pure Clay Mask" uses kaolin clay to purify and mattify the skin while minimizing the appearance of pores.[81] GlamGlow's "SUPERMUD® Clearing Treatment" combines activated charcoal and kaolin clay to unclog pores, exfoliate the skin, and reduce the appearance of acne and blemishes.[77] Mary Kay's "Clear Proof Deep-Cleansing Charcoal Mask" also uses activated charcoal and kaolin clay to absorb excess oil, impurities, and environmental

**Table 11.2** Formulations and marketed products that use clay minerals.

| Product name (manufacturer) | Clay mineral(s) | Description | Reference |
|---|---|---|---|
| Indian healing clay (Aztec Secret Health & Beauty) | Bentonite | A facial mask made from 100% natural calcium bentonite clay to deep cleanse the skin | 73 |
| Charcoal pore pudding intensive wash-off treatment (Boscia) | Activated charcoal, kaolin clay | A face mask that claims to detoxify and balance the skin while reducing the appearance of pores and blackheads | 74 |
| Goddess skin clay mask (Charlotte Tilbury) | Spanish clay | A face mask that claims to smooth, brighten, and hydrate the skin while reducing the appearance of pores and imperfections | 75 |
| Umbrian clay pore purifying face mask (Fresh) | Umbrian clay | A face mask made from Umbrian clay to absorb impurities and clarify the skin | 76 |
| SUPERMUD® clearing treatment (GlamGlow) | Activated charcoal, kaolin clay | A face mask that claims to unclog pores, exfoliate the skin, and reduce the appearance of acne and blemishes | 77 |
| Jeju volcanic pore clay mask (Innisfree) | Volcanic ash | A face mask made from Jeju volcanic ash to absorb sebum and purify pores | 78 |
| Rare earth deep pore cleansing masque (Kiehl's) | Amazonian white clay | A face mask containing Amazonian white clay to detoxify and minimize pores | 79 |
| Effaclar clay mask (La Roche-Posay) | Kaolin | A face mask containing kaolin clay to absorb excess sebum and unclog pores | 80 |
| Pure clay cleanser (L'Oréal Paris) | Kaolin, montmorillonite, Moroccan lava clay | A facial cleanser containing three different types of clay minerals to detoxify and clarify the skin | 81 |
| Clear Proof® deep-cleansing charcoal mask (Mary Kay) | Activated charcoal, kaolin and bentonite clays | A face mask that claims to absorb excess oil, impurities, and environmental pollutants from the skin while minimizing the appearance of pores | 82 |
| Irish moor mud purifying black mask (Peter Thomas Roth) | Irish moor mud | A face mask that claims to purify, detoxify, and hydrate the skin while minimizing the appearance of fine lines and wrinkles | 83 |
| Ceramic slip cleanser (Sunday Riley) | French green clay | A facial cleanser containing French green clay to gently exfoliate and remove impurities | 84 |
| Purifying mask (Tata Harper) | Kaolin clay | A face mask containing Kaolin clay to detoxify and clarify the skin | 85 |

pollutants from the skin while minimizing the appearance of pores.[82] Boscia's "Charcoal Pore Pudding Intensive Wash-Off Treatment" uses activated charcoal and kaolin clay to detoxify and balance the skin while reducing the appearance of pores and blackheads.[74]

Other examples of cosmeceutical products that use clay minerals include Fresh's "Umbrian Clay Pore Purifying Face Mask" that uses Umbrian clay to purify and balance the skin while reducing the appearance of pores and blemishes.[76] Kiehl's "Rare Earth Deep Pore Cleansing Masque" uses Amazonian white clay to detoxify and exfoliate the skin while minimizing the appearance of pores and oiliness.[79] Peter Thomas Roth's "Irish Moor Mud Purifying Black Mask" uses Irish moor mud to purify, detoxify, and hydrate the skin while minimizing the appearance of fine lines and wrinkles.[83] Charlotte Tilbury's "Goddess Skin Clay Mask" uses Spanish clay to smooth, brighten, and hydrate the skin while reducing the appearance of pores and imperfections.[75] Overall, these cosmeceutical products that use clay minerals provide a range of benefits for the skin, from detoxification and exfoliation to purifying and balancing, and are popular among those seeking natural and effective skincare solutions.

## 11.5 Conclusions and Future Perspectives

Clay minerals, particularly montmorillonite and kaolin, have been commonly used in cosmetic and personal care products due to their unique properties, such as adsorption, absorption, and ability to regulate oil production. These properties make them effective in various skin care applications, including detoxification, exfoliation, and treatment of various skin conditions such as acne, rosacea, and eczema. Clay minerals have also been found to have anti-inflammatory and antioxidant effects, making them beneficial for skin health. However, more research is needed to fully understand their mechanisms of action and potential side effects. It is important to choose a high-quality and pure product, and to patch test before using it on a larger area of skin. Overall, the use of clay minerals in cosmetics has shown promising results and can be a useful addition to a skin care routine.

The use of clay minerals as cosmetics is expected to continue to grow in the future, due to the increasing demand for natural and environmentally-friendly cosmetic products, as well as the growing awareness of the benefits of clay minerals for the skin. One of the key future perspectives of clay mineral use in cosmetics include increased demand for natural cosmetics. As consumers become more conscious of the ingredients used in cosmetics, there is expected to be increased demand for natural and organic products that are free from harmful chemicals and preservatives. Clay minerals are a natural and safe ingredient that can be used in cosmetics, and they are expected to continue to be in high demand. Meanwhile, the development of new cosmetic products that utilize clay minerals is expected to continue, as manufacturers look for ways to improve the benefits and performance of their products. Clay minerals can be used in a range of cosmetic products, including face masks, soaps, lotions, and gels, and there is potential for new products to be developed that incorporate these minerals in innovative ways. However, the use of clay minerals in cosmetics is expected to expand into new markets, as consumers become more aware of the benefits of these minerals and their unique properties. Clay minerals are becoming increasingly popular in developing countries, and there is potential for growth in these markets in the future. As consumers become more conscious of the environmental impact of cosmetics, there is expected to be a greater focus on sustainability practices in the industry. Clay minerals are a sustainable and environmentally-friendly ingredient that can be used in cosmetics, and their use is expected to increase as the industry moves towards more sustainable practices. Therefore, the use of clay minerals as cosmetics is expected to continue to grow in the future, due to the increasing demand for natural and environmentally-friendly products, the development of new cosmetic products, the expansion into new markets, and the improved sustainability practices of the industry. Clay minerals have unique properties and benefits for the skin, and they are expected to play an important role in the future of cosmetics.

## References


1. M. Nadziakiewicza, S. Kehoe and P. Micek, Physico-chemical properties of clay minerals and their use as a health promoting feed additive, *Animals*, 2019, **9**, 714.
2. L. Dzene, E. Ferrage and J. C. Viennet, *et al.*, Crystal structure control of aluminized clay minerals on the mobility of caesium in contaminated soil environments, *Sci. Rep.*, 2017, **7**, 1–12.
3. G. O. Akintola, F. Amponsah-Dacosta and S. E. Mhlongo, Geotechnical evaluation of clayey materials for quality burnt bricks, *Heliyon*, 2020, **6**, e05626.
4. S. Ghosh, J. Mukhopadhyay and A. Chakraborty, Clay mineral and geochemical proxies for intense climate change in the permian gondwana rock record from eastern India, *Research*, 2019, **2019**, 8974075.
5. F. Bergaya and G. Lagaly, Chapter 1 General Introduction: Clays, Clay Minerals, and Clay Science, *Dev. Clay Sci.*, 2006, **1**, 1–18.
6. H. H. Murray, Traditional and new applications for kaolin, smectite, and palygorskite: a general overview, *Appl. Clay Sci.*, 2000, **17**, 207–221.
7. C. Viseras, E. Carazo and A. Borrego-Sánchez, *et al.*, Clay minerals in skin drug delivery, *Clays Clay Miner.*, 2019, **67**, 59–71.
8. I. S. Khurana, S. Kaur, H. Kaur and R. K. Khurana, Multifaceted role of clay minerals in pharmaceuticals, *Future Sci. OA*, 2015, **1**, FSO6.
9. T. A. Ho, L. J. Criscenti and J. A. Greathouse, Revealing Transition States during the Hydration of Clay Minerals, *J. Phys. Chem. Lett.*, 2019, **10**, 3704–3709.
10. N. Kumari and C. Mohan, Basics of clay minerals and their characteristic properties, *Clays Clay Miner.*, 2021, **24**, 1–29.

11. Q. Wang and Y. Song, Clay minerals and major elements concentrations of Zhuanglang Miocene red clay in Longzhong Basin, China, *Data Brief*, 2018, **17**, 297–304.
12. F. Deon, F. van Ruitenbeek and H. van der Werff, *et al.*, Detection of Interlayered Illite/Smectite Clay Minerals with XRD, SEM Analyses and Reflectance Spectroscopy, *Sensors*, 2022, **22**, 3602.
13. E. Maris, P. Botané, P. Wavrer and D. Froelich, Characterizing plastics originating from WEEE: A case study in France, *Miner. Eng.*, 2015, **76**, 28–37.
14. J. Dong, Z. Cheng, S. Tan and Q. Zhu, Clay nanoparticles as pharmaceutical carriers in drug delivery systems, *Expert Opin. Drug Delivery*, 2021, **18**, 695–714.
15. M. Isah, M. H. Asraf and N. A. N. N. Malek, *et al.*, Preparation and characterization of chlorhexidine modified zinc-kaolinite and its antibacterial activity against bacteria isolated from water vending machine, *J. Environ. Chem. Eng.*, 2020, **8**, 103545.
16. S. N. Ishak, N. A. N. N. Malek and Z. Yusop, *et al.*, Evaluation of phase transformation behaviors of zeolite and antibacterial properties against Gram-positive and -negative bacteria, *J. Chin. Chem. Soc.*, 2020, **67**, 2042–2049.
17. M. Hovaneţ, R. Viorel Ancuceanu and M. Dinu, *et al.*, Toxicity and anti-inflammatory activity of Ziziphus jujuba mill. Leaves, *Farmacia*, 2016, **64**, 5.
18. I. Yellapurkar, S. Shaikh and G. Pavale, *et al.*, Kaolin-catalysed one-pot synthesis of thiophene containing aminonaphthols under solvent-free condition and their *in vitro* anticancer and antioxidant activity, *Res. Chem. Intermed.*, 2021, **47**, 4067–4082.
19. B. N. Hupp and J. J. Donovan, Quantitative mineralogy for facies definition in the Marcellus Shale (Appalachian Basin, USA) using XRD-XRF integration, *Sediment. Geol.*, 2018, **371**, 16–31.
20. M. A. Nilforoushzadeh, M. A. Amirkhani and P. Zarrintaj, *et al.*, Skin care and rejuvenation by cosmeceutical facial mask, *J. Cosmet., Dermatol.*, 2018, **17**, 693–702.
21. H. Oh, J. S. Lee and H. S. Lee, *et al.*, A Novel Polyvinylpyrrolidone-Stabilized Illite Microparticle with Enhanced Antioxidant and Antibacterial Effect, *Polymers*, 2021, **13**, 4275.
22. J. D. D. Moraes, S. R. A. Bertolino and S. L. Cuffini, *et al.*, Clay minerals: Properties and applications to dermocosmetic products and perspectives of natural raw materials for therapeutic purposes-A review, *Int. J. Pharm.*, 2017, **534**, 213–219.
23. C. dos Santos and A. de J. Santos, Chemical, Mineralogical, and Rheological Characterization of Regional Clays for Potential Use in Cosmetics and Pharmaceutical Products, *Clays Clay Miner.*, 2021, 61–77.
24. M. K. Uddin, A review on the adsorption of heavy metals by clay minerals, with special focus on the past decade, *Chem. Eng. J.*, 2017, **308**, 438–462.

25. L. Cao, W. Xie and H. Cui, *et al.*, Fibrous Clays in Dermopharmaceutical and Cosmetic Applications: Traditional and Emerging Perspectives, *Int. J. Pharm.*, 2022, **625**, 122097.
26. C. Viseras, C. Aguzzi and P. Cerezo, Medical and health applications of natural mineral nanotubes, *Natural Mineral Nanotubes: Properties and Applications*, 2015.
27. J. Cervini-Silva, A. Nieto-Camacho and M. T. Ramírez-Apan, *et al.*, Anti-inflammatory, anti-bacterial, and cytotoxic activity of fibrous clays, *Colloids Surf., B*, 2015, **129**, 1–6.
28. K. L. Konkol and S. C. Rasmussen, An ancient cleanser: Soap production and use in antiquity, *ACS Symp. Ser.*, 2015, **1211**, 245–266.
29. C. S. F. Gomes and M. Rautureau, in *Historical Evolution of the Use of Minerals in Human Health BT - Minerals latu sensu and Human Health: Benefits, Toxicity and Pathologies*, ed. C. Gomes and M. Rautureau, Springer International Publishing, Cham, 2021, pp. 43–79.
30. R. Ion, R. Fierăscu and S. Teodorescu, *et al.*, Ceramic materials based on clay minerals in cultural heritage study, *Clays, Clay Miner Ceram Mater Based Clay Miner*, 2016.
31. C. Viseras, R. Sánchez-Espejo and R. Palumbo, *et al.*, Clays in cosmetics and personal-care products, *Clays Clay Miner.*, 2021, **69**, 561–575.
32. S. M. Auerbach, K. A. Carrado and P. K. Dutta, *Handbook of Zeolite Science and Technology*, 2003.
33. C. Viseras, C. Aguzzi, P. Cerezo and A. Lopez-Galindo, Uses of clay minerals in semisolid health care and therapeutic products, *Appl. Clay Sci.*, 2007, **36**, 37–50.
34. A. López-Galindo, C. Viseras and P. Cerezo, Compositional, technical and safety specifications of clays to be used as pharmaceutical and cosmetic products, *Appl. Clay Sci.*, 2007, **36**, 51–63.
35. M. I. Carretero and M. Pozo, Clay and non-clay minerals in the pharmaceutical and cosmetic industries Part II. Active ingredients, *Appl. Clay Sci.*, 2010, **47**, 171–181.
36. S. Guggenheim, J. M. Adams and D. C. Bain, *et al.*, Summary of recommendations of nomenclature committees relevant to clay mineralogy: Report of the Association Internacionale pour l'Etude des Argiles (AIPEA) nomenclature committee for 2006, *Clays Clay Miner.*, 2006, **54**, 761–772.
37. C. de S. F. Gomes and J. B. P. Silva, Minerals and clay minerals in medical geology, *Appl. Clay Sci.*, 2007, **36**, 4–21.
38. C. Dumousseaux, Powdered anhydrous composition, useful formakeup and/or care of keratinmaterial e.g. skin, comprises solid ground cosmetic active agent e.g. moisturizing agent, effervescent system and mineral filler e.g. talc, *FR pat.*, FR2958156A1(B1), 2011.
39. G. Meli and L. Pagis, Cosmetic compositions comprising a talc particulate, *EP pat.*, EP3270871A1(B1), European Patent Office, 2018.
40. S. Kang, C. Park, M. Park and H. Yun, Cosmetic composition comprising talc and butane for moisturizing and refreshing, *KR pat.*, KR100602144B1(A), Korean Intellectual Property Office, 2004.

41. T. Kaida, Y. Murui, Y. Osawa and F. Suzuki, Interferencecolor flaky talc and cosmetic prepared by compounding it, *JP pat.*, JP2004339185A, Korean Intellectual Property Office, 2004.
42. H. Albrecht and R. Stelzer, Aqueous cosmetic composition useful as a shaving or cleansing agent contains fatty acids or salts, emulsifiers, complexers, waxes and talc, *DE pat.*, DE0221812A1, Danish Patent and Trademark Office, 2003.
43. M. Kohut and S. Ruppert, Talc-containing cosmetic cleansing emulsion, *EP pat.*, EP1458350A2, European Patent Office, 2004.
44. R. Wittmer, Production of nanoparticulate synthetic talc, useful as a filler for polymers and in pharmaceutical and cosmetic compositions, comprises using nanoparticulate silica, *DE pat.*, DE10125879A1, The German Patent and Trade Mark Office, 2002.
45. D. Arseguel, H. Balard and F. Clauss, Attaching quinone or indole compounds to talc, useful e.g. as ultra-violet absorbers in cosmetics, by reductive solubilization, application to talc and reoxidation, *FR pat.*, FR2792322A1, French Patent Office, 2000.
46. C. Hechavarria, Cosmetic powder containing talc and mica, *NZ pat.*, NZ248534A, Intellectual Property Office of New Zealand, 1994.
47. M. Iida and H. Nishimura, Modified talc and cosmetic, *JP pat.*, JPH10114514A, Japan Patent Office, 1998.
48. Y. Kim, The cosmetics materials which uses the kaoline and manufacturing method of skin beauty care kaolin-powder pack, kaolin-powder mask pack, *KR pat.*, KR20070083251A, Korean Intellectual Property Office, 2007.
49. C. H. Kim and J. G. Kim, Cosmetic composition based on kaolin, *KR pat.*, KR20120008659A, Korean Intellectual Property Office, 2012.
50. S. Koely-Therouin and J.-L. Mattei, Solid, warming, kaolin containing cosmetic composition, and use thereof for make-up removal, *EP pat.*, EP1106164A1(B1), European Patent Office, 2001.
51. T. Nishi, Synthetic colored kaolin, production method therefor, and cosmetics using same, *Japan pat.*, JP2001172529A, Japan Patent Office, 2001.
52. A. R. Hamilton, M. Roberts, G. A. Hutcheon and E. E. Gaskell, Formulation and antibacterial properties of clay mineral-tetracycline and -doxycycline composites, *Appl. Clay Sci.*, 2019, **179**, 105148.
53. A. O. R. Beringhs, J. M. Rosa and H. K. Stulzer, *et al.*, Green Clay and Aloe Vera Peel-Off Facial Masks: Response Surface Methodology Applied to the Formulation Design, *AAPS PharmSciTech*, 2013, **14**, 445.
54. N. Asthana, K. Pal and A. A. Aljabali, *et al.*, Polyvinyl alcohol (PVA) mixed green-clay and aloe vera based polymeric membrane optimization: Peel-off mask formulation for skin care cosmeceuticals in green nanotechnology, *J. Mol. Struct.*, 2021, **1229**, 129592.
55. L. C. Thiesen, P. E. Bretzke and C. M. d. S. Bittencourt, *et al.*, Litchi chinensis leaf extract provides high *in vitro* photoprotection associated to a natural mineral clay, *Photodermatol., Photoimmunol. Photomed.*, 2020, **36**, 61–62.
56. M. Ghadiri, W. Chrzanowski, W. H. Lee and R. Rohanizadeh, Layered silicate clay functionalized with amino acids: wound healing application, *RSC Adv.*, 2014, **4**, 35332–35343.

57. M. Mattioli, L. Giardini, C. Roselli and D. Desideri, Mineralogical characterization of commercial clays used in cosmetics and possible risk for health, *Appl. Clay Sci.*, 2016, **119**, 449–454.
58. R. Mendoza, Cosmetic compositions, *EP pat.*, EP3166574A1(A4,B1), European Patent Office, 2015.
59. S. Gamoudi and E. Srasra, Green synthesis and characterization of colored Tunisian clays: Cosmetic applications, *Appl. Clay Sci.*, 2018, **165**, 17–21.
60. R. K. B. C. Cavalcanti, C. T. Brasileiro, R. O. Macedo and H. S. Ferreira, Mineral make up developed from natural and organophilic bentonite clays, *Ceramica*, 2018, **64**, 266–275.
61. S. K. Niazi, *Handbook of Pharmaceutical Manufacturing Formulations*, 2019.
62. S. Chakraborty, V. Anoop and N. George, *et al.*, Physicochemical stability evaluation of cosmetic formulations of PVA, starch and MMT clay nanocomposites, *SN Appl. Sci.*, 2019, **1**, 1–13.
63. K. Bae, B. Choi, Y. J. Jang, D. Kim and E. Lee, Cosmetic composition for preventing or improving skin anti-aging containing bentonite extract for increasing telomerase activity, *Korean pat.*, KR102228930B1, 2021.
64. Y. J. Suh and K. Cho, Immobilization of nanoscale sunscreening agents onto natural halloysite micropowder, *Mater. Trans.*, 2015, **56**, 899–904.
65. C. del Hoyo, M. A. Vicente and V. Rives, Preparation of drug-montmorillonite UV-radiation protection compounds by gas-solid adsorption, *Clay Miner.*, 2001, **36**, 541–546.
66. G. Cao, Y. Deng and R. Feng, *et al.*, Application of surface organic modified attapulgite emulsion in preparation of traditional Chinese medicine cosmetics, *Chinese Pat.*, CN111467250A, 2020.
67. S. Cho, M. Ju and D. S. Kil, Hybrid powder of halloysite nanotube-light scattering nanoparticles, preparation method thereof, and UV screening cosmetic composition containing same as active ingredient, International Patent System, *WO pat.*, WO2013047915A1, Patent Cooperation Treaty (PTC), 2013.
68. D. G. Shchukin, G. B. Sukhorukov, R. R. Price and Y. M. Lvov, Halloysite nanotubes as biomimetic nanoreactors, *Small*, 2005, **1**, 510–513.
69. Y. J. Suh, D. S. Kil and K. S. Chung, *et al.*, Natural nanocontainer for the controlled delivery of glycerol as a moisturizing agent, *J. Nanosci. Nanotechnol.*, 2011, **11**, 661–665.
70. S. Sadjadi, Halloysite-based hybrids/composites in catalysis, *Appl. Clay Sci.*, 2020, **189**, 105537.
71. N. Deleo, A. Howell and S. Patel, *et al.*, Laponite clay in cosmetic and personal care products, *US pat.*, US2015335543A1, The United States Patent and Trademark Office, 2015.
72. A. L. Howell, M. A. Lull and C. D. Novack, Laponite clay in cosmetic and personal care products, *US pat.*, US2015335544A1(B2), The United States Patent and Trademark Office, 2015.

73. Aztec Secret. https://aztecsecret.in/, accessed 10 March 2023.
74. Charcoal Pore Pudding – Boscia. https://www.boscia.com/products/charcoal-pore-pudding, accessed 10 March 2023.
75. Goddess Skin Clay Mask – Purifying Clay Face Mask | Charlotte Tilbury. https://www.charlottetilbury.com/us/product/goddess-skin-clay-mask, accessed 10 March 2023.
76. Umbrian Clay Pore-Purifying Face Mask, 100Ml | Skincare | Fresh Beauty US. https://www.fresh.com/us/skincare/categories/masks/umbrian-clay-pore-purifying-face-mask-H00006628.html, accessed 10 March 2023.
77. SUPERMUD® Charcoal Mud Mask | GLAMGLOW. https://www.glamglow.com/product/15132/39399/face-masks/supermud/instant-clearing-treatment-mask, accessed 10 March 2023.
78. Super Volcanic Pore Clay Mask 2X 100ml – Innisfree India. https://in.innisfree.com/products/super-volcanic-pore-clay-mask-2x-100ml?_pos=1&_sid=6bd626ce9&_ss=r, accessed 10 March 2023.
79. Clay Mask – Kiehl's Rare Earth Deep Pore Cleansing Mask. https://kiehls.in/rare-earth-deep-pore-cleansing-mask.html, accessed 10 March 2023.
80. Effaclar Clay Mask for Oily Skin | La Roche-Posay. https://www.laroche-posay.us/our-products/acne-oily-skin/face-mask/effaclar-shine-control-clay-mask-3337875533317.html, accessed 10 March 2023.
81. Facial Cleanser Detox & Brighten Cleanser | L'Oréal Paris. https://www.lorealparisusa.com/skin-care/facial-cleansers/pure-clay-cleanser-detox-and-brighten, accessed 10 March 2023.
82. ClearProof® Deep-Cleansing Charcoal Mask | Mary Kay. https://www.marykay.com/en-us/products/skincare/concern/blemishes-acne/clear-proof-deepcleansing-charcoal-mask-301029#tabbed3, accessed 10 March 2023.
83. Irish Moor Mud Mask | Masks & Peels | Peter Thomas Roth. https://www.peterthomasroth.com/irish-moor-mud-mask-1301015.html, accessed 10 March 2023.
84. Sunday Riley Ceramic Slip Cleanser. https://sundayriley.com/products/ceramic-slip-cleanser, accessed 11 March 2023.
85. Purifying Mask | Cleansing Mask Treatment | 100% Natural Antiaging Skincare | Tata Harper Skin Care. https://www.tataharperskincare.com/purifying-mask, accessed 11 March 2023.

CHAPTER 12

# *Nanotechnology for Cosmetics*

GAGANDEEP KAUR[a], MANDEEP KAUR[a], KIRTANJOT KAUR[b], SHWETA SAREEN[c] AND VISHAL MUTREJA[d*]

[a]PG Department of Chemistry, Sri Guru Teg Bahadur Khalsa College, Sri Anandpur Sahib, Punjab 140118, India; [b]PG Department of Chemistry, Sri Guru Gobind Singh College, Sector 26, Chandigarh 160019, India; [c]University Centre for Research and Development, Chandigarh University, Mohali, Punjab 140413, India; [d]Department of Chemistry, University Institute of Science, Chandigarh University, Mohali, Punjab 140413, India
*E-mail: vishal.mutreja@gmail.com

## 12.1 Introduction

Nanotechnology is a cutting-edge scientific field that constitutes the development, fabrication, characterization, and utilization of substances and systems in the nanoscale range (1–100 nm). It is considered the most impending technology of the twenty-first century owing to its applications in chemistry, physics, materials science and engineering. It has also benefited health and cosmetics science through various interesting applications, such as effective delivery, drug loading, anti-bacterial activity, UV-filters *etc.*, which could meet the expectations of end users.

Cosmetics are substances that have been employed by humans for a very long time and are mainly utilized for regeneration purposes. They are designed for cleansing, grooming, enhancing beauty, or changing the look without altering the layout of the body in accordance with the definition by the US Food and Drug Administration. They are often applied externally and can be prepared from a single substance or a mixture of compounds from either natural or artificial sources.[1] In 1961, Raymond Reed, a co-founder of the US Society of

Bioprospecting of Natural Sources for Cosmeceuticals
Edited by Deepika Kathuria, Ajay Sharma, Meenakshi Verma and Gulzar Ahmad Nayik

Published by the Royal Society of Chemistry, www.rsc.org

Cosmetic Chemists, invented the term "cosmetics".[2] Cosmetics comprise various personal care products such as lip gloss, lipsticks, balms for lips, lotions, moisturizers, and sunscreens for skin *etc.* Another segment of cosmetics is "cosmeceuticals" which consists of products that can be used as cosmetics as well as having health benefits.[3] The problems of hair damage, acne, wrinkles, dryness and photoaging have been resolved using such cosmeceutical products.[4] Nowadays, the development of cosmetic products involves the substantial usage of nanomaterials and nanoparticles. Numerous cosmetic firms are making investments in the manufacturing of cosmetics *via* new concepts of nanotechnology. The preparation of cosmetics, as well as cosmeceutical products, *via* the application of nanomaterials and nanoparticles has generated the new terms, nanocosmetics and nanocosmeceuticals, which are used when such nanomaterials are employed in the cosmetics industry owing to their distinct characteristics, such as small size, large surface area, and effective penetrating power.[5,6] There are numerous distinctive features of nanocosmetics in comparison to the conventional cosmetics.

Nanocosmetics have better texture, improved performance, effective entrapment, controlled and focused delivery, and occlusive properties.[7] The cosmetics industries are using a wide range of nanoparticles for the preparation of cosmetic products depending upon the type of application. For instance, ZnO nanoparticles act as UV filters and are employed in skin care products such as sunscreens to provide better protection from UV radiation. There are many more advantages of nanomaterials that can be used as active components, formulation supplements and carriers in cosmetic products. There are two different types of nanoformulations that have been mostly used in cosmetics: (i) the first nanoformulation system involves disintegration to their molecular form when applied on the skin and (ii) the second system doesn't disintegrate after their application to the skin. The first type of nanoformulation system comprises nanocarriers like liposomes, nanopigments and niosomes *etc.*, while metal-based nanoparticles are considered in the second group. Various personal care products, such as moisturizing creams, hair shampoos, anti-aging creams, conditioners, and skin-whitening creams, are using such nanomaterials.[8] Furthermore, the objective of slow transport of active and specific ingredients to particular parts of the human body could also be achieved by nanomaterials.

## 12.2 Nanomaterials in Cosmetics

The cosmetics sector is utilizing nanotechnology to its advantage by developing nanomaterials for the improved functionality and bioavailability of active constituents in various cosmetic products. They are numerous nanomaterials/nanocarriers that are being used in personal health care products. The various nanomaterials used in cosmetics are summarized in Figure 12.1. They are broadly divided into two types: organic nanomaterials and inorganic nanomaterials. Organic nanomaterials are further categorized into polymeric nanoparticles, vesicular system nanoparticles, non-vesicular

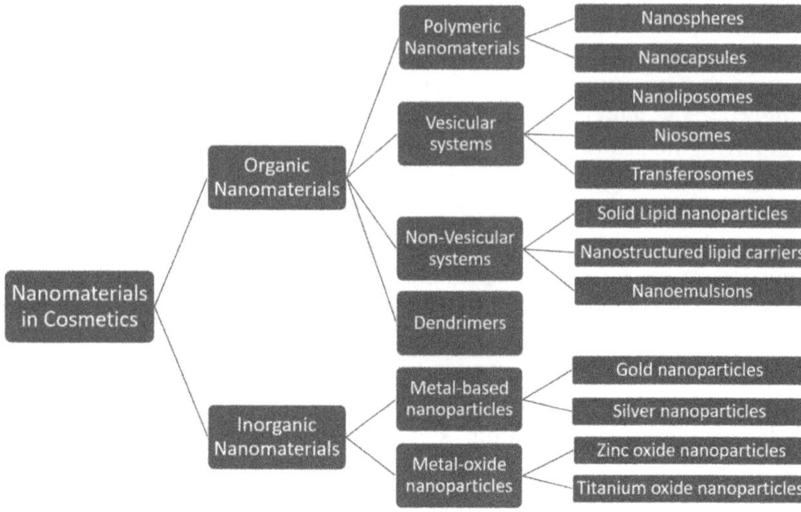


**Figure 12.1** Types of nanomaterials in cosmetics.

system nanoparticles and dendrimers. Inorganic nanomaterials consists of metal nanoparticles and metal oxide nanoparticles.

## 12.2.1 Inorganic Nanomaterials

These include nanoparticles of metals such as silver, gold and metal oxide, *viz.* CuO and FeO *etc.*, which are discussed below:

### *12.2.1.1 Metallic Nanoparticles*

Some of the metallic nanoparticles, such as gold, copper and silver, being biocompatible, hydrophilic, and remarkably stable, are used in different products. Moreover, the antimicrobial properties of gold and silver nanoparticles make them the best candidate for use in cosmetic formulations. Specifically, silver nanoparticles act as preservatives in various cosmetic products due to their antimicrobial properties. These have been discussed below:

**12.2.1.1.1 Gold-based Cosmetics.** Gold has always been a center of attraction in cosmetic products because of its substantial advantages, including potential to freshen the skin, better elasticity, enhanced circulation of blood, imparting stiffness to the skin, and impeding the ageing of the skin. It also has antibacterial, antiseptic and antifungal properties.[9] Furthermore, it is non-reactive, non-cytotoxic, and highly stable, and thus is considered appropriate for cosmetic formulations.[10,11] Gold nanostructures can be prepared in different size ranges with various shapes, such as nanorods, nanoclusters, nanospheres, nanocubes, nanostars, nanotriangles and nanoshells.

They can be employed as drug delivery carriers to target cells that are located deep inside the skin. They are used to repair skin damage and to enhance flexibility, elegance, *etc.* The generation of collagen can be boosted by the application of gold nanoparticles several times.[12] Cosmetics companies like L'Core Paris and L'Oréal utilize gold nanoparticles in their products to enhance their efficiency. Sunburn, skin inflammation and hypersensitivity can be cured by the soothing properties of gold nanoparticles and thus they are successfully employed in face masks and various types of cosmetic products.

**12.2.1.1.2 Silver-based Cosmetics.** Silver nanoparticles have been employed in cosmetic formulations and oral health care products, owing to their antifungal, antibacterial, antiviral, oral health care, and wound healing, properties. They are preferable in contrast to bulk silver in cosmetic products because bulk silver could precipitate out in the mixture.[13] Because of their antibacterial properties, silver nanoparticles act as preservatives in numerous cosmetic products such as soaps, lip care products, shampoos, toothpastes, wet wipes, deodorants *etc.* Silver nanoparticles in skin gel are considered a better option than silver sulfadiazine to treat skin infections in burn victims.[14] In a recent study, it was found that silver nanoparticles have antifungal properties, and are capable of treating fungal infections of the toe and finger nails.[15]

#### *12.2.1.2 Metal Oxide Nanoparticles*

$TiO_2$ is one of the best-known metal oxide nanoparticles used in sunscreens with high sun protection factor (SPF) and transparency. Furthermore, $TiO_2$ plays a vital role in protecting the skin from UV as it absorbs ultraviolet B (UVB) and ZnO absorbs ultraviolet A (UVA).[16] The proficiency of $TiO_2$ and ZnO nanoparticles to behave as UV filters is due to their small size (from 20 nm to 200 nm).[17] Additionally, they can create a thin transparent coating on the external skin with uniformity, which hinders the interaction of radiation with the skin and thus protects it. When using ZnO nanoparticles, short-lived oxygen species can be generated with the release of $Zn^{2+}$ that initiates the antibacterial action of ZnO by disturbing the bacteria's cell wall structure.[18]

### 12.2.2 Organic Nanomaterials

Organic nanomaterials include polymer-based nanostructures, vesicular nanoparticles and non-vesicular nanoparticles. These nanomaterial systems are used in different cosmetic products,[19] and have been discussed in detail in the following sections.

#### *12.2.2.1 Polymeric Nanoparticles*

Polymeric nanoparticles can be prepared by using both natural and synthetic polymers. The synthetic polymers consist of poly caprolactone (PCL), derivatives of cellulose and polyacrylate, whereas gelatine, chitosan and alginate fall in the category of natural polymers.[20] Further, depending upon the shape, morphology and structure, these are divided into types: nanocapsules and nanospheres. Nanocapsules are particles in which the main ingredient or active component can be loaded in the inner core of the polymeric material or is a liquid that is bound with a polymeric membrane, while nanospheres are systems in which the active substance is trapped in the polymeric core of the spheres.

**12.2.2.1.1 Nanocapsules.** Nanocapsules are polymeric nanoparticles that enclose a liquid core of oil or water within them.[21] These materials can be used in formulations utilized as carriers or applied directly to the skin after being merged into semi-solid systems. The use of surfactants and polymers regulates the skin penetration level.[22] They are used in beauty care formulations to safeguard chemicals, and eliminate offensive scents. Generally, the bioactive agent is present in the inner core and the outer polymeric part regulates its discharge. These nanocapsules are used in antiperspirants to improve their potency.[17] Recent reports suggest the synthesis of unique stimuli-responsive polymeric nanocapsules to transport vitamins and extracts, which can then be used in semi-solid compounds like creams. When these compounds are used on the skin, stimuli brought on by the presence of the skin's enzymes and pH shift prompted the nanocapsules to secrete their active components at the specific skin location. Benzophenone-3-loaded nanocapsules with poly (ε-caprolactone) from carrot oil and Pluronic F68 were synthesized by an emulsification–diffusion method and then were employed in sunscreen products with SPF activity, they were also nontoxic.[23] Nanocapsules were used also as nanocarriers to release the trapped fragrance molecules present in the inner core, which is coated with a polymeric wall made up of poly-L-lactic acid by using the nano precipitation method.[24]

**12.2.2.1.2 Nanospheres.** Nanospheres comprise a polymeric framework, which allows for the entrapment or adsorption of a variety of active substances.[18] The polymeric framework of the nanospheres protects the drug from enzymatic and chemical degradation. Furthermore, the nanospheres are divided into two types depending on the use of polymer, one is biodegradable nanospheres and the other is non-biodegradable nanospheres. These nanospheres can hold a wide range of both hydrophobic and hydrophilic compounds, but their primary use is to enhance the performance of volatile, weakly water-soluble, and chemically labile agents.[25] In addition to this, these nanocarriers deliver persistent release of the trapped molecules, thus extending their effectivity.[26]

**12.2.2.1.3 Dendrimers.** Dendrimers are a category of polymeric nanoparticle with a central core and a network of long branches with functional moieties located at their ends.[27–29] They can also be used to efficiently supply cosmetic products into the skin because of their size in the nanoscale range (2–20 nm). Furthermore, they are widely used in other cosmetic products such as deodorants and shampoo.[30] Additionally, they are used as a carrier to improve the penetration of vitamins into the skin's inner layers. They are considered perfect molecules for the delivery of drugs because of their special features, like polyvalency, stability and monodispersity. Moreover, dendrimers have found wide applications in the manufacture of sunscreens, lotions, anti-acne products, hairgels, and spraygels.[20] The development of a dendrimer–resveratrol formulation helped to enhance the solubility, stability and skin penetration of resveratrol, which had anti-aging properties by entrapment of resveratrol within the dendrimer, and was then successfully commercialized.[31]

### *12.2.2.2 Vesicular Nanoparticle Systems*

Depending upon the size and number of bilayers, vesicular nanoparticles are categorized as small unilamellar vesicles (SUVs), large unilamellar vesicles (LUVs) and multilamellar vesicles (MLVs).[32] The small and large unilamellar vesicles are synthesized from multilamellar vesicles by sonication, extrusion *etc.* while multilamellar vesicles can be prepared from raw materials.[33] These vesicle systems are used to improve the penetration and to transport ingredients deep into the skin layer.[34]

**12.2.2.2.1 Niosomes.** Niosomes are vesicles made up of a non-ionic surfactant bilayer that offer better drug encapsulation efficiency and stability over traditional liposomes. They are divided into three types on the basis of size, namely multilamellar vesicles with a diameter in the range 0.5–10 μm, large unilamellar vesicles with a diameter of 0.1–1 μm and small unilamellar vesicles with a diameter of 25–500 nm.[35] Most of the niosomes are synthesized by using cholesterol and non-ionic surfactants such as Tween, Brij, Span, sorbitan ester, alkyl amides, and polyoxyethylene alkyl ethers.[36] The manufacturing of such nanocarriers is cost-effective and their scaling-up is also easy to handle when compared with liposomes.[37] Niosomes are frequently used in cosmetics because they improve skin permeation by decreasing the *stratum corneum* (horny layer) barrier resistance and permitting the active components to enter into the tissues at faster speeds . The synthesis of niosomes depends on various factors, which include the surfactant's structure and nature, the nature of the embedded drug, and the composition of the membrane. In 1970, L'Oréal developed the first niosomes and in 1987 they patented them under the brand name Lancôme.[36]

**12.2.2.2.2 Nanoliposomes.** Liposomes in the nano range are known as nanoliposomes. Liposomes are spherical vesicular nanocarriers with an aqueous core that is wrapped by a phospholipid bilayer to prevent any hazardous side effects.[38] They can be classified as unilamellar vesicles (ULVs), MLVs, *etc.* depending on how the lipid outer coating is arranged.[39] The biodegradability and biocompatibility of nanoliposomes make them unique nanocarriers in nanocosmetics.[30] They exhibit good skin penetrability and therefore are applicable in moisturizing formulations. They also have the potential to fix and transfer nutrients, as well as to give attractive fragrances to body washes, lipsticks, and deodorants.[18] Owing to their smaller particle size, they are extensively used in cosmetics, as they provide better hydration to the skin, resulting in smoother, and more elastic skin. Nanoliposomes can be prepared by using micro fluidization and sonication techniques. They help in the protection of active components from UV and due to the presence of lipids they increase the product's shelf life. The release of the active ingredients at a particular site is also possible with the help of liposomes. Although they have attractive qualities, their commercialization in the cosmetics market is an issue of concern due to physicochemical instability.[40]

**12.2.2.2.3 Transferosomes.** Transferosomes are liposome-based formulations with phospholipids and an edge activator, which is a long chain surfactant such as Span 80, Tween 20 *etc.* that gives ultra-deforming capacity to the transferosome and permits greater penetration into the skin's transdermal layer. Due to the presence of an edge activator molecule in transferosomes, they have "self-optimizing deformable" ability that enables them to alter the structure of their membrane and fit through extremely narrow holes.[41] These systems can also be used for acne treatment and have utility in UV protection.[42]

### *12.2.2.3 Non-vesicular Nanoparticle Systems*

This category includes a variety of heterogeneous systems stabilized by surfactants, and made up of a dispersion of lipophilic and aqueous immiscible phases.

**12.2.2.3.1 Solid Lipid Nanoparticles (SLNs).** These nanoparticles have an unusual exterior shell with an inner core consisting of lipid or oily components.[3] They have a polymeric character and contain various fatty acids, including stearic acid or a combination of fatty acids. Due to the hydrophobic nature of the SLN particles, the skin is protected from drying. These nanoparticles are therefore more effective carriers, are non-toxic, and have superior skin penetration. The formulations' small size allows for direct penetration into the corneum layer, which improves the uptake of active substances into the skin.[43] In cosmetics, they exhibit better biocompatibility and work well as a transport supply system.[18] SLNs are frequently utilized in sunscreen formulations, serving as active transporters for molecular sunscreen compounds. Compared to commonly used formulations, they require lower amounts of the active species; however, they could provide the same level of protection. SLNs are also used in perfume products owing to their slow release of the fragrance bearing species over a longer length of time.

**12.2.2.3.2 Nanostructured Lipid Carriers (NLCs).** These nanoparticles are believed to be lipid nanoparticles of second generation. To overcome the shortcomings of SLNs, they are developed with distinctive features. There are three types of NLC based on synthesis methodology and composition employed in formulations, namely imperfect type, multiple type and amorphous type. The particle size of NLCs lies within the range 10–1000 nm.[44] Due to the decreased risk of toxicities, NLC has received more scientific and commercial attention in recent years. Because of the deformed structure that helps to provide additional space, NLCs exhibit greater drug-carrying capability for enclosed bioactive molecules as compared to SLNs. The synthesis of NLCs addresses some of the other SLN issues, such as lowering the particle concentration and preventing drug ejection during storage. They are prepared from physiological and biodegradable lipids, which are comparatively less toxic.[45]

12.2.2.3.3 **Nanoemulsions.** Nanoemulsions are formed when a water phase, an oil phase and a surfactant are combined to form a liquid dispersion that is thermodynamically or kinetically stable. Water-in-oil (W/O), oil-in-water (O/W), and bi-continuous layered systems (O/W/O or W/O/W) are three different types of nanoemulsions based on their composition. Furthermore, they are also categorized based on their preparation method, which results in various formulations, such as cream-like, water-like or gel-like.[46,47] Nanoemulsions are generally synthesized using various techniques, such as sonication, phase inversion, high-pressure homogenization, and microfluidization during the incorporation of surfactant. Nanoemulsions consist of small-sized particles ranging from a few nanometers to 200 nm and have much lower interfacial tension between the oil and water phases. They are considered as a suitable candidate for the distribution of lipophilic molecules because of the presence of a core that is lipophilic and is enclosed by a layer of phospholipids. They are clear or translucent and have some good features, which include little viscosity, large interfacial area, high kinetic stability, and good solubilization ability.[48] Nanoemulsions are employed in cosmetic goods as they afford quick diffusion, effective transport of active substances, and moisturization of the skin.

## 12.3 Nanomaterials Used in Commercialized Products

Nanotechnology has recently offered creative answers to a number of issues in the health care and pharmaceutical fields. The same idea has been used in cosmetics (like nanoliposomes, SLNs, nanospheres, nanocapsules, nanoemulsions, dendrimers, nanostructured lipid nanoparticles and metallic nanoparticles), leading to cutting-edge formulations known as nanocosmeceuticals that offer specialized treatments for cosmeceutical issues. A smaller size may be crucial for the unique features as it aids in the development of new characteristics, like improved solubility, chemical reactivity, transparency and durability. Various nanomaterials have been used in different cosmetic products for different applications, so that they can transfer various active molecules into the skin (see Table 12.1).

**Table 12.1** Various nanomaterials used in different cosmetic products.[a]

| S. no. | Nanomaterials in cosmetics | Functions of cosmetic products | Name of products | Manufacturer | Reference |
|---|---|---|---|---|---|
| 1 | Nanoliposomes | Moisturizer | Dermosome™ | Croda Canada | 3 and 49 |
| | | | Rehydrating liposome day crème | Kerstin Florian | 3 and 49 |
| | | Anti-wrinkle | Capture totale | Dior | |
| | | Skin repair whitening | Azelac RU liposomal serum | Sesderma | 50 |
| 2 | Nanoemulsions | Moisturizer | NanoVital vitanics crystal moisture cream | VITACOS Cosmetics | 18 |
| | | Anti-aging | Bepanthol ultra facial protect cream | Bayer HealthCare | |
| 3 | Niosomes | Anti-wrinkle | Niosome+ perfected age treatment | Lancôme | |
| | | Anti-aging and skin repair | Mayu niosome base cream | Laon Cosmetics | 3 and 49 |
| 4 | Dendrimers | Sunscreen | Topical resveratrol formulations | — | 13 |
| 5 | Solid lipid nanoparticles (SLNs) | Body moisturizer | Allure body lotion | Chanel | |
| | | Perfume | Allure parfum bottle | Chanel | |
| 6 | Nanostructured lipid nanoparticles | Moisturizer, skin repair | Phyto NLC active cell repair | Sirech Emas | 18 and 49 |
| | | Anti-aging | Regeneration cream intensive ampoules | Scholl | 51 |
| 7 | Nanocapsules | Anti-aging | Nano vita C | Eccos | 18 |
| | | Anti-aging and skin repair/moisturizing | Super aqua skin cream line | ENPRANI | 52 |
| 8 | Gold nanoparticles | Anti-aging | Nano gold BB Cream SPF 50 PA+++ | Tony Moly | 3 |
| | | | Nano gold mask pack | Lexon NanoTech | 9 |
| 9 | Silver nanoparticles | Skin balance and hydration | Microsilk white lotus intensive lotion & mask | Joyona International Marketing Ltd. | |
| | | Skin cleanser | Cosil nanobeauty soap | Natural Korea | 9 |
| 10 | ZnO | Skin care | Sunscreen, skincare and lip balms | Dermatone | 38 |
| 11 | $TiO_2$ | Skin care | Dior snow pure UV base SPF 50 | Christian Dior | 9 |

[a]S. no. – serial number

The market for personal care products such as cosmetics is considered to have expanded exponentially in the last two decades. Numerous nanocosmeceuticals are often used in nail, lip, hair and skin care products. The categorization of nanocosmetics is shown in Figure 12.2.

*Nail Care*. Nanocosmetic nail care products provide a greater benefit than traditional products. Nanocosmetic nail paints provide benefits like longevity, quick-drying, enhanced strength, chip resistance, and ease of application due to their elastic nature. The utilization of innovative methods like mixing metal oxide and silver nanoparticles in nail paints for the treatment of toenail fungal infections is another feature of nanotechnology in cosmetics.

*Lip Care*. Lipsticks and glosses can be manufactured from nanoparticles and these nanocosmetic products prevent transepidermal moisture loss, to keep the lips smooth. In addition, they avoid any transfer of color from the lips and retain their color for a prolonged duration. Liposome-containing lip

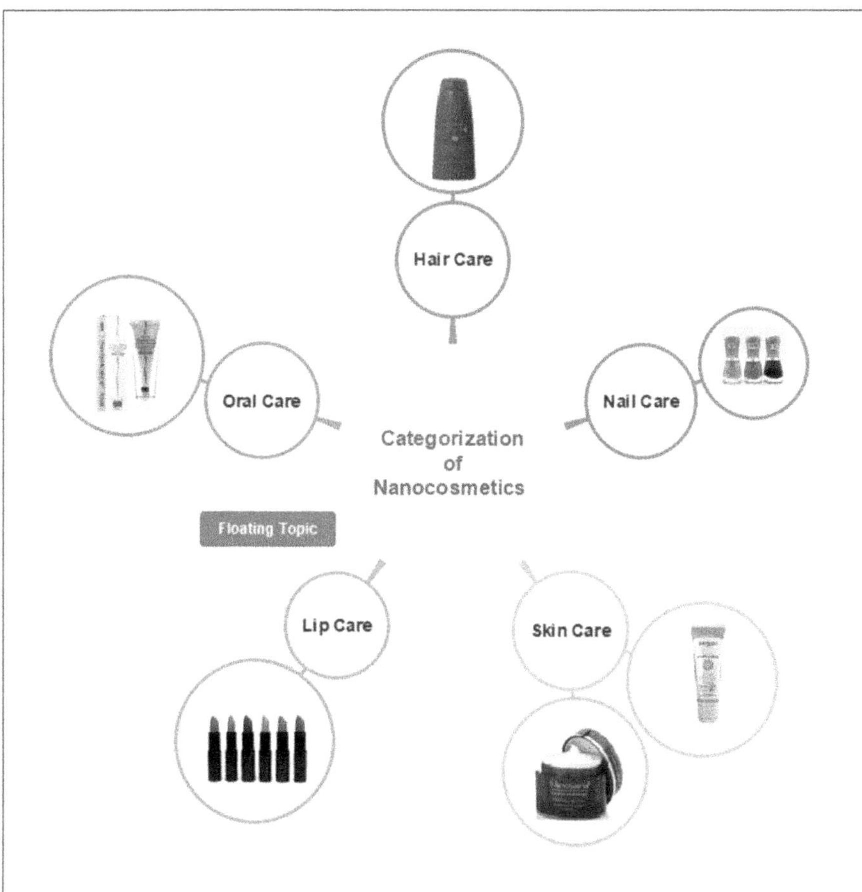


**Figure 12.2** Categories of different products that could employ nanocosmetics.

volumizer moisturizes intensify, accentuate and highlight the lip contours and clears wrinkles.

*Hair Care*. Nanotechnology-based hair care products include shampoos, conditioners, styling products, hair growth stimulants and coloring agents. The unique features and distinct size of the nanoparticles enable them to target the hair follicles and shafts, and boost the amount of active substance. Shampoos containing nanoparticles trap the moisture by improving resident interaction with the scalp, and within cuticles and hair follicles by generating a barrier film. Nanocosmeceutical conditioners are designed to add softness, sheen, silky smoothness, and luster, while improving hair detangling. The primary role of unique carriers such as niosomes, microemulsions, nanoemulsions, nanospheres, and liposomes is to heal destroyed cuticles, restore texture and make the hair glossy and non-oily.

*Oral Care*. Toothpaste and mouthwash are products that are used to maintain oral hygiene. The practice of manufacturing these items with nanotechnology has become extremely prevalent in an effort to increase their performance and efficacy. The majority of products for oral care use polymeric nanoparticles, *viz.* dendrimers and hydrogels, as nanocarriers and gold and silver nanoparticles as antibacterial agents. Nanotechnology-enhanced toothpastes are designed with the objective of promoting self-healing of the teeth, regenerating enamel, and preventing bacterial infections. For improved whitening and protection, toothpastes made with nanosized hydroxyapatite and $TiO_2$ are also available. Toothpastes containing silver nanoparticles are considered to be effective microbicides.

*Skin Care*. Cosmetics for skin care products improve skin function and texture by promoting collagen formation and fending off the damaging effects of free radicals. Furthermore, they are designed and formulated to nourish and give protection to keratin's structure to improve the overall health of the skin. The most powerful elements of sunscreen lotions are ZnO and $TiO_2$ nanoparticles, which penetrate deep into the skin to protect it and to keep the product less oily and transparent. Nanocarriers such as nanoemulsions, nanoliposomes, SLNs and niosomes used in skin care products create a thin layer of humectants and hold onto moisture for an extended period of time.

## 12.4 Conclusions

Despite their numerous performance boosting benefits (see Figure 12.3), there are also some challenges associated with nanoparticles, such as toxicity, stability, and high cost. The nanoparticles used in the fabrication of cosmetics could be toxic and can lead to oxidative stress. For instance, $TiO_2$ nanoparticles, which act as a UV protector, have been reported to cause damage to the RNA and DNA within the cells. Therefore, the regulatory bodies need to do rigorous scrutiny before approving any product.

Furthermore, these products may cause adverse effects on the environment. However, greener nanostructures prepared using plant-based extracts

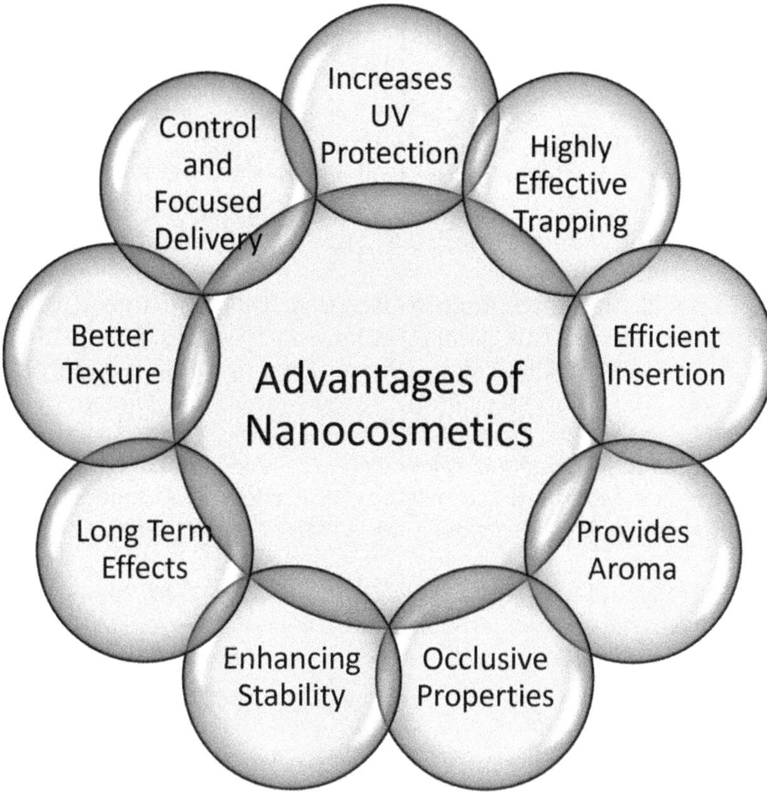


**Figure 12.3** Advantages of nanocosmetics.

as reducing and stabilizing agents could be a viable solution to the aforementioned problems. Such nanostructures have shown promising results for different applications (antibacterial, antifungal and drug delivery) due to the cooperative or synergetic effect among different phytoconstituents and metallic structures.

## References

1. P. Shende, D. Patel and A. Takke, *Nanomaterial-based Cosmeceuticals*, INC, 2020.
2. S. Dhawan, P. Sharma and S. Nanda, *Nanocosmetics*, 2020, 141–169.
3. S. Kaul, N. Gulati, D. Verma, S. Mukherjee and U. Nagaich, *J. Pharm.*, 2018, **2018**, 1–19.
4. K. Srinivas, *J. Chem. Pharm. Res.*, 2016, **8**, 906–914.
5. M. D. Gianeti and P. M. B. G. Maia Campos, *Molecules*, 2014, **19**, 18268–18282.

6. U. Ahmad, Z. Ahmad, A. A. Khan, J. Akhtar, S. P. Singh and F. J. Ahmad, *Drug Res.*, 2018, **68**, 545–552.
7. B. H. Abbasi, H. Fazal, N. Ahmad, M. Ali, N. Giglioli-Guivarch and C. Hano, *Nanomaterials for Cosmeceuticals: Nanomaterials-Induced Advancement in Cosmetics, Challenges, and Opportunities*, INC, 2020.
8. T. P. Vinod and R. Jelinek, *Nanocosmetics from Ideas to Products*, 2019, pp. 29–46.
9. A. Lohani, A. Verma, H. Joshi, N. Yadav and N. Karki, *ISRN Dermatol.*, 2014, **2014**, 1–14.
10. A. Gupta, R. Mathur, S. Singh, N. Bag, U. A. Khan, F. J. Ahmad, G. A. Gabr, P. Kesharwani and G. K. Jain, *J. Pharm. Sci.*, 2021, **110**, 888–897.
11. R. A. Bapat, T. V. Chaubal, S. Dharmadhikari, A. M. Abdulla, P. Bapat, A. Alexander, S. K. Dubey and P. Kesharwani, *Int. J. Pharm.*, 2020, **586**, 119596.
12. K. Singh and R. Jain, *Anal. Chem. Lett.*, 2021, **11**, 1–14.
13. V. Gupta, S. Mohapatra, H. Mishra, U. Farooq, K. Kumar, M. J. Ansari, M. F. Aldawsari, A. S. Alalaiwe, M. A. Mirza and Z. Iqbal, *Gels*, 2022, **8**, 173.
14. W. T. J. Ong and K. L. Nyam, *Saudi J. Biol. Sci.*, 2022, **29**, 2085–2094.
15. M. Cobos, I. De-La-pinta, G. Quindós, M. D. Fernández and M. J. Fernández, *Nanomaterials*, 2020, **10**, 376.
16. H. Yang, S. Zhu and N. Pan, *J. Appl. Polym. Sci.*, 2004, **92**, 3201–3210.
17. G. Fytianos, A. Rahdar and G. Z. Kyzas, *Nanomaterials*, 2020, **10**, 1–16.
18. A. C. Santos, F. Morais, A. Simões, I. Pereira, J. A. D. Sequeira, M. Pereira-Silva, F. Veiga and A. Ribeiro, *Expert Opin. Drug Delivery*, 2019, **16**, 313–330.
19. K. Khezri, M. Saeedi and S. Maleki Dizaj, *Biomed. Pharmacother.*, 2018, **106**, 1499–1505.
20. E. B. Souto, A. R. Fernandes, C. Martins-Gomes, T. E. Coutinho, A. Durazzo, M. Lucarini, S. B. Souto, A. M. Silva and A. Santini, *Appl. Sci.*, 2020, **10**, 1–24.
21. S. Nafisi and H. I. Maibach, *Nanotechnology in Cosmetics*, 2017.
22. L. Vinet and A. Zhedanov, *A 'Missing' Family of Classical Orthogonal Polynomials*, 2011, vol. 44.
23. T. C. Barbosa, L. É. D. Nascimento, C. Bani, T. Almeida, M. Nery, R. S. Santos, L. R. de Oliveira Menezes, A. Zielinska, A. R. Fernandes, J. C. Cardoso, A. Jäger, E. Jäger, E. Sanchez-Lopez, L. Nalone, E. B. Souto and P. Severino, *Toxics*, 2019, 7, 51.
24. L. A. Frank, R. V. Contri, R. C. R. Beck, A. R. Pohlmann and S. S. Guterres, *Wiley Interdiscip. Rev.: Nanomed. Nanobiotechnol.*, 2015, **7**, 623–639.
25. F. S. Poletto, R. C. R. Beck, S. S. Guterres and A. R. Pohlmann, *Nanocosmetics and Nanomedicines*, 2011, pp. 49–68.
26. B. Begines, T. Ortiz, M. Pérez-Aranda, G. Martínez, M. Merinero, F. Argüelles-Arias and A. Alcudia, *Nanomaterials*, 2020, **10**, 1–41.
27. B. Surekha, N. S. Kommana, S. K. Dubey, A. V. P. Kumar, R. Shukla and P. Kesharwani, *Colloids Surf., B*, 2021, **204**, 111837.

28. A. Sheikh, Md. Shadab and P. Kesharwani, *J. Controlled Release*, 2021, **340**, 221–242.
29. A. Sheikh and P. Kesharwani, *Eur. Polym. J.*, 2021, **159**, 110746.
30. L. Rigano and N. Lionetti, in *Nanobiomaterials in Galenic Formulations and Cosmetics: Applications of Nanobiomaterials*, 2016, pp. 121–148.
31. T. Pentek, E. Newenhouse, B. O'Brien and A. Singh Chauhan, *Molecules*, 2017, **22**, 137.
32. P. van Hoogevest and A. Fahr, in *Nanocosmetics*, 2019.
33. C. Marianecci, L. Di Marzio, F. Rinaldi, C. Celia, D. Paolino, F. Alhaique, S. Esposito and M. Carafa, *Adv. Colloid Interface Sci.*, 2014, **205**, 187–206.
34. A. C. Carita, J. O. Eloy, M. Chorilli, R. J. Lee and G. R. Leonardi, *Curr. Med. Chem.*, 2017, **25**, 606–635.
35. A. Sankhyan and P. Pawar, *J. Appl. Pharm. Sci.*, 2012, **2**, 20–32.
36. N. Ali, S. L. Harikumar and A. Kaur, *Int. J. Res. Pharm. Chem.*, 2012, **2**, 479–487.
37. X. Ge, M. Wei, S. He and W. E. Yuan, *Pharmaceutics*, 2019, **11**, 55.
38. S. Raj, S. Jose, U. S. Sumod and M. Sabitha, *J. Pharm. BioAllied Sci.*, 2012, **4**, 186–193.
39. S. K. Dubey, A. Dey, G. Singhvi, M. M. Pandey, V. Singh and P. Kesharwani, *Colloids Surf., B*, 2022, **214**, 112440.
40. J. Joseph, B. N. Vedha Hari and D. Ramya Devi, *Eur. J. Pharm. Sci.*, 2018, **112**, 38–51.
41. J. R. Walve, S. R. Bakliwal, B. R. Rane and S. P. Pawar, *Int. J. Appl. Biol. Pharm. Technol.*, 2011, **2**, 204–213.
42. R. Pahwa, S. Pal, K. Saroha, P. Waliyan and M. Kumar, *J. Appl. Pharm. Sci.*, 2021, **11**, 1–8.
43. M. Singh, S. Mohapatra, Sanskriti, N. Kaur, A. Mushtaq, S. Zahid, A. A. Pandith, S. Mansoor and Z. Iqbal, in *Dietary Phytochemicals*, 2021, pp. 223–251.
44. S. L. Patwekar, S. R. Pedewad, A. P. Potulwar, M. S. Gaikwad, A. B. Suryawanshi and S. A. Khan, *Int. J. Pharm. Pharm. Res.*, 2016, **6**, 168–180.
45. D. K. Purohit, T. D. Nandgude and S. S. Poddar, *Asian J. Pharm.*, 2016, **10**, S1–S9.
46. B. Gorain, H. Choudhury, A. B. Nair, S. K. Dubey and P. Kesharwani, *Drug Discovery Today*, 2020, **25**, 1174–1188.
47. H. Choudhury, N. F. B. Zakaria, P. A. B. Tilang, A. S. Tzeyung, M. Pandey, B. Chatterjee, N. A. Alhakamy, S. K. Bhattamishra, P. Kesharwani, B. Gorain and Md. Shadab, *J. Drug Delivery Sci. Technol.*, 2019, **54**, 101301.
48. R. P. Patel and J. R. Joshi, *Int. J. Pharm. Sci. Res.*, 2012, **3**, 4640–4650.
49. V. Dhapte-Pawar, S. Kadam, S. Saptarsi and P. P. Kenjale, *Future Sci. OA*, 2020, **6**, FSO613.
50. H. Zhou, D. Luo, D. Chen, X. Tan, X. Bai, Z. Liu, X. Yang and W. Liu, *Clin., Cosmet. Invest. Dermatol.*, 2021, **14**, 867–887.
51. L. Battaglia and E. Ugazio, *J. Nanomater.*, 2019, 1–22.
52. B. Jose, E. J. Jesy and R. J. Nedumpara, *World J. Pharm. Res.*, 2014, **3**, 5041–5048.

# Subject Index


algae, 145, 165
algal secondary metabolites, 147
algal-based raw materials (biometabolites), 152
    instrumental extraction methods, 152
        conventional methods, 152, 154
        enzyme-assisted extraction (EAE), 154–155
        microwave-assisted extraction (MAE), 155
        pressurized liquid extraction (PLE), 157
        subcritical water extraction (SWE), 157
        supercritical fluid extraction (SFE), 156
        ultrasound-assisted extraction (UAE), 156–157
alkaloid secondary metabolites, 86–91
    berberine, 89
    caffeine, 86–88
    capsaicin, 88–89
    piperine, 90
    pyrrolizidine alkaloids, 90–91
*Aloe vera,* 123
anthocyanins, 10, 12
anti-aging, 158–159
antibacterial properties, 178–179
antifungal properties, 179–180
anti-inflammatory agents, 186–188
antimicrobial agents, 176–177
antioxidants, 3, 160–161, 180–183
astaxanthin, 145

bacteria, 200
bacterial secondary metabolites, 203
    bacterial cellulose, 203–205
    bacterial polysaccharides, 206–209
    bio-surfactants, 209
    cyanobacteria, 205–206
    oligosaccharides, 209
bee products, 94
berberine, 14, 89
bimatoprost, 10
bioactive cosmeceuticals, uses, 150, 152, 209–210
biochemical metabolites, 175–176
biological active ingredients, 2
biopterin glucose, 214
bioresource marine natural products, 224–225
bisabolol, 80–81

caffeine, 14, 86–88
capsaicin, 88–89
carminic acid, 93–94
caryophyllene, 78
chitosan, 8
cinnamic acid, 14, 82–83
clay minerals, 276–278

cosmeceutical-based formulations, 283–285
history, use as cosmetics, 278–279
marketed products, 283–285
uses, cosmeceuticals, 280–283
coco beans, 3
Consumer Packaging and Labeling Act (CPLA), 26
cosmeceutical activities, 158
anti-aging, 158–159
antioxidant, 160–161
healing and therapeutic agent, 161
moisturizing, 160
skin whitening, 159–160
cosmeceutical properties, 3–5
cosmeceutical-based formulations, 161–162, 216
Cosmetic Regulation Act, 26
coumarin, 83–85
cultural and historical background, cosmetics, 146, 200–203
cyclodextrin glucanotransferase, 212

daphnoretin, 84
dermatocosmetic products, 9
dioctahedral smectites, 282
drug repositioning, 10
Drugs and Cosmetics Act, 1940, 40

EcoCert, 27
ectoines, 215
emulsions, 6
enzyme-assisted extraction (EAE), 154–155
essential oils (EOs), 54–55, 122
application, cosmetics, 65–67
authentication methods, 64–65
characterization methods, 63
chemistry of, 55–56
isolation methods, 56–63
exfoliants, 5

ferulic acid, 83
Food, Drug, and Cosmetic Act (FD&C Act), 19, 22–24, 54

gellan, 209, 212
geraniol, 78–79
glycolipids, 11
glycyrrhizic acid, 3
green cosmetics, economic overview, 5

hair care, 9–11, 302
mushrooms, 140
natural cosmetic-based formulations, 9–11
Health Canada, 26, 27
humectants, 4
hydrodistillation (HD) method, 58
4-hydroxybenzoate, 214

*Inocybe,* 131

kaempferol, 14, 85–86
kaolinite, 277

lactate dehydrogenase (LDH), 213
lactic acid, 215
Laponite®, 282
latanoprost, 10
lichens, 172, 173
antibacterial properties, 178–179
antifungal properties, 179–180
anti-inflammatory agents, 186–188
antimicrobial agents, 176–177
antioxidants, 180–183
biochemical metabolites in, 175–176
moisturizers, 189–190
morphology, symbiosis, and diversity, 173–175
in perfume and other products, 190–191
secondary metabolites, 176

lichens (*continued*)
skin lightening and depigmenting agents, 188–189
sunscreen agents, 183–186
lignin, 9
limitations, natural cosmeceuticals, 14
limonene, 14, 76–78
linalool, 14, 76–78
lip care, 12–13, 301–302

marine environments, bioresources, 230–231
bacteria-derived cosmeceuticals, 231–232
coral-derived cosmeceuticals, 235
fungi-derived cosmeceuticals, 231
macroalgal-derived cosmeceuticals, 232–233
marine sponge-derived cosmeceuticals, 234–235
microalgal-derived cosmeceuticals, 233–234
marine natural products, 225–226
applications, cosmeceuticals, 243
anti-acne activities, 245–246
antiaging activities, 243–244
antimelanogenic activities, 244–245
antioxidant activities, 246
antiviral activities, 253
moisturizing activities, 248–249
repairing activities, 250
skin lipid activities, 250
sunscreens, 246–247
wound healing, anti-inflammatory, and analgesic activities, 247–248
biodiversity challenges, 226–227
cosmeceutical applications, properties, 235
market challenges, 229–230
side effects, cosmeceuticals, 254
supply and technical challenges, 227–229
market products, 216
marketed products, 161–162
mass-market cosmetics, 216
melanogenesis, 9
metal oxide nanoparticles, 295
metallic nanoparticles, 294–295
gold-based cosmetics, 294–295
silver-based cosmetics, 295
microalgal secondary metabolites, 147, 150
microwave-assisted extraction (MAE), 61, 155
moisturizers, 189–190
Mordor Intelligence, 20
mushrooms, 129, 130
constituents, 134
carbohydrates, 134–135
lipids, 136
minerals, 135–136
proteins, 135
vitamins, 136
cosmetics applications, 137–138
hair care, 140
skin care, 138–139
edible, 133–134
medicinal, 132–133
poisonous, 131–132

nail care, 12, 301
nanocosmetics, 293
nanomaterials, 293–294
commercialized products, 299–302
hair care, 302
lip care, 301–302

nail care, 301
oral care, 302
skin care, 302
inorganic nanomaterials
metal oxide nanoparticles, 295
metallic nanoparticles, 294–295
organic nanomaterials, 295
non-vesicular nanoparticle systems, 298–299
polymeric nanoparticles, 295–297
vesicular nanoparticle systems, 297–298
nanotechnology, 12, 292
natural bioactive substances, 224
natural cosmetic-based formulations, 6
natural cosmetics, history, 4
natural product-based cosmeceuticals, 3
cultural history of, 106
natural products, 1
natural tri-glycerides, 6
non-vesicular nanoparticle systems, 298–299
nanoemulsions, 299
nanostructured lipid carriers (NLCs), 298–299
solid lipid nanoparticles (SLNs), 298
nutraceuticals, 130

Ohmic-assisted hydrodistillation (OAHD), 58
oral care, 302
organic triglycerides, 7
over-the-counter (OTC) medicines, 19, 20

parabens, 214
perfumes, 14
pharmacological activities, 210–216
phenolic secondary metabolites, 81–86
cinnamic acid, 82–83
coumarin, 83–85
ferulic acid, 83
kaempferol, 85–86
psoralens, 85
phenylpropanoids, 56
structure, 57
phytochemicals, 136–137
phytosphingosine (PS), 7
piperine, 90
polymeric nanoparticles, 295–297
dendrimers, 296–297
nanocapsules, 296
nanospheres, 296
polyphenol, 8
polysaccharides, 135
potassium cocoate, 5
pressurized liquid extraction (PLE), 157
primary metabolites, 73, 74
propolis, 14
psoralens, 85
puerarin, 60
pullulan films, 8
pyrrolizidine alkaloids, 90–91

regulations, 21
Canada, 26–27
China, 29–31
European Union, 31–34
Fair Trade Standard, 35
Natural and Organic Cosmetics Standards, 34–35
quality management, 35–36
India, 36–40
ISO 16128-1:2016 and ISO 16128-2:2017/Amd 1:2022(en), 41
Japan, 28–29
United States of America, 21–22

regulations (*continued*)
Food, Drug, and Cosmetic Act (FD&C Act), 22–24
New Laws, 24–25
USDA (United States Department of Agriculture), 25–26
regulatory guidelines, 18

Safe Cosmetics and Personal Care Products Act, 24
scytonemin, 215
secondary metabolite marine natural products
carbohydrates, 236
carrageenans, 237
chitosan, 238–239
exopolysaccharides, 238
fucoidans, 237
cosmeceutical applications, 235–236
marine phytochemical compounds, 239
astaxanthin, 240–241
phlorotannins, 239–240
pseudopterosins, 241
mycosporines, 241–242
neo fatty (carboxylic) acid derivatives, 242–243
secondary metabolites, 73–75, 176
alkaloid secondary metabolites, 86–91
cosmetic applications, 106–107
bakuchiol, 118, 120
boswellic acids (BAs), 114–116
caffeine, 109, 111
capsaicin, 111, 114
curcuminoids, 107–109
polyphenolics, 120
resveratrol, 116, 118
microorganisms and animals, 91–93
bee products, 94
carminic acid, 93–94
toxicological profile, 95
phenolic secondary metabolites, 81–86
terpene secondary metabolites, 76–81
skin aging, 7
skin care, 6–9, 54, 302
mushrooms, 138–139
anti-aging effect, 139
moisturizing ability/hydrating agent, 139
skin whitening effect, 139
natural cosmetic-based formulations, 6–9
skin depigmenting agents, 188–189
skin lightening agents, 188–189
skin whitening, 159–160
agent, 3
spongothymidine nucleoside, 224
spongouridine nucleoside, 224
steam distillation (SD) method, 57
subcritical water extraction (SWE), 157
sunscreen agents, 183–186
supercritical fluid extraction (SFE), 59, 156
synthetic products, 2

tea tree oil, 79–80
terpene secondary metabolites, 76–81
bisabolol, 80–81
caryophyllene, 78
geraniol, 78–79
limonene and linalool, 76–78
tea tree oil, 79–80
terpenes, structures, 56
terpenoids, 55
toiletries, 11–12
toxicological profile, 95
tretinoin, 200

ultrasound-assisted extraction (UAE), 156–157
ultrasound-assisted extraction method, 61

vesicular nanoparticle systems, 297–298
  nanoliposomes, 297–298
  niosomes, 297
  transferosomes, 298
volatile oils, 55

xanthan gum, 211

Yuki JAS, 29